Math Girls⁵

Galois Theory

Hiroshi Yuki

Translated by Tony Gonzalez

BENTO BOOKS

http://bentobooks.com

MATH GIRLS 5: GALOIS THEORY by Hiroshi Yuki

Originally published as *Sūgaku Gāru Garoa Riron*
Copyright © 2012 Hiroshi Yuki
Softbank Creative Corp., Tokyo

English translation © 2021 by Tony Gonzalez
Edited by M.D. Hendon and David Slutzky
Cover design by Kasia Bytnerowicz

All rights reserved. No portion of this book in excess of fair use considerations may be reproduced or transmitted in any form or by any means without written permission from the copyright holders.

Published 2021 by

Bento Books, Inc.
Houston, TX 77043

bentobooks.com

ISBN 978-1-939326-48-5 (hardcover)
ISBN 978-1-939326-46-1 (trade paperback)
ISBN 978-1-939326-47-8 (case laminate)
Library of Congress Control Number: 2021950570

Printed in the United States of America
First edition, November 2021

Math Girls^5:
Galois Theory

Contents

1 Ladder Diagrams   5
   1.1 Lots of Ladders   5
      1.1.1 Exchanging Ends   5
   1.2 More Ladders   8
      1.2.1 Counting   8
      1.2.2 Yuri's Question   10
   1.3 Common-Sense Ladders   11
      1.3.1 Smoothies   11
      1.3.2 That Which is Irreplaceable   12
      1.3.3 Creating all Patterns   12
   1.4 The Ladders You Love   16
      1.4.1 Three Vertical Lines   16
      1.4.2 Squared ladders   19
      1.4.3 Cubed ladders   21
      1.4.4 The Diagram   22
      1.4.5 Searching for Deeper Mysteries   25

2 Quadratic Equations and Sleeping Beauty   27
   2.1 Square Roots   27
      2.1.1 Yuri   27
      2.1.2 Negative × Negative   28
      2.1.3 The Complex Plane   29
   2.2 The Quadratic Formula   31

      2.2.1   Quadratic Equations   31
      2.2.2   Equations and Polynomials   33
      2.2.3   Deriving the Quadratic Formula   34
      2.2.4   Getting Your Feelings Across   37
  2.3   Solutions and Coefficients   37
      2.3.1   Tetra   37
      2.3.2   Solutions and Coefficients   38
      2.3.3   Getting your Head Straight   41
  2.4   Symmetric Polynomials and the Field Perspective   42
      2.4.1   Miruka   42
      2.4.2   Relations Between Roots and Coefficients, Revisited   43
      2.4.3   Solutions to Equations, Revisited   48
      2.4.4   On the Way Home   54

3   Searching for Form   59
  3.1   The Form of Equilateral Triangles   59
      3.1.1   At the Hospital   59
      3.1.2   Fever, Revisited   67
      3.1.3   Awakening from the Dream   68
  3.2   Symmetric Groups   69
      3.2.1   In the Library   69
      3.2.2   Group Axioms   70
      3.2.3   Axioms and Definitions   78
  3.3   Cyclic Groups   80
      3.3.1   Taking a Break   80
      3.3.2   Structure   81
      3.3.3   Subgroups   82
      3.3.4   Order   85
      3.3.5   Cyclic Groups   86
      3.3.6   Abelian Groups   90

4   In a Yoke with You   95
  4.1   In the Library   95
      4.1.1   Tetra   95
      4.1.2   Factorization   96

- 4.1.3 Scope of Numbers 97
- 4.1.4 Polynomial Division 99
- 4.1.5 The Twelfth Root of 1 101
- 4.1.6 Regular n-gons 103
- 4.1.7 Trigonometric Functions 104
- 4.1.8 Onward 107
- 4.2 Cyclic Groups 108
  - 4.2.1 Miruka 108
  - 4.2.2 Twelve Complex Numbers 109
  - 4.2.3 Making a Table 111
  - 4.2.4 Polygons with Shared Vertices 112
  - 4.2.5 Primitive 12th Roots of Unity 114
  - 4.2.6 Cyclotomic Polynomials 116
  - 4.2.7 Cyclotomic Equations 121
  - 4.2.8 The Yoke's on You 124
  - 4.2.9 Cyclic Groups and Generating Elements 125
- 4.3 Mock Exams 127
  - 4.3.1 At the Testing Site 127

5 Trisected Angles 131
- 5.1 The World of Graphs 131
  - 5.1.1 Yuri 131
  - 5.1.2 Trisecting Angles 132
  - 5.1.3 Misunderstandings about Angle Trisection 136
  - 5.1.4 Straightedges and Compasses 137
  - 5.1.5 Constructibility 138
- 5.2 The World of Numbers 139
  - 5.2.1 A Concrete Example 139
  - 5.2.2 Arithmetic by Construction 143
  - 5.2.3 Extracting Square Roots by Construction 146
- 5.3 The World of Trigonometric Functions 149
  - 5.3.1 At the Narabikura Library 149
  - 5.3.2 Lisa 150
  - 5.3.3 Parting Words 153

  5.4 The World of Equations 154
    5.4.1 Seeing Structures 154
    5.4.2 Warming Up with Rational Numbers 158
    5.4.3 One Step at a Time 161
    5.4.4 The Next Step? 162
    5.4.5 Discovery? 164
    5.4.6 Predictions and Theorems 166
    5.4.7 Where to Next 168

6 Pillars of the Heavens 173
  6.1 Dimensions 173
    6.1.1 Neighborhood Festival 173
    6.1.2 Fourth Dimension 174
    6.1.3 Octopus Balls 176
    6.1.4 Supports 177
  6.2 Linear Space 180
    6.2.1 In the Library 180
    6.2.2 Coordinate Planes 181
    6.2.3 Vector Spaces 183
    6.2.4 $\mathbb{C}$ as a Vector Space over $\mathbb{R}$ 186
    6.2.5 $\mathbb{Q}(\sqrt{2})$ as a Vector Space over $\mathbb{Q}$ 188
    6.2.6 Expanse 192
  6.3 Linear Independence 196
    6.3.1 The Basics 196
    6.3.2 Dimensional Invariance 199
    6.3.3 Degree of Extension 200

7 The Secrets of Lagrange Resolvents 203
  7.1 The Cubic Formula 203
    7.1.1 Tetra 203
    7.1.2 Red Card: The Tschirnhaus Transformation 204
    7.1.3 Orange Card: Relations Between Roots and Coefficients 207
    7.1.4 Yellow Card: Lagrange Resolvents 208
    7.1.5 The Green Card: Sums of Cubes 212

- 7.1.6 The Blue Card: Products of Cubes 216
- 7.1.7 The Indigo Card: From Coefficients to Roots 218
- 7.1.8 The Violet Card: The Cubic Formula 221
- 7.1.9 A Map of Our Journey 223
- 7.2 Lagrange Resolvents 227
  - 7.2.1 Miruka 227
  - 7.2.2 Properties of Lagrange Resolvents 231
  - 7.2.3 Applications 235
- 7.3 The Quadratic Equation 236
  - 7.3.1 Lagrange Resolvents for Quadratic Equations 236
  - 7.3.2 Discriminants 238
- 7.4 Roots of Fifth-degree Polynomials 240
  - 7.4.1 A Quintic Formula? 240
  - 7.4.2 The Significance of "5" 241

8 Building Towers 243

- 8.1 Music 243
  - 8.1.1 One Note 243
  - 8.1.2 One Encounter 245
- 8.2 A Lecture 246
  - 8.2.1 In the Library 246
  - 8.2.2 Degrees of Field Extensions 246
  - 8.2.3 Extension Fields and Subfields 247
  - 8.2.4 $\mathbb{Q}(\sqrt{2})/\mathbb{Q}$ 248
  - 8.2.5 A Problem 250
  - 8.2.6 $\mathbb{Q}(\sqrt{2}, \sqrt{3})/\mathbb{Q}$ 251
  - 8.2.7 Products of Degrees of Extension 253
  - 8.2.8 $\mathbb{Q}(\sqrt{2}+\sqrt{3})/\mathbb{Q}$ 256
  - 8.2.9 Minimal Polynomials 258
  - 8.2.10 A Discovery? 262
- 8.3 A Letter 267
  - 8.3.1 Going Home 267
  - 8.3.2 At Home 268
  - 8.3.3 The Letter 268

- 8.3.4 Constructible Numbers 268
- 8.3.5 Dinner 270
- 8.3.6 Solvability of Equations 271
- 8.3.7 Splitting Fields 272
- 8.3.8 Normal Extensions 273
- 8.3.9 Tackling Truth 276

9 The Form of Feelings 279
- 9.1 The Symmetric Group S_3 279
  - 9.1.1 At the Narabikura Library 279
  - 9.1.2 Classifications 284
  - 9.1.3 Cosets 286
  - 9.1.4 Clean Forms 289
  - 9.1.5 Creating Groups 291
- 9.2 Notational Form 297
  - 9.2.1 At Oxygen 297
  - 9.2.2 Swapping Notation 298
  - 9.2.3 Lagrange's Theorem 300
  - 9.2.4 Notation for Normal Subgroups 305
- 9.3 The Form of Parts 305
  - 9.3.1 $\sqrt[3]{2}$, All Alone 305
  - 9.3.2 Pursuing Form 306
  - 9.3.3 "Proper" Decompositions 307
  - 9.3.4 Further Dividing C_3 308
  - 9.3.5 Equivalence with Division 310
- 9.4 The Form of Symmetric Group S_4 314
  - 9.4.1 In Beryllium 314
- 9.5 The Form of Feelings 317
  - 9.5.1 Iodine 317
  - 9.5.2 Lights Out 317

10 Galois Theory 319
- 10.1 The Galois Festival 319
  - 10.1.1 A Chronology 319
  - 10.1.2 Galois's First Paper 322
- 10.2 Definitions 325
  - 10.2.1 Definition of Reducibility and Irreducibility 325

- 10.2.2 Definition of Substitution Groups 327
- 10.2.3 Two Worlds 329
- 10.3 Lemmas 330
  - 10.3.1 Lemma 1: Properties of Irreducible Polynomials 330
  - 10.3.2 Lemma 2: Creating V from Roots 333
  - 10.3.3 Lemma 3: Representing Roots in terms of V 335
  - 10.3.4 Lemma 4: Conjugates of V 337
- 10.4 Theorems 341
  - 10.4.1 Theorem 1: Definition of the Galois group of a Polynomial 341
  - 10.4.2 The Galois Group for $x^2 - 3x + 2 = 0$ 342
  - 10.4.3 The Galois Group for $ax^2 + bx + c = 0$ 345
  - 10.4.4 Creating Galois Groups 349
  - 10.4.5 The Galois Group for $x^3 - 2x = 0$ 352
  - 10.4.6 Theorem 2: Reduction of the Galois group of a Polynomial 355
  - 10.4.7 Galois's Error 359
  - 10.4.8 Theorem 3 (Adjoining all Roots of an Auxiliary Equation) 361
  - 10.4.9 Repeated Reductions 363
  - 10.4.10 Theorem 4: Properties of Reduced Galois Groups 364
- 10.5 Theorem 5: Necessary and Sufficient Conditions for an Equation to be Algebraically Solved 365
  - 10.5.1 Galois's Question 365
  - 10.5.2 What Does it Mean to Algebraically Solve an Equation? 368
  - 10.5.3 Tetra's Question 369
  - 10.5.4 Adjoining pth Roots 370
  - 10.5.5 Galois's Adjunction Elements 374
  - 10.5.6 Yuri's Response 378
- 10.6 Two Towers 378
  - 10.6.1 The General Cubic Equation 378

        10.6.2   The General Quartic Equation   380
        10.6.3   General Quadratic Equations   385
        10.6.4   No Formula for the Roots of Quintic Equations   386
   10.7   Summer's End   387
        10.7.1   The Fundamental Theorem of Galois Theory   387
        10.7.2   Browsing the Exhibits   391
        10.7.3   Oxygen at Night   392
        10.7.4   Irreplaceable   393

Epilogue   395

Afterword   401

Recommended Reading   405

Index   417

To my readers

This book contains math problems covering a wide range of difficulty. Some will be approachable by elementary school students, while others will challenge even college students.

The characters often use words and diagrams to express their thoughts, but in some places equations tell the tale. If you find yourself faced with math you don't understand, feel free to just browse through it before continuing with the story. Tetra and Yuri will be there to keep you company.

If you have some skill at mathematics, then please follow not only the story, but also the math. Doing so is the best way to fully engage in the tale that is being told.

—Hiroshi Yuki

Prologue

> I love the nights in summertime.
> Moonlit nights are wonderful, of course,
> but even the darkest nights are
> adorned with flittering fireflies.
>
> Sei Shonagon
The Pillow Book

There are nights I cannot forget. Nights of stars, nights of storms. Nights with many friends, nights with one, nights alone. So many nights.

Let me speak of her. I connected with mathematics, and with her through mathematics. Through her, I connected with myself.

What is so important I could never replace it?

Who is so important as to never be replaced?

There are things I will never relinquish, no matter the price. Things I would never trade.

But what?

Her words, her form, her smile. That moment that had held me up throughout my life, that moment with form and scale that cannot be expressed in words.

Let me speak of him. He encountered math by being left behind. He grew, but twice he failed. He encountered mathematics, and several years later solved the hardest of problems, producing a new mathematics.

And yet...

> He was blessed with genius, but not with fortune.
>
> He was blessed with teachers, but not by the age
> in which he lived.

He is no longer with us, having fallen while still young. His life burned bright. The day before his duel, he wrote a letter. His words, written so hastily, passed on a message that became a mathematics for a new age.

That thing which for him there was no replacement was a thing with no replacement in mathematics.

Let me speak of myself. I am here. I am here, now. My past is now gone, my future has yet to arrive. It is thus today that I must live.

> As he lived his life, I will live mine, no matter the
> differences in our time, in our abilities.

I will use my life to pass on that irreplaceable thing.

The greatest things begin from tiny seeds. Like problems posed by my annoying cousin—

CHAPTER 1

Ladder Diagrams

> True, sometimes giving names to things can help by leading us to focus on some mystery. It's harmful, though, when naming leads the mind to think that names alone bring meaning close.
>
> MARVIN MINSKY
> *The Society of Mind*

1.1 LOTS OF LADDERS

1.1.1 Exchanging Ends

"Hey, cuz! Think you can set up a ladder diagram[1] that works like this?" Yuri said, slapping down her notebook to an open page.

I looked at what she had drawn. "Let's see..."

---

[1] *Amidakuji* (lit., "the Buddha's lots"), a method for "randomly" mapping the elements of one set to those of another, frequently used in Japan to, e.g., assign tasks or roles among group members. To create a ladder diagram, write the elements of one set in an upper row and elements of the other set in a lower row, then use vertical lines to connect elements in the two sets. Next, randomly add some number of short horizontal lines ("rungs") connecting pairs of adjacent vertical lines, ensuring that no two adjacent rungs touch. Finally, start tracing down the vertical lines from elements in the upper row, crossing to adjacent vertical lines each time a horizontal rung is encountered, until you arrive at one of the elements in the lower row. When this is performed for all upper-row elements, each should be uniquely mapped to a lower-row element. See the Wikipedia entry for "Ghost Leg" (the Chinese name) for a more detailed description.

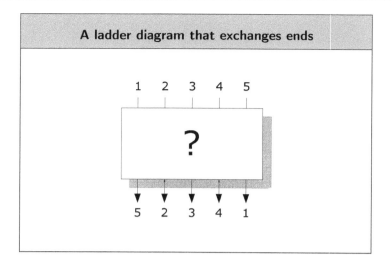

"You have to fill in lines so that the top gets arranged like the bottom," she said.

"Right. So you want the 1 to move all the way to the right, the 5 to move all the way to the left, and you want the 2, 3, and 4 to come straight down. So we're just swapping the ends, in other words."

"Yep! You do know how ladder diagrams work, right?"

"C'mon, Yuri. I'm almost a high school graduate. How could I not know?"

"Just checking."

Yuri is my cousin, a third-year junior high school student, now preparing for her high school entrance exams. She lived in my neighborhood and most weekends came to hang out in my room, where we would talk, do math puzzles, read books, do our homework, whatever. We were practically siblings, and best of friends.

It was nearly summer vacation, and the last Saturday before finals. Yuri had spread her notebook on my desk, where I had been studying. She was wearing jeans and a T-shirt. Her hair was done in pigtails instead of her usual ponytail, making her look younger than she was.

"What's with the braids?" I asked.

She twirled one in response. "It's a retro twentieth-century thing. Twintails!"

"What's a twintail?"
"Nothing you'd be interested in. Focus on the problem instead!"
"Doesn't require much focus, honestly. Here."

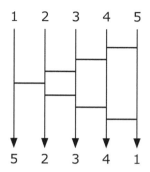

A ladder diagram that swaps ends.

"Well that was too fast to be any fun," Yuri said.

"Think of it like this. You can draw four rungs to make something like steps going down and to the left to move the 5 to the leftmost position, then three more going down and to the right to move the 1. Then you've swapped the 1 and the 5, without changing the others."

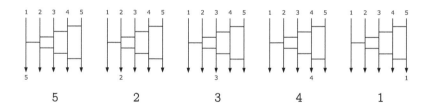

"Okay, that works. It's not the answer I came up with, though." Yuri turned the page to show me what she had done.

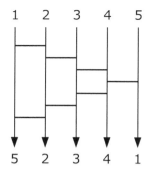

Another ladder diagram that swaps ends.

"Interesting," I said.

"It works, right?" She turned the page again. "Here's another solution."

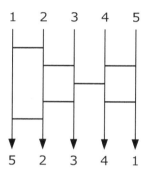

Yet another ladder diagram that swaps ends.

1.2 More Ladders

1.2.1 Counting

"My turn," I said.

> **(Counting ladder diagrams)**
>
> How many ladder diagrams can be created using five vertical lines?

"I don't get it," Yuri said.

"Counting is fundamental for a math lover. We're talking about ladder diagrams, so it's only natural to wonder how many there are."

"But you can make as many as you want! I mean, all you have to do is add more rungs. You could add thousands, tens of thousands!"

I chuckled. "Sure, you can always add more lines, but if you just end up with the same result as some other configuration we don't count it. Take a look at these from before, for example."

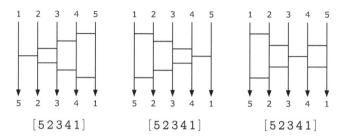

"All three have lines drawn in different ways, but we consider them to be the same since each one just shuffles $1, 2, 3, 4, 5$ into $5, 2, 3, 4, 1$. We can give all ladder diagrams like this the same name."

[5 2 3 4 1]

"Clever name," Yuri snarked.

"Anyway, we'll consider all ladders that produce the same pattern as one type. So it isn't the way we hang rungs that's important, it's the pattern they produce."

"Yeah, I get it. I'm sure this will be a huge help on your entrance exams, by the way."

"Don't remind me that I should be studying."

As a third-year senior high school student, college entrance exams were rapidly approaching, and summer vacation was peak study time for kids my age. I loved math, so I was okay there, but over the summer I really needed to focus on other subjects. It was really annoying that entrance exams would get in the way of what I really wanted to study, but I figured there wasn't much I could do about it until exams were done. Still... annoying.

Summer vacation would start as soon as my spring semester finals were done. To make sure I didn't fall behind in my studies, I was planning to enroll in a summer course at a prep school, one designed for current high school students. I would still be able to go to my school library to work on problems after my daily classes. But the prep school also conducted mock exams, so it would be a busy summer.

Yuri interrupted my thoughts. "So isn't counting ladders just another 'number of cases' problem, then?"

"Well, yeah."

"Then this is easy! The answer is 120."

"That's right, there are 120 different ladders you can make using five vertical lines. Well done! Can you explain how you got that?"

"No prob. We just have to think about how the numbers are arranged after the ladders are done. One of the numbers 1 through 5 will fall down into the far left position. For each of those five possibilities, there are four other numbers that could fall into the second position, any number except for the one that's in the first position. And so on—three possibilities for the third position, two for the fourth, and only one for the far right position. Multiply those together and you get five factorial."

$$5! = 5 \times 4 \times 3 \times 2 \times 1 = 120$$

"Exactly. You even said 'for each possibility,' so bonus points."

"I'll take that. Not sure how much I deserve it, though. You've taught me all this before[2]. You called it permutations, I think?"

"I did indeed."

1.2.2 Yuri's Question

"Hang on a second, though," Yuri said. Sunlight gleamed off of a chestnut braid. "I'm not convinced there's definitely 120 solutions."

"Why not?" I asked. "We just talked about how there should be 5! cases if you have five vertical lines, and that's 120."

"Yeah, but how can we be sure a ladder diagram can make *all* possible permutations? What if there's a pattern that it can't make?"

---

[2]See *Math Girls 4: Randomized Algorithms*, Chapter 3.

I slowly nodded. *Well played, Yuri.* She had always been good at finding overlooked conditions and gaps in my logic.

"Okay, good point. You're right, that is something we need to pay attention to. Our answer only makes sense if all patterns are possible. One constraint is that we can only create rungs between two adjacent vertical lines. We need to make sure we can still make 120 patterns even under that constraint."

"Right? I mean, I think we can, but still. I just want to be sure. Doesn't feel right, otherwise."

"Actually, I don't think it will be all that hard to set your mind at ease."

"Seriously? I sure don't see how."

"You will. Let's think through it together."

Yuri pulled out her metal-framed glasses. "You bet!"

My mother called from the kitchen. "Hey, kids! Anybody want a cold drink?"

"I do!" Yuri called back. She stood and tugged on my hand. "C'mon! Let's do this while we have some juice!"

1.3 COMMON-SENSE LADDERS

1.3.1 Smoothies

Yuri and I went to the living room, where my mother entered bearing colorful drinks with thick straws.

"What's this?" I asked.

"Smoothies!" she said. "You just put frozen fruit and yogurt and ice in a blender and mix away. I added bananas, blueberries, raspberries, and strawberries."

"Cold and delicious!" Yuri chirped.

"Always such a dear child," Mother said.

I took a sip of the sweet concoction. "Wow, this actually is good."

"Did you decide where you want to go to school?" my mother asked Yuri.

"Same school he's at," she said, pointing at me.

"Is that so? Well you're an excellent student, so I'm sure you'll get in."

"I wish I was so sure..."

"You'd better teach her what she needs to know," Mom said to me.

"Doing my best," I said.

"I certainly hope so," she said, heading back into the kitchen.

1.3.2   That Which is Irreplaceable

"Speaking of which, where's your boyfriend going to school?" I asked, referring to a boy from a different school she'd recently met.

"Somewhere else," she said.

"Yeah? You okay with that?"

"Doesn't have anything to do with me. And he's not my boyfriend, by the way."

"Are you still trading math problems with each other?" I knew they had a couple of shared notebooks they passed back and forth, in which they posed problems, discussed solutions, compared answers, and teased each other. That's what Yuri had told me, at least.

"I guess." She was using her straw to poke at stray bits of ice at the bottom of her glass. "Tell me something, is there anything you consider irreplaceable?"

"Time, I guess. Time is precious." I had been thinking about that recently. I resented the time that passed while I was studying for college entrance exams, knowing there was no way to ever get that time back. No method, nothing I could exchange... "Time is the very definition of irreplaceable. It's so precious there's nothing that can replace it."

Yuri was thinking in silence, a serious expression clouding her face.

1.3.3   Creating all Patterns

"Oh, right, we were working on a problem," Yuri said. "Let's finish."

"Back to it, then," I said. "First off, let's be sure we know exactly the problem we want to solve."

Patterns ladder diagrams can create

Can all 120 possible patterns be created using a ladder diagram with five vertical lines?

"That doesn't really help me much," she said.
"I'm just asking if we can devise some rung pattern that creates any given number pattern."
"Ah. Well I think we can do that. We could use, say, a pattern that changes $1, 2, 3, 4, 5$ to $3, 5, 1, 4, 2$. But still, all 120 patterns? Seems like it would be a pain to test every one."
"It would be. Sounds like the approach Tetra might take, though."
"Tetra I am not."
"Nor I. What we really want is some way to say that we can make any kind of ladders without having to actually do so. In other words, we want to think about *how* to make them. Whatcha think?"
"You already know the answer, don't you."
"I think so. But your explanation gave me a big hint, when you said one of the five numbers $1, 2, 3, 4, 5$ will fall down to the leftmost position."
"That was a hint?"
"If we can actually do it. Think you can show how to move each number there?"
"Move each number 1 through 5 to the leftmost position? Sure, that's easy."

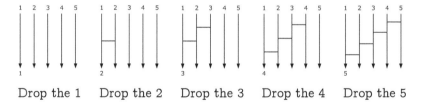

Drop the 1   Drop the 2   Drop the 3   Drop the 4   Drop the 5

Dropping each number to the leftmost position.

"That works. You can just make staircases that go down to the left."
"Yep. What's next?"
"Just more of the same, actually. Next is to find some way of moving any number to down to the second-to-left position without disturbing the leftmost number. Think you can do that?"

"Sure, just make another down–left staircase, right?"
"Sounds good. Let's try it for [3 5 1 4 2]."
I picked up my mechanical pencil and sketched out a diagram.

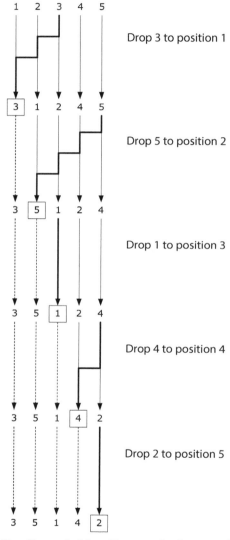

Creating a ladder diagram for [3 5 1 4 2].

"You did it!"
"Yeah, if you look closely you can see how—by chaining three ladders together. I connected [3 1 2 4 5] and [1 5 2 3 4] and [1 2 3 5 4] to make [3 5 1 4 2]."

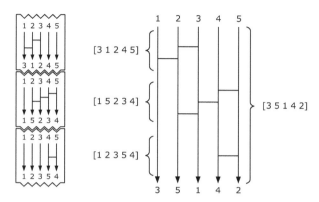

Creating a ladder diagram for [3 5 1 4 2].

Yuri cocked her head. "How's that? Oh, I see. You connected three down–left staircases, right?"
"Right."
"So you didn't connect three of them, you connected five."
"Huh?"
"Well look at what you drew. There's five of them!"

[3 1 2 4 5], [1 5 2 3 4], [1 2 3 4 5], [1 2 3 5 4], [1 2 3 4 5]

"Oh, you're counting the ones that don't change anything, which would be [1 2 3 4 5]."
"Of course. Precision is vital to mathematics, Dr. Watson." Yuri puffed on an imaginary pipe.
"It is indeed, Ms. Holmes. Even so, nothing changes, no matter how many [1 2 3 4 5]'s you add."
"Yeah, everything just plops straight down."
" 'Plops' ?"
"Yup. So let's call a [1 2 3 4 5] a 'plopper'!"

"Uh, sure. Anyway, with this we've shown that we can create any pattern, just by swapping adjacent numbers. So now—"

A flash of insight left me frozen, though I could feel my pulse quickening with excitement.

"You okay?" Yuri asked.

"Yuri! This is the reverse of a bubble sort!"

"That sounds familiar..."

"You know, that sorting algorithm Tetra gave a presentation on[3]! It compares adjacent numbers, and swaps them if they're out of order. It's an algorithm that takes a given pattern and puts it in order. We're doing exactly the opposite—taking a sorted array of numbers and swapping adjacent pairs to put them in a given pattern."

Bubble sort: Takes a given arrangement of numbers and puts them in order

Ladder diagram: Takes an ordered arrangement of numbers and puts them in a given arrangement.

"Ah, right. I remember now," Yuri said. Clearly my enthusiasm wasn't infectious.

"Anyway, I'm starting to think that the ladder diagram might be a lot deeper than we first thought. I'd like to play with them some more, but 120 patterns from just five lines? Wow..."

"So simplify."

"Good idea. Let's see what we can do with just three lines."

1.4 THE LADDERS YOU LOVE

1.4.1 Three Vertical Lines

"So can you whip up some three-line ladders?" I asked.

"Sure!" Yuri said. "There's only six possibilities, you know."

---

[3]See *Math Girls 4: Randomized Algorithms*, Chapter 6.

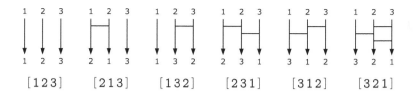

Three-line ladder diagrams.

I nodded. "That's right. $3! = 3 \times 2 \times 1 = 6$."

"Kinda neat to line them up like this. It looks like we can make groups of friends, depending on how many rungs they have."

"Friends?"

"Like you said, counting is fundamental for a math lover! Check it out."

· 0 rungs: [1 2 3]

· 1 rung: [2 1 3] and [1 3 2]

· 2 rungs: [2 3 1] and [3 1 2]

· 3 rungs: [3 2 1]

"Doesn't it look like something's hiding in that?" Yuri asked. "Any ideas, Dr. Watson?"

"Enough with the detective roleplay. More importantly, rather than think of [3 2 1] as having three rungs, I think we should make it friends with [2 1 3] and [1 3 2]."

"You want to stick the three-rung ladders in with the one-rung ladders? Why?"

"Well, the number of rungs is one way to look at things, but I think there's a deeper structure to perceive."

· [2 1 3] is an exchange of 1 and 2

· [1 3 2] is an exchange of 2 and 3

But also...

· [3 2 1] is an exchange of 3 and 1

"See? Each of these is a single exchange of two elements, so maybe they belong together?"

"Sure, I'll buy that. They're swappers."

"Another of your new math terms?"

Yuri nodded. "When you just exchange two elements, that's a swapper."

"So what are [2 3 1] and [3 1 2]?"

"Hmm... Those would be scramblers. See how all three numbers get scrambled into new positions?"

I rolled my eyes, but couldn't hold back a smile.

"So there are three kinds of three-line ladder diagrams," she said, making a list.

- [1 2 3] is a plopper; it just drops numbers straight down.

- [2 1 3], [1 3 2], [3 2 1] are swappers; they exchange two elements.

- [2 3 1], [3 1 2] are scramblers; every element moves to a different position.

"Ploppers, swappers, and scramblers, huh? Fair enough. Now that we have all this, we don't really need all the lines in a ladder diagram diagram—they just get in the way. Instead, we can just use arrows to show where each element ends up. Like this."

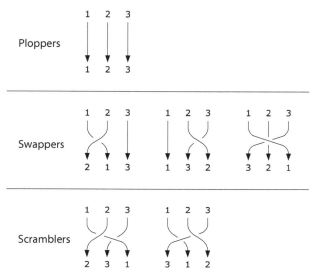

Categories for three-line ladder diagrams.

1.4.2 Squared ladders

Yuri's interesting naming gave me a new idea.

"Hey, let's see what happens if we take two ladders that just exchange two elements—"

"Swappers," she said.

"—swappers, right—and join them together. When we join two of the same ladders, let's call that 'squaring.'"

"Squaring, like when we multiply two of the same number?"

"Exactly."

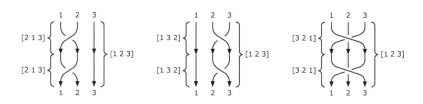

· Joining two [2 1 3]'s gives [1 2 3].

- Joining two [1 3 2]'s gives [1 2 3].

- Joining two [3 2 1]'s gives [1 2 3].

"Check it out. Joining two of the same swapper gives you a plopper!"

"That's so totally amazing! *Not.* If you swap two elements, then swap them back, of course you're going to get what you started with."

"Well, yeah. Not a surprising result, I guess. But still, it's kinda cool, if only because it lets us write things like an equation. Since we're calling joining two of the same ladders 'squaring,' we can write that like this, using an equals sign to show same patterns."

$$[2\,1\,3]^2 = [1\,2\,3]$$
$$[1\,3\,2]^2 = [1\,2\,3]$$
$$[3\,2\,1]^2 = [1\,2\,3]$$

"Neat," Yuri said.
"Wow, this feels so much better."
"Because you're an equation freak."
"I wonder what happens when we square other kinds of ladders?"
"You'll find out soon!" Yuri said, writing in the notebook. She soon held up the page. "Interesting! When we square a scrambler, we get a different scrambler!"

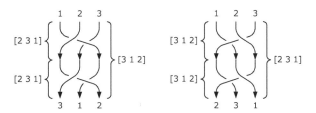

A squared scrambler is another scrambler.

"We sure do. A squared [2 3 1] gives [3 1 2], for example, another scrambler."

"Right! And when we square [3 1 2], we get [2 3 1]."

$$[2\,3\,1]^2 = [3\,1\,2]$$
$$[3\,1\,2]^2 = [2\,3\,1]$$

"So a squared swapper is a plopper, and a squared scrambler is a different scrambler. Hmm..." Yuri played with one of her braids as she thought. "Ooh! Ooh! So cool!"

"What is?"

"What happens when you *cube* a scrambler!"

1.4.3 Cubed ladders

"A cubed scrambler is a plopper!" Yuri yelled.

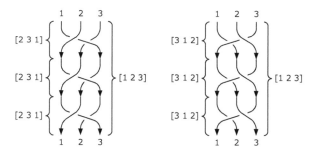

A cubed scrambler is a plopper.

$$[2\,3\,1]^3 = [1\,2\,3]$$
$$[3\,1\,2]^3 = [1\,2\,3]$$

"Huh, it sure is. So there's one type we can square to get a plopper, and one we can cube to get a plopper. Actually, I guess we could also say a plopper to the first power is a plopper."

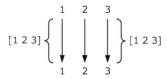

A plopper to the first power is a plopper.

$$[1\,2\,3]^1 = [1\,2\,3]$$

"Hang on," Yuri said. "There's something I want to try."

She began writing a large graph in the notebook. I tried to get a peek, but she blocked my view and told me to wait until it was done. I waited for a good while, but she remained out in math world. Knowing quite well how important silence could be for focused concentration, I took the opportunity to quietly clear away our empty glasses, and opened up my world history textbook to get in a little studying while I waited.

1.4.4 The Diagram

"Done!" Yuri finally said after nearly an hour.

I flipped my textbook over. "Welcome back."

"Look what I did! I got the ploppers and the swappers and the scramblers all in one graph! Now I'm satisfied." With a flourish, she showed me what she'd been writing.

Ladder Diagrams 23

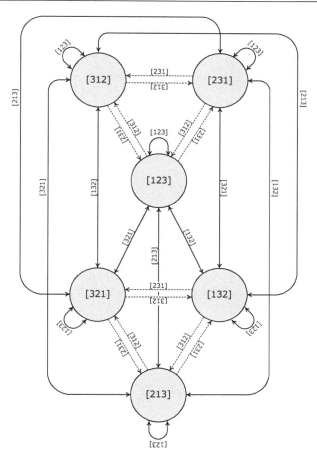

Three-lined ladder diagram.

"What on earth..."

"See? This is a swapper, and these two are scramblers. Oh, this is so exciting. Hey, I know! I should use this for my summer research project!"

"Maybe you can tell me what it is first?"

"What, you don't get it? Okay, let me explain. Each of these circles is a ladder diagram. We were using chained ladders to make new ladders, right? That's what the lines show. Like, if you join [3 1 2] with [3 2 1] it becomes [2 1 3], while if you go the other way, joining [2 1 3] with [3 2 1], you get [3 1 2]. It's a round trip starting from [3 2 1]. I'm just using one line there to cut down on the clutter."

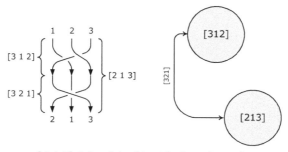

[3 1 2] joined to [3 2 1] gives [2 1 3].

Yuri continued. "Then if you join [3 2 1] with [2 3 1], that's a scrambler, you get [2 1 3], right? That's the dashed line, here. Solid lines for swappers, dashed lines for scramblers."

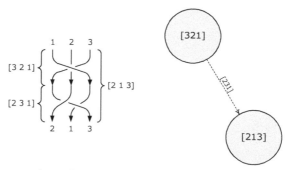

[3 2 1] joined to [2 3 1] gives [2 1 3].

"And look! Remember we talked about how joining three scramblers in a row brings us back to where we started? You can follow the [2 3 1] scrambler to make a triangle!"

Ladder Diagrams

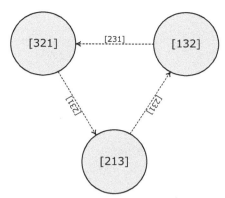

A triangle formed by the scrambler [2 3 1].

"Isn't this cool?"

1.4.5 Searching for Deeper Mysteries

I was honestly impressed with Yuri's graph.

"This is excellent work, Yuri. Truly excellent."

"I'd say that drawing diagrams is also fundamental for a math lover," she teased.

"I know what Tetra would say if she were here: 'Now I feel like I really get it.'"

"Enough about Tetra."

"Wow, there's so much we can read from this diagram. For example—"

"Ooh!" Yuri held up a hand to stop me. "What about this! Every three-lined ladder can be transformed into a plopper, either by squaring it, cubing it, or just leaving it alone, right? I wonder if we could do the same thing with five-line ladders, by taking them up to the fifth power?"

"Why don't you try and see?"

"Because there's 120 of them."

"How about four-line ladders, then? There's only 4! = 24 of those."

"Still too much work."

"I'll bet Tetra would do it."

Yuri made a face. "Why do you keep going on about Tetra? Sheesh!"

"Hence we can go backward and obtain any desired permutation, by starting with all elements in order and then exchanging appropriate pairs of adjacent elements."

DONALD KNUTH
The Art of Computer Programming,
Vol. 4, Fascicle 2

CHAPTER 2

Quadratic Equations and Sleeping Beauty

> The Princess shall indeed pierce her hand with a spindle; but instead of dying, she shall only fall into a profound sleep, which shall last a hundred years; at the expiration of which a king's son shall come and waken her.
>
> CHARLES PERRAULT,
> TRANS. BY A.E. JOHNSON
> *The Sleeping Beauty in the Wood*

2.1 SQUARE ROOTS

2.1.1 Yuri

"Welcome home!" Yuri shouted.

"Yuri? What are you doing here?" For some reason Yuri had come to my front door to welcome me home from school. Wearing an apron, no less. "What's going on?"

It was evening on a Thursday, the last day of my final exams. There was an end-of-semester assembly scheduled for the following day, then summer vacation would officially begin.

Yuri smiled. "My parents had to go out of town until tomorrow, so they left their darling daughter in the care of her favorite aunt, who I am currently helping with dinner."

My mother entered the hallway. "I've got it from here, Yuri. Thanks for your help. We'll have dinner when your uncle gets home. You two go play until then."

"Will do!" Yuri said, then turned to me. "I have a question."

2.1.2 Negative × Negative

Yuri followed me to my room, still wearing her apron.

"So when you square i, you get -1, right?" she said.

"Sure, i is called the imaginary unit, and $i^2 = -1$ is pretty much its definition."

"Okay, and if you square -3 you get 9, right?"

"Of course. Because $(-3)^2 = (-3) \times (-3) = 9$."

"And when you multiply two positives you get a positive, but you also get a positive when you multiply two negatives. In other words, you never get a negative by squaring a number. So how is it that i squared is negative? I've been meaning to ask you this for ages."

I nodded slowly as I considered how to answer.

"Well, everything you've said is correct. You get a positive result both when you multiply two positive numbers and when you multiply two negatives, so the square of that kind of number will always be positive. You're good so far. But that only holds for real numbers."

"Oh yeah?"

"Yeah. So technically, it isn't correct to say that the square of a number is always positive. Instead, you should say the square of a *real* number is always positive. Or zero."

"And I suppose i isn't a real number?"

"Exactly. It's explicitly defined outside of the real numbers, as a new kind of number whose square is negative."

"One of those 'let's assume that such-and-such exists' things you math people like to pull?"

"Yes, which is fine—the existence of such a number wouldn't cause contradictions anywhere else. It lets us create a new class of numbers, which we call the complex numbers."

"Ah, I've heard that before."

"Complex numbers are formed by combining the imaginary unit with the reals. We can still perform the standard arithmetic operations on them: addition, subtraction, multiplication, and division. But the fact that they include this new number i expands the world of numbers."

"Expands how?"

"Well, you can add $i+i$ to get $2i$, for example. You can also add a real number to that, like $3+2i$. You can also perform multiplication and division, all the usual stuff. Doing so gives you lots of new numbers you couldn't have created otherwise, all complex numbers."

"Hmm."

"When we want to treat all the real numbers as a set, we usually name it $\mathbb{R}$, and we usually call the set of complex numbers $\mathbb{C}$. It's actually kinda cool—you have this vast ocean called $\mathbb{R}$, into which you add just a tiny drop called i, and poof, you get a new ocean named $\mathbb{C}$. When you mix in the arithmetic operations, you get a far larger ocean."

"Just a drop of i, huh?" She looked very serious, despite the apron. She was plenty fun when she was joking or pouting, but I liked this side of her too.

"Tell you what. I think you'll get a much better handle on all this if we talk about the complex plane."

2.1.3 The Complex Plane

"You know about the number line, right?" I said.

Yuri rolled her eyes. "Since, like, first grade."

"Then you know that every real number is a single point on the number line. We designate one point as 0, and call every real number to the left of 0 negative, which we indicate with a minus sign, and everything to the right positive. Let's call this the real axis."

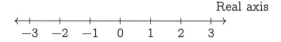

"Now we want to put imaginary numbers on the vertical axis, which we'll call the imaginary axis."

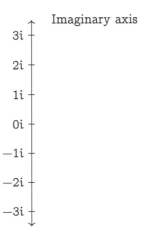

"When we combine the two, we get what's called the complex plane."

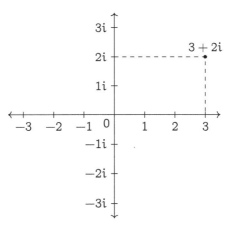

"Just like we can represent any real number as a point on the real number line, we can plot any complex number as a point on the complex plane."

"Real numbers on the number line, complex numbers in the complex plane," Yuri said. "Got it."

"$\mathbb{C}$, the set of complex numbers, is a really wonderful set. You know what a quadratic equation is, right? And how sometimes they're unsolvable?"

"Unsolvable because they're hard, or because solving them is impossible?"

"Because it's impossible. Using only real numbers, at least. But things are different if we expand our horizons—in the complex plane, there's always a solution to any quadratic equation. Doesn't that feel great? So when you're doing math, you should always keep in mind whether you're dealing with just real numbers, or if you've expanded your scope to the complex numbers."

"Scope, huh."

"Hey kids!" Mom called from downstairs. "Time for dinner!"

2.2 The Quadratic Formula

2.2.1 Quadratic Equations

After dinner, my father retreated into his study. I took a bath, then headed to the living room to kick back for a while. Now that finals were over with, I felt like I could finally take a breather. But still... entrance exams.

Yuri appeared in her pajamas as I was flipping through English vocabulary flashcards. *Abandon... ancient... determined... individual...*

"Nothing like a hot bath," she said, toweling her hair.

"Want some tea?" my mother asked her.

"Yes, please!" She turned to me. "Okay, back to what we were talking about. Quadratic equations."

I opened my notebook on our dining room table and sat next to Yuri, who smelled of soap and shampoo.

"Okay, here's the basic form of a quadratic equation."

$$ax^2 + bx + c = 0 \qquad (a \neq 0)$$

"What's with the $a \neq 0$?" Yuri asked. "Is that really necessary?"

"It is. If a were 0, the x^2 term would disappear and it wouldn't be a quadratic equation, because it wouldn't have a squared term. It would be a linear equation instead."

"That's all?"

"That's all. You know that the a, b, c here are called 'coefficients,' right?"

"Yep, learned that."

"Great. Technically, we call $ax^2 + bx + c = 0$ 'a quadratic equation in x,' because x is the unknown variable. Unknown for now, at least—it's what we want to find to solve the equation. We can call x a 'variable' when we think in terms of functions, or an 'indeterminate' when we're thinking in terms of polynomials."

"Got it," Yuri said. She was listening with her chin propped on a hand.

"Okay. Solving a quadratic means finding a value, or values, of x that makes $ax^2 + bx + c = 0$ a true statement. If some number α satisfies a quadratic equation, we can call α a 'solution to that equation.' If α makes the polynomial equal 0, we can call alpha a 'root' of that polynomial. Good so far?"

"I think so. This α represents a number, right?"

"Right. If we're given a specific quadratic equation, a solution for it will be some specific number, like 3 or 7.5. Right now we're talking about quadratics in general, though, so we can't use concrete numbers for their solutions. That's why I'm using α instead."

Yuri nodded. "Okay, I'm good, then."

"And you're sure you're good with what it means for an equation to be satisfied, right? That if some number α satisfies the quadratic equation $ax^2 + bx + c = 0$, then this equation will be true."

$$a\alpha^2 + b\alpha + c = 0$$

"I'm sure. Pretty straightforward stuff."

"Okay. Quadratic equations have two solutions, so we'll often use two letters to represent them, α and β."

"And this β is just some other number, right?"

"Right, except that there can also be something called 'repeated roots,' where $\alpha = \beta$. In any case, this β makes the quadratic true, just like α does."

$$a\beta^2 + b\beta + c = 0$$

"But anyway, yes, all of these letters—coefficients a, b, c and solutions α, β just stand for numbers."

2.2.2 Equations and Polynomials

"Okay, pop quiz, then," I said. "How about $x^2 - 3x + 2$? Is this—"

"Yes, I can solve that," Yuri said.

"You're jumping the gun. I was going to say, is this an equation?"

"Oh. Well you aren't going to trick me that easily. No, it isn't."

"That's right. This isn't an equation, it's a polynomial. And I wasn't trying to trick you, I just want to be sure you're paying attention to the difference between the two. The polynomial $x^2 - 3x + 2$ only becomes an equation when you set it equal to something, often 0 in the case of a quadratic."

$x^2 - 3x + 2$ a *polynomial* in x

$x^2 - 3x + 2 = 0$ an *equation* in x

"All this is still old news, you know," Yuri said.

"Okay, okay. Try solving this equation, then."

$$x^2 - 3x + 2 = 0$$

Yuri sat silently, looking at me.

"What's wrong?"

She grinned. "Just waiting to make sure you're done giving the problem."

"I'm done."

"Okay, easy then. The answer is $x = 1, 2$."

"That was fast. How'd you solve it?"

"Factorization, of course."

$$x^2 - 3x + 2 = (x-1)(x-2)$$

"Then you just solve $(x-1)(x-2) = 0$, and there's the answer."

"Well done. So $x = 1$ and $x = 2$ are both solutions. This is a simple quadratic, so you can factor the polynomial on the left like you did. But sometimes the factorization doesn't come so easily. In that case, we use—"

"The quadratic formula."

"Exactly."

2.2.3 Deriving the Quadratic Formula

"Let's find the quadratic formula for ourselves. Think you can do it?"

"Nope," she answered without hesitation. "My math teacher did it on the board once, but wow, what a mess. I couldn't follow it at all. So I just memorized it instead—negative b plus or minus the square root of b squared minus four ac over 2a," she rattled off. "Er, I think that's right..."

I laughed. "Hard to be sure when you're just reciting from memory isn't it?"

"Very funny. It's that square root that always gets me."

"Yeah, you can't avoid square roots when you're dealing with quadratics, so dealing with them is key to deriving the quadratic formula. To derive it, just be sure to remember that your goal is to get things into this form."

$$<\text{Part including x}>^2 = <\text{Part not including x}>$$

Yuri sat up in her seat. "Okay, now you have my interest. So I want everything including x on the left and squared, and everything else on the right?"

"That's right. That's the form you want to be moving toward as you move stuff around. Once you get there you can take the square root of both sides, then the rest is just cleanup. Let's take a shot at doing that."

"I'm actually looking forward to it."

"Okay, here's our quadratic equation."

$$ax^2 + bx + c = 0 \qquad (a \neq 0)$$

"Good start," Yuri said.

"Keep everything with an x on the left, and move everything else to the right."

$$ax^2 + bx = -c \qquad \text{move c to the right}$$

"Easy enough."

Quadratic Equations and Sleeping Beauty

"Now if we multiply both sides by $4a$, the x^2 term becomes $4a^2x^2$, in other words $(2ax)^2$."

$$4a^2x^2 + 4abx = -4ac \qquad \text{multiply by } 4a$$
$$(2ax)^2 + 4abx = -4ac \qquad \text{because } 4a^2x^2 = (2ax)^2$$

"Yeah, but—"

"Then we can add b^2 to both sides, which leaves us just one step from our goal."

$$(2ax)^2 + 4abx + b^2 = b^2 - 4ac \qquad \text{add } b^2 \text{ to both sides}$$

"Hang on, hang on. Where are you getting these things from? Multiply by $4a$, add a b^2, all that?"

"Just look carefully at the left side," I said.

$$\underbrace{(2ax)^2 + 4abx + b^2} = b^2 - 4ac \qquad \text{one step from our goal}$$

"What am I looking for?"

"Remember our target form?"

"Sure, <Part including x>^{2} = <Part not including x>, right?"

"Right. So all I've been doing is getting everything including an x on the left in a form where it can be squared. And look! What's there now can be factored!"

"We can factor $(2ax)^2 + 4abx + b^2$?"

"Sure. It's in the form $A^2 + 2AB + B^2 = (A+B)^2$, right?"

"Now that you mention it... Oh, that's $(2ax + b)^2$, isn't it!"

"It sure is. So now we're ready for our final push."

$$\underbrace{(2ax)^2 + 4abx + b^2} = b^2 - 4ac \qquad \text{one step from our goal}$$
$$\underbrace{(2ax+b)^2} = b^2 - 4ac \qquad \text{factor the left side}$$
$$\underbrace{(2ax+b)^2}_{\text{<part including x>}^2} = \underbrace{b^2 - 4ac}_{\text{<part not including x>}}$$

"Hey! That was our goal!" Yuri said. "Let me find the square root!"

$(2ax + b)^2 = b^2 - 4ac$ our goal
$2ax + b = \pm\sqrt{b^2 - 4ac}$ take square root
$2ax = -b \pm \sqrt{b^2 - 4ac}$ move b to the right
$x = \dfrac{-b \pm \sqrt{b^2 - 4ac}}{2a}$ divide both sides by 2a (recall, $a \neq 0$)

"And there it is," I said.
"Sure is!"

The quadratic equation

The solutions (roots) of a quadratic equation $ax^2 + bx + c = 0$ can be obtained as

$$x = \frac{-b \pm \sqrt{b^2 - 4ac}}{2a}.$$

"Well done, Yuri."

"I guess, but still... No way I'd ever notice how to factor into a $(2ax + b)^2$ like that. I can already feel myself forgetting everything you did."

"That's normal. I doubt I ever would have stumbled across it myself. I had to learn it too. The key is to remember the $b^2 - 4ac$ that pops up. You know you want it to be on the right when you take the square root, so you want to multiply by $4a$ and add b^2 to get things in that form."

"So that's where those come from! Okay, remembering $b^2 - 4ac$ should be easier than memorizing the whole quadratic equation, I guess."

"Yeah, but you should probably memorize it anyway."

"Whaaa—? I thought the whole point was to learn how to derive it so I don't have to memorize."

"Actually, the point is to understand what you're doing so well that you can derive the equation yourself. But you'll use the quadratic equation so often that memorizing it can be a great time saver."

I suddenly sneezed.

"Ah! That means someone's talking about you!" Yuri said.

"I think it just means the air conditioner is set too low," I replied.

2.2.4 Getting Your Feelings Across

Late that night, I was still at the dining room table, flipping through my English vocabulary cards. *Permanent... significant... traditional...*

"Hey," Yuri said. She was lounging on the sofa.

"Hmm?"

"Why is it so hard to express your feelings to other people?"

"What's this all about?"

"Just wondering why some people can say mean things, even at important times. I mean, the less time you have, the more important it is, right? But that's when I find it hardest to say the right thing. I'm such an idiot..."

"Yuri, are you okay?"

My mother walked into the room. "What, you're still up? It's late! Off to bed!"

"Okay," Yuri said.

"I laid out a futon in the guest room."

"Great, thanks!"

"It's been so long since you spent the night!" My mother was beaming. "You came for sleepovers all the time when you were in elementary school."

"And always cried the next morning when you had to go home," I added.

"Never have I done such a thing." Yuri stuck her tongue out at me.

2.3 SOLUTIONS AND COEFFICIENTS

2.3.1 Tetra

The next day, I gave Tetra a rundown of what Yuri and I had talked about. We were in the school library, where we had headed after the end-of-semester assembly. The library was empty except for us. I heard the sounds of some sports team practicing outside, hidden by the sycamore that grew just outside the window.

"Sounds like you guys had fun!" Tetra said.

Tetra was a second-year student at my school, one year behind me. She was small, with short hair and large eyes. The way she moved quickly about reminded me of a squirrel. She had been having problems with math when she first came to our school, but had come to enjoy it through working with Miruka and me. She approached mathematics with vim and vigor, and I was still helping her out with new topics and study methods.

"Sure, it's always fun to teach math to Yuri."

"I'm so jealous. I wish I had been there."

"For a sleepover?"

"Er, not— I mean, uh— Hey, can you believe it's already summer vacation tomorrow!"

2.3.2 Solutions and Coefficients

"So you were talking about quadratic equations?" Tetra asked.

I nodded. "Along with imaginary numbers, the complex plane, formulas for finding roots, that kind of thing."

"I think I understand all that stuff, but no way I could explain it like you do. I think there's a big gap between understanding something and being able to teach it." Tetra threw her arms out wide, suggesting more of a gulf than a gap.

"I guess. But explaining something to someone is a good way to make sure you really understand it, so when I help you and Yuri with math, I'm learning a lot myself."

"Well I'm glad you get something out of it too!"

"Come to think of it, last night I never got to the relation between roots and coefficients. You know about that, right?"

"Sure. I think so, at least. Oh! Let me explain it to you so I can make sure!"

"An excellent idea."

"So let's see... Okay, my lecture today is 'The Relation between Roots and Coefficients in Quadratic Equations.'" Tetra took out a

Quadratic Equations and Sleeping Beauty

notebook and started writing as she explained. "Say you're given a quadratic equation in x, like this."

$$ax^2 + bx + c = 0 \quad (a \neq 0)$$

"This quadratic has two roots, which we'll call α and β. There's a relation between these roots and the equation's coefficients a, b, c, namely..."

$$\alpha + \beta = -\frac{b}{a}, \quad \alpha\beta = \frac{c}{a}$$

"And that's it! These two equations show the relation between roots and coefficients for a quadratic! How'd I do?"

"Very well," I said. "One more thing, though. Can you briefly explain the difference between the quadratic formula and the relation between the roots and coefficients in a quadratic?"

"Briefly, huh? I guess the quadratic formula is just a formula for findings roots, while the other is just equations for showing how those roots are related to the coefficients. Is that wrong?"

"It isn't, but maybe it's cleaner to describe it like this."

- The quadratic formula uses coefficients to represent the roots.

- The relational equations use coefficients to represent sums and products of the roots.

"Hey, you're right! We use the quadratic formula to represent the roots like this."

$$\alpha = \frac{-b + \sqrt{b^2 - 4ac}}{2a}, \quad \beta = \frac{-b - \sqrt{b^2 - 4ac}}{2a}$$

"So yes, we're using the coefficients a, b, c to describe the roots α, β. And the relational equations for the roots and coefficients are this."

$$\alpha + \beta = -\frac{b}{a}, \quad \alpha\beta = \frac{c}{a}$$

"So we're using a, b, c to show the sum and the product of the solutions. Neat!"

"I think so too."

"Okay, I'll remember this. The quadratic formula expresses the roots using coefficients, while the relational equations express the sums and products of roots in terms of coefficients. Makes perfect sense when you write it that way."

"By the way, it looks like you've memorized the relational equations, but can you derive them?"

"Uh, maybe? Maybe not?"

"It's not that hard. We say the roots of a quadratic $ax^2 + bx + c$ are $x = \alpha, \beta$, but maybe it's more precise to say the roots are $x = \alpha \overset{or}{\vee} x = \beta$. So..."

$$x = \alpha, \beta$$
$$\iff x = \alpha \vee x = \beta$$
$$\iff x - \alpha = 0 \vee x - \beta = 0$$
$$\iff (x - \alpha)(x - \beta) = 0$$
$$\iff a(x - \alpha)(x - \beta) = 0$$
$$\iff a\bigl(x^2 - (\alpha + \beta)x + \alpha\beta\bigr) = 0$$
$$\iff ax^2 - a(\alpha + \beta)x + a\alpha\beta = 0$$

"So we know this is true."

$$ax^2 - a(\alpha + \beta)x + a\alpha\beta = 0 \iff ax^2 + bx + c = 0$$

"Now we just have to compare the coefficients to derive the relational equations."

Relation between quadratic roots and coefficients

Letting α, β be the roots of a quadratic $ax^2 + bx + c$, the following relations hold:

$$\alpha + \beta = -\frac{b}{a}, \quad \alpha\beta = \frac{c}{a}$$

"Of course, since we can use the quadratic formula to find the two roots, we can also directly calculate their sum and product."

$$\begin{aligned}
\alpha + \beta &= \frac{-b + \sqrt{b^2 - 4ac}}{2a} + \frac{-b - \sqrt{b^2 - 4ac}}{2a} \\
&= \frac{(-b + \sqrt{b^2 - 4ac}) + (-b - \sqrt{b^2 - 4ac})}{2a} \\
&= -\frac{2b}{2a} \\
&= -\frac{b}{a} \\
\alpha\beta &= \frac{-b + \sqrt{b^2 - 4ac}}{2a} \cdot \frac{-b - \sqrt{b^2 - 4ac}}{2a} \\
&= \frac{(-b)^2 - (\sqrt{b^2 - 4ac})^2}{(2a)^2} \\
&= \frac{b^2 - (b^2 - 4ac)}{4a^2} \\
&= \frac{4ac}{4a^2} \\
&= \frac{c}{a}
\end{aligned}$$

"Got it!" Tetra said.

2.3.3 Getting your Head Straight

"Every time I talk to you about math it feels like I'm cleaning up some of the mess in my head," Tetra said.

"Yeah?"

"Sure! I've studied all this in class, and I can solve the problems we're assigned, but still, deep down I've always felt like there was something I was missing. When you lay things out nice and neat like this..." She pointed to what she'd written in her notebook.

- The quadratic formula uses coefficients to represent the roots.

- The relational equations use coefficients to represent sums and products of the roots.

"...it feels like everything I've crammed in my head has been reorganized and put in place. You show me what's important." Tetra looked at me with her huge eyes. "I wish I had a book that could do that. When I read something I don't understand, a finger would emerge from the page and point at the place I need to focus on."

I looked her in the eyes and laughed. "That's...kinda' creepy."

"Speaking of which, one more thing to organize in my head. What's the big deal about $\alpha + \beta$ and $\alpha\beta$, the sum and product of the roots? Why are they important?"

"Huh?" I was taken aback by this unexpected query. "Well, they're important because, uh..."

A crisp voice came from behind me.

"Relations between roots and coefficients? Interesting."

Miruka...

2.4 Symmetric Polynomials and the Field Perspective

2.4.1 Miruka

Ah, Miruka. Where to begin?

Miruka was another third-year student, and my classmate. She had long black hair, elegant poise, and a citrus scent. She was good at math in ways I could only dream of, and when she got in the mood she would impart bits of her knowledge through lectures she delivered to us. Sharp-tongued, genius, and just a touch pugnacious. She loved math, books, chocolate, and spinning her pen. She hated weakness. No matter how I tried to describe her, however, I could never get across her true form, in the same way you can't truly describe a color in words.

Miruka had guided me on many excursions into the world of mathematics, from right here in this library. In that realm there was no dragon fierce enough to give her pause, no jungle dense enough for her to lose her way. Indeed, she usually emerged from such encounters bearing wondrous treasures.

Ah, Miruka...

2.4.2 Relations Between Roots and Coefficients, Revisited

"Relations between roots and coefficients? Interesting."

Miruka was looking at the equations Tetra and I had written.

"It is!" Tetra said. "We were just talking about how the quadratic formula shows roots as coefficients, and how relational equations use the coefficients to show the sum and product of the roots."

"And how to derive those," I said.

"Hmph." Miruka closed her eyes and turned her face slightly upward. "Sums and products of roots, huh?"

Her black hair slipped back, revealing the graceful lines of her cheek and jaw. Tetra and I sat in silence for some three seconds before she spoke again.

"Let's talk about symmetric polynomials," she said. She sat next to me and commandeered my notebook and pencil. "A symmetric polynomial in α and β is one in which the polynomial doesn't change value when you swap α and β. For example, the sum $\alpha + \beta$ is a symmetric polynomial, because we can swap α and β to make $\beta + \alpha$, which is equal to the original $\alpha + \beta$. It's invariant, in other words."

$$\alpha + \beta \qquad \text{(a symmetric polynomial in } \alpha \text{ and } \beta)$$

"The difference $\alpha - \beta$ isn't symmetric, though. If we swap the α and β we get $\beta - \alpha$, which isn't the same as $\alpha - \beta$. So in this case the polynomial isn't invariant."

$$\alpha - \beta \qquad \text{(a non-symmetric polynomial in } \alpha \text{ and } \beta)$$

"Interestingly, however, if we square $\alpha - \beta$ it becomes symmetric. $(\beta - \alpha)^2 = (\alpha - \beta)^2$, so the polynomial is invariant when we swap α and β."

$$(\alpha - \beta)^2 \qquad \text{(a symmetric polynomial in } \alpha \text{ and } \beta)$$

"But that's too easy. Let's try something a little more complex."

$$\alpha\beta + (\alpha - \beta)^2 + 2\alpha^3\beta^2 + 2\alpha^2\beta^3 \qquad \text{(a symmetric polynomial in } \alpha \text{ and } \beta)$$

"The $\alpha + \beta$ and $\alpha\beta$ that popped up in the relational equations you were talking about are both symmetric polynomials in α and β.

In fact, these two polynomials in particular are called *elementary symmetric polynomials*."

$\alpha + \beta$, $\alpha\beta$ (elementary symmetric polynomials in α and β)

"So we could also say that the relational equations you were talking about use coefficients to represent elementary symmetric polynomials. Yes, Tetra?"

I turned to see Tetra with her hand in the air.

"Why are these called 'symmetric'? When I think of symmetry it's about figures, point symmetry and line symmetry and things like that. I'm not sure what swapping variables has to do with symmetry."

Miruka smiled. "Always concerned with the words, aren't you. But you're right, 'symmetry' is a word we use to describe forms. But polynomials have forms too, so it isn't so strange to describe them as being symmetric."

Tetra made a dubious face. "Polynomials have forms?"

"Sure," Miruka said. "After all, doesn't symmetry suggest that something doesn't change? It's a property related to invariance. You could even say that symmetry is just one kind of invariance."

"I guess so, but I'm still not sure I understand. When I say something is symmetric, I mean something like it being the same on the right and left. If something is invariant, it sounds more like it will never ever change. But here both words mean the same thing?"

"Okay, then. Let's consider shapes that are left–right symmetric. Isosceles triangles, for example."

"Swap the left and right sides in this left–right symmetric shape, and nothing changes. It's invariant. *That's* what I mean when I say symmetry is one kind of invariance."

"Ah, okay. I'm starting to get this. So a symmetric shape is one that's invariant when you do some kind of swapping, right?"

Miruka nodded. "Right. But it doesn't necessarily have to be some kind of swap. There can also be symmetries under permutations, rotations, all kinds of operations. In any case, symmetry often refers to a geometric form, but that's not the only thing it applies to."

"I see," Tetra said, nodding as she jotted something down in her notebook. "Okay, so swapping the α and β in $\alpha + \beta$ gives $\beta + \alpha$, which is the same as the original $\alpha + \beta$, so it's invariant. Got it."

$$\alpha + \beta \xrightarrow{\text{swap } \alpha \text{ and } \beta} \beta + \alpha$$

"Very good," Miruka said.

Listening to the two of them talking, I felt something like a flame igniting within me. That phrase Miruka had used—symmetry is just one kind of invariance. I felt sure there was more to that.

"Back to symmetry," Miruka said. She waved an index finger like a conductor's baton and resumed her lecture. "We can always use elementary symmetric polynomials to write a symmetric polynomial, which is where the 'elementary' part comes from. For example, here's the symmetric polynomial $(\alpha - \beta)^2$."

$$\underbrace{(\alpha - \beta)^2}_{\substack{\text{symmetric} \\ \text{polynomial}}} = (\underbrace{\alpha + \beta}_{\substack{\text{elementary} \\ \text{symmetric} \\ \text{polynomial}}})^2 - 4(\underbrace{\alpha \times \beta}_{\substack{\text{elementary} \\ \text{symmetric} \\ \text{polynomial}}})$$

"Here's how we get that."

$$\begin{aligned}
(\alpha - \beta)^2 &= \alpha^2 - 2\alpha\beta + \beta^2 && \text{expand} \\
&= \alpha^2 + 2\alpha\beta + \beta^2 - 4\alpha\beta && \text{prepare to create square} \\
&= (\alpha^2 + 2\alpha\beta + \beta^2) - 4\alpha\beta && \text{parenthesize squared part} \\
&= (\alpha + \beta)^2 - 4\alpha\beta && \text{as a square}
\end{aligned}$$

"The fact that any symmetric polynomial can be written using elementary symmetric polynomials is called the 'fundamental theorem of symmetric polynomials.' Let's see how a symmetric polynomial in the roots can always be written using the elementary symmetric polynomials $\alpha + \beta$ and $\alpha\beta$ of those roots."

Miruka wrote this in the notebook:

> A symmetric polynomial in the roots can be expressed using the elementary symmetric polynomials of those roots (from the fundamental theorem of symmetric polynomials).

"As you saw when looking at the relation between roots and coefficients, you can write the elementary symmetric polynomials of the roots using coefficients."

She wrote this too:

> An elementary symmetric polynomials of the roots can be expressed using coefficients (from the relation between roots and coefficients).

"Putting them together, we can say this."

> A symmetric polynomial in the roots can be expressed using coefficients.

"Taking it further, since we know a symmetric polynomial in the roots of a quadratic is invariant when we swap roots, we get this."

> Polynomials in roots α and β that are invariant when α and β are swapped can be expressed using coefficients.

"So what's next, Tetra?" Miruka asked.

"Who, me? I...I'm not sure I'm following this, what it means, even. An example, maybe...?"

"I'm sure he can come up with one," Miruka said, pointing at me.

"It's like this, Tetra," I said. "Say you have α and β as the solutions to a quadratic equation $ax^2 + bx + c = 0$, and think about a symmetric polynomial made from them. I'll use Miruka's example."

$$\alpha\beta + (\alpha - \beta)^2 + 2\alpha^3\beta^2 + 2\alpha^2\beta^3 \quad \text{(symmetric polynomial in } \alpha \text{ and } \beta\text{)}$$

"What Miruka's saying is we can use coefficients a, b, c to write this symmetric polynomial. Here, I'll write it out."

$\alpha\beta + (\alpha - \beta)^2 + 2\alpha^3\beta^2 + 2\alpha^2\beta^3$	example symmetric polynomial
$= \alpha\beta + \alpha^2 - 2\alpha\beta + \beta^2 + 2\alpha^3\beta^2 + 2\alpha^2\beta^3$	expand
$= \alpha\beta + (\alpha^2 + 2\alpha\beta + \beta^2) - 4\alpha\beta + 2\alpha^3\beta^2 + 2\alpha^2\beta^3$	prepare to write as square
$= \alpha\beta + (\alpha + \beta)^2 - 4\alpha\beta + 2\alpha^3\beta^2 + 2\alpha^2\beta^3$	as $\alpha + \beta$ and $\alpha\beta$
$= (\alpha + \beta)^2 - 3\alpha\beta + 2\alpha^3\beta^2 + 2\alpha^2\beta^3$	bring together $\alpha\beta$ terms
$= (\alpha + \beta)^2 - 3\alpha\beta + 2(\alpha\beta)^2(\alpha + \beta)$	factor out $2(\alpha\beta)^2$
$= \left(-\dfrac{b}{a}\right)^2 - 3\left(\dfrac{c}{a}\right) + 2\left(\dfrac{c}{a}\right)^2\left(-\dfrac{b}{a}\right)$	relation between roots and coefficients

"Oh, I get it!" Tetra said, running a finger down the equations. "You're right! You wrote the symmetric polynomial using the coefficients! And Miruka was saying we can do this for *any* symmetric polynomial." She paused and frowned. "It's just, I can follow along with what you've written here, but I would never come up with it on my own. How can I get to that point?"

"With practice," Miruka said.

"Ah, of course. But in particular, how would I ever think to write $\alpha^2 - 2\alpha\beta + \beta^2$ as $(\alpha^2 + 2\alpha\beta + \beta^2) - 4\alpha\beta$?"

"I guess the important thing is learning to see the direction you want to head in," I said. "In this case, we want to head toward a representation using elementary symmetric polynomials."

"The direction I want to head in..." Tetra said, taking notes.

"It's like..." I started, but paused, again feeling my previous frustration. What was it? This feeling that seemingly unconnected math was coming together to form some new melody under the direction of Miruka's baton. For some reason, though, I could only barely hear the tune. All I was picking up was something connecting solving equations with swapping solutions... "So frustrating..." I muttered.

"Equations, coefficients, solutions, symmetric polynomials..." Miruka chanted. "They're all related. The mathematician Joseph-Louis Lagrange went deep into research on solutions to equations, and discovered how those solutions are related to swapping roots. Évariste Galois studied what Lagrange had found, and from that revealed far more about equations."

"That name sounds familiar," Tetra said.

Miruka started speaking faster. "Equations are related to a mathematical concept called fields, which are quite a beast to tackle. But Galois found a way to associate hard-to-handle fields with groups, which are much more manageable. This connection between fields and groups is called a Galois correspondence." Miruka stood and softly placed a hand on Tetra's shoulder, then one on mine. "It's a bridge between two worlds, the world of fields and the world of groups. It's also the foundation of one of the most beautiful of all mathematical theories, Galois theory."

2.4.3 Solutions to Equations, Revisited

Tetra raised a hand. "Um, fields? Groups? I know we've talked about this before,[1] but..."

"A quick review, then," Miruka said. "You can think of a field as a set of numbers where the four basic arithmetic operations are defined, so it makes sense to say $x + y, x - y, x \times y,$ and $x \div y$ for elements x, y in the field. So the set of all real numbers is a field, as is the set of all complex numbers."

"So I can just think of a field as a set of numbers that I can calculate with."

"That's fine, but you have to be sure you clearly understand what you mean by 'calculate.' Only the four basic arithmetic operators are defined—addition, subtraction, multiplication, and division. You can't rely on there being square roots in a field, for example. There can be, but that has nothing to do with the thing being a field. So you can't be sure you can go from 9 to finding $+\sqrt{9}$ and $-\sqrt{9}$, in other words ± 3. That's pertinent to what we're talking about, because a square root appears in the quadratic formula."

---

[1] See *Math Girls 2: Fermat's Last Theorem*, Chapter 7.

The quadratic formula

The roots to a quadratic formula $ax^2 + bx + c = 0$ are

$$\frac{-b \pm \sqrt{b^2 - 4ac}}{2a}.$$

"Of course, the $\pm\sqrt{b^2 - 4ac}$ part," Tetra said.

"Let's take a closer look at all the operations the quadratic formula uses," Miruka said as she carefully rewrote the formula.

$$\left(0 - b \pm \sqrt{b \times b - 4 \times a \times c}\right) \div (2 \times a)$$

Interesting. This new format for such a familiar formula, explicit multiplication signs between multiplied terms and division signs in place of fractions, was a refreshing take on it. Written like this, it was impossible to miss any of the arithmetic that was going on.

Miruka tapped what she had written. "The quadratic formula uses all four basic arithmetic operators—addition, subtraction, multiplication, and division—but also uses a square root operation. Put another way, so long as you can use the basic arithmetic operators and square roots, you can use the coefficients in any quadratic equation you come across to solve it."

"Well yeah," Tetra said, "but isn't that kind of—"

Miruka raised a finger. "Obvious? It is, just from reading the quadratic formula. So let's start from that obviousness and take another look at quadratic equations from the perspective of fields."

I saw that twinkle in Miruka's eye that always meant she was having fun.

"As an example," she continued, "let's solve this quadratic."

$$x^2 - 2x - 4 = 0$$

"The coefficients here are $a = 1, b = -2, c = -4$, so here's the solution."

$$x = \frac{-b \pm \sqrt{b^2 - 4ac}}{2a}$$
$$= \left(0 - b \pm \sqrt{b \times b - 4 \times a \times c}\right) \div (2 \times a)$$
$$= \left(0 - (-2) \pm \sqrt{(-2) \times (-2) - 4 \times 1 \times (-4)}\right) \div (2 \times 1)$$
$$= \left(2 \pm \sqrt{4 + 16}\right) \div 2$$
$$= \left(2 \pm \sqrt{20}\right) \div 2$$
$$= \left(2 \pm \sqrt{2^2 \times 5}\right) \div 2$$
$$= \left(2 \pm 2\sqrt{5}\right) \div 2$$
$$= 1 \pm \sqrt{5}$$

"Okay, so the roots of this quadratic are $1 + \sqrt{5}$ and $1 - \sqrt{5}$. No problems so far, right?" Miruka paused, then pointed at Tetra. "What are the coefficients for $x^2 - 2x - 4$?"

"That would be, uh, $1, -2, -4$."

"That's right. So let's find a field that contains these coefficients."

"Like, the integers, maybe?"

"No. The set of all integers isn't a field, because for example $1 \div (-2)$ isn't an integer."

"Oh, right. When we perform arithmetic, the answer has to be in the set too."

"Correct. Another way to put it is to say that the field must be 'closed' under the arithmetic operations."

"Okay, this is all coming back to me."

"The smallest field that coefficients $1, -2, -4$ all belong to is the rational numbers, $\mathbb{Q}$. So let's consider the field of coefficients to be the field of rational numbers $\mathbb{Q}$."

"What's the field of coefficients?"

"The field to which the coefficients of a quadratic equation belong. So in this case we're saying that coefficients $1, -2, -4$ all belong to $\mathbb{Q}$."

$$1 \in \mathbb{Q}, \quad -2 \in \mathbb{Q}, \quad -4 \in \mathbb{Q} \quad \text{(all coefficients belong to } \mathbb{Q}\text{)}$$

"Each of $1, -2$, and -4 are integers, but they're also rational numbers," I added.

"So the coefficients for $x^2 - 2x - 4 = 0$ are all in $\mathbb{Q}$," Miruka said. "However, the roots for this quadratic, $1 + \sqrt{5}$ and $1 - \sqrt{5}$, are not."

$$1 + \sqrt{5} \notin \mathbb{Q}, \quad 1 - \sqrt{5} \notin \mathbb{Q} \qquad \text{(neither root belongs to } \mathbb{Q}\text{)}$$

Tetra nodded. "Okay, the coefficients for this quadratic are rational numbers, but its roots aren't. Got it."

"But look what that means—so long as we stay within the scope of $\mathbb{Q}$, the polynomial $x^2 - 2x - 4 = 0$ doesn't have a solution."

"Huh," I grunted. "Yeah, looks like that's the case. So what do we do?"

"Make ourselves a new playground," Miruka said, lowering her voice as if passing on a secret. I couldn't help but lean in closer. "We have all the rationals in $\mathbb{Q}$, but we also need $\sqrt{5}$ to do all the arithmetic we'd like. So let's just add $\sqrt{5}$ to $\mathbb{Q}$ and use the resulting set to make a new field, which we'll call $\mathbb{Q}(\sqrt{5})$. Also, 'add' isn't quite right. Instead I'm going to call this 'adjoining' $\sqrt{5}$ to $\mathbb{Q}$, and the result is an 'adjunction.'"

$\mathbb{Q}(\sqrt{5})$ The field resulting from adjoining $\sqrt{5}$ to $\mathbb{Q}$

"So what kind of numbers might reside in $\mathbb{Q}(\sqrt{5})$? All of the rational numbers, of course, so for example $1, 0, -1, 0.5$, and $\frac{1}{3}$ are all in there. So is $\sqrt{5}$, and the results of arithmetic between a rational number and $\sqrt{5}$. Examples would be $1 - \sqrt{5}$ or $\frac{1}{3} + \sqrt{5}$, even numbers like $2 - 7\sqrt{5}$ and $\frac{1+\sqrt{5}}{3}$ and $\frac{1+\sqrt{5}}{1-\sqrt{5}}$. Generally, we can write any element in $\mathbb{Q}(\sqrt{5})$ as $\frac{p+q\sqrt{5}}{r+s\sqrt{5}}$, where p, q, r, s are rational numbers. With some rationalization, we can write these numbers as $P + Q\sqrt{5}$."

$$\frac{p+q\sqrt{5}}{r+s\sqrt{5}}$$

$$=\frac{p+q\sqrt{5}}{r+s\sqrt{5}}\cdot\frac{r-s\sqrt{5}}{r-s\sqrt{5}} \qquad \text{multiply by } \frac{r-s\sqrt{5}}{r-s\sqrt{5}} \text{ to rationalize denominator}$$

$$=\frac{(p+q\sqrt{5})(r-s\sqrt{5})}{(r+s\sqrt{5})(r-s\sqrt{5})}$$

$$=\frac{pr-ps\sqrt{5}+qr\sqrt{5}-qs\sqrt{5}\sqrt{5}}{r^2-s^2\sqrt{5}\sqrt{5}} \qquad \text{calculate numerator and denominator}$$

$$=\frac{pr-5qs+(qr-ps)\sqrt{5}}{r^2-5s^2} \qquad \sqrt{5} \text{ gone from denominator}$$

$$=\frac{pr-5qs}{r^2-5s^2}+\frac{qr-ps}{r^2-5s^2}\sqrt{5} \qquad \text{bring out } \sqrt{5}$$

"Since $\frac{pr-5qs}{r^2-5s^2} \in \mathbb{Q}$ and $\frac{qr-ps}{r^2-5s^2} \in \mathbb{Q}$, if we let $P = \frac{pr-5qs}{r^2-5s^2}$ and $Q = \frac{qr-ps}{r^2-5s^2}$, then P and Q will both be rational numbers, and we can write elements of $\mathbb{Q}(\sqrt{5})$ in the form $P + Q\sqrt{5}$, with the condition that P and Q are rationals."

Tetra was frantically scribbling notes. "So the entire field of rational numbers $\mathbb{Q}$ is included in this new field that we created by adding, er, *adjoining* $\sqrt{5}$ to $\mathbb{Q}$... right?"

"Right. You can also use set notation and say $\mathbb{Q} \subset \mathbb{Q}(\sqrt{5})$. In this case, we call $\mathbb{Q}(\sqrt{5})$ a 'field extension' of $\mathbb{Q}$."

Tetra nodded and added this term to her notes.

"Back to the subject at hand. To summarize what we've said so far, the roots of $x^2 - 2x - 4 = 0$ are $1 \pm \sqrt{5}$, which are not rational numbers, so this equation does not have a solution within the scope of the field of rational numbers $\mathbb{Q}$. We can, however, find a solution in this new field $\mathbb{Q}(\sqrt{5})$ that we created by adjoining $\sqrt{5}$ to $\mathbb{Q}$, since $\mathbb{Q}(\sqrt{5})$ includes both $1+\sqrt{5}$ and $1-\sqrt{5}$."

Quadratic Equations and Sleeping Beauty

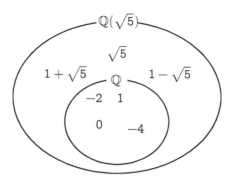

"So here's what we've found. The polynomial $x^2 - 2x - 4$—"

- has no roots in the field $\mathbb{Q}$, but
- has roots in the field $\mathbb{Q}(\sqrt{5})$.

"Put another way, the polynomial equation $x^2 - 2x - 4 = 0$—"

- cannot be solved in the field $\mathbb{Q}$, but
- can be solved in the field $\mathbb{Q}(\sqrt{5})$.

"Does this help you get a feel for polynomials from the perspective of fields?"

"Slowly but surely," Tetra said.

"Then let's stop talking about a specific polynomial like $x^2 - 2x - 4 = 0$, looking instead at the general case $ax^2 + bx + c = 0$. First, we have to think about where we got this number $\sqrt{5}$ that we adjoined to $\mathbb{Q}$ so we could avoid unsolvability issues. As you might guess, this $\sqrt{5}$ came from the $\sqrt{b^2 - 4ac}$ in the quadratic equation. If we let our field of coefficients be K, we won't always be able to solve $ax^2 + bx + c = 0$ within the scope of K. Maybe we can, maybe we can't—it all depends on whether $\sqrt{b^2 - 4ac}$ is an element of K. The key, in other words, is whether $b^2 - 4ac$ is the square of some number that's an element of our field of coefficients. This key, $b^2 - 4ac$, is called the 'discriminant' of the polynomial.

"Another quick summary. The polynomial $ax^2 + bx + c$—"

- *Might* have no roots within the scope of field K, but

- it *will* have roots within the scope of field $K(\sqrt{b^2-4ac})$.

"Put another way—"

- The polynomial equation $ax^2+bx+c=0$ *might* be unsolvable within the scope of field K, but
- it *will* be solvable within the scope of the field $K(\sqrt{b^2-4ac})$.

"We know where the solutions to the polynomial will lie, namely in this field."

$$K(\sqrt{\text{discriminant}})$$

"So there you have it, the relation between fields and solving polynomials."

"A tiny drop, and poof," I said.

"What's that?" Tetra asked.

"Something I said to Yuri yesterday. We were talking about how you can just add a new number i to the reals, and poof, you get the complex numbers. This stuff about solutions to quadratics feels the same way. Just by adding this new number $\sqrt{b^2-4ac}$ to a field of coefficients K, we get $K(\sqrt{b^2-4ac})$, a whole new field that's guaranteed to include the solution to any quadratic."

"The library is *closed*," came the announcement by Ms. Mizutani, bringing our mathematical excursion to an end.

2.4.4 On the Way Home

Miruka, Tetra, and I headed toward the train station. Everything we had talked about was still buzzing in my brain.

Before today, I had only thought of the relation between roots and coefficients to be the fact that I could represent roots as sums and products of the coefficients. That wasn't wrong, but Miruka had taken the same thing and shown us how to use coefficients to represent elementary symmetric polynomials for the roots. Not only that, she tied together representations using coefficients with swapped roots to show the relation between swapping roots and solving polynomials.

So sure, I could consider the quadratic formula as a description of roots using coefficients. But Miruka had drawn a completely new picture from the perspective of fields, sets of numbers that allowed arithmetic, of all things. Simply being able to perform arithmetic using a quadratic's coefficients isn't enough to guarantee finding its solutions, since it might be impossible to create $\sqrt{b^2 - 4ac}$. That depends on one fact, whether $\sqrt{b^2 - 4ac}$ is an element in the field of coefficients. If so, the solutions would lie in the field of coefficients, and we could solve the quadratic within that scope.

Miruka had also talked about how we can add (*no, adjoin!*) $\sqrt{b^2 - 4ac}$ to the field of coefficients K to create $K(\sqrt{b^2 - 4ac})$, a new field in which the solutions would definitely lie. In other words, a quadratic equation could always be solved within the scope of $K(\sqrt{b^2 - 4ac})\ldots$

I was still far from completely understanding this new "field-based perspective," but I could see enough to know it was a different approach from the method I was used to, just grinding through calculations. The relation between roots and coefficients, the quadratic equation—these were the most basic of tools I used to manipulate polynomials. I had thought I knew all there was to know about them. But now I saw that things weren't quite so simple. It was like I had climbed to higher ground and discovered some new territory, and a path leading to deeper understanding.

Tetra, who was walking in front of me, turned and said, "So, do you have any plans for summer vacation?"

Ah, right. Summer vacation starts tomorrow. I really need to study for college entrance exams, but this new math... So tempting...

"Just studying for exams," I said. "I signed up for a summer seminar, so I guess that's where I'll be spending my mornings. I'll probably go to the school library in the afternoon to work on practice problems."

"Sounds like you'll be busy."

"How about you, Miruka?" I asked. "What will you be up to?"

"Oh, this and that."

"I know what I'm going to do!" Tetra said. "I'm going to dig through my notebooks and find my notes from when we talked about groups and fields and all that! I need to review!"

I let out a sudden sneeze.

"Oops, did you catch cold?" Tetra asked, looking at me with genuine concern.

"Nah, I'm fine."

> We define the discriminant as the square of a product of differences, because squaring makes the discriminant a symmetric polynomial... allowing us to represent it using the polynomial's coefficients.
>
> SHOICHI NAKAJIMA
> *Algebraic Equations and Galois Theory* [24]

My notes

Completing the Square

The method for deriving the quadratic equation that I showed to Yuri avoids fractions and is easy to follow, but it requires keeping the form of $b^2 - 4ac$ in mind.

Another method called "completing the square" can be a little bit more involved, but more naturally leads to the <polynomial in x>^{2} form that we're after.

A quadratic comes in this form:

$$ax^2 + bx + c = 0$$

Divide both sides by a to make the coefficient for x^2 become 1.

$$x^2 + \frac{b}{a}x + \frac{c}{a} = 0$$

We want to get the left side into the form <bits with x>^{2} + <bits with no x>, something like this:

$$x^2 + \frac{b}{a}x + \frac{c}{a} = \underbrace{\left(x + \blacksquare\right)^2}_{\text{(bits with } x)^2} + \underbrace{\frac{c}{a} - \blacksquare^2}_{\text{(bits without } x)}$$

So the question is, what should go in the black boxes?

In $(x + \blacksquare)^2 = x^2 + 2\blacksquare x + \blacksquare^2$ the coefficient of x equals b/a, so we should have $\blacksquare = \frac{b}{2a}$

$$x^2 + \frac{b}{a}x + \frac{c}{a} = \left(x + \frac{b}{2a}\right)^2 + \frac{c}{a} - \blacksquare^2$$

from $\blacksquare = \frac{b}{2a}$ we can also get the final term.

$$x^2 + \frac{b}{a}x + \frac{c}{a} = \left(x + \frac{b}{2a}\right)^2 + \frac{c}{a} - \left(\frac{b}{2a}\right)^2$$

Now all that's left is calculations. The second term first:

$$= \left(x + \frac{b}{2a}\right)^2 - \frac{b^2 - 4ac}{4a^2}$$

This should equal zero, giving us

$$\left(x + \frac{b}{2a}\right)^2 - \frac{b^2 - 4ac}{4a^2} = 0$$

and thus

$$\left(x + \frac{b}{2a}\right)^2 = \frac{b^2 - 4ac}{4a^2}$$

Solving for x gives the quadratic equation:

$$x = \frac{-b \pm \sqrt{b^2 - 4ac}}{2a}$$

Completing the square is thus a way of naturally deriving the quadratic equation by focusing on the form $(x + \blacksquare)^2$ as your goal, then subtracting out the $\blacksquare^2$ term that arises to put the formula in the proper form.

CHAPTER 3

Searching for Form

> What is Life? One dissects a body but finds no life inside. What is Mind? One dissects a brain but finds no mind therein. Are life and mind so much more than the "sum of their parts" that it is useless to search for them?
>
> MARVIN MINSKY
> *Society of Mind*

3.1 THE FORM OF EQUILATERAL TRIANGLES

3.1.1 At the Hospital

I wasn't fine.

That night I broke out in a fever, one so serious that my parents rushed me to the hospital. My fever had broken by the next day, but apparently I was still at risk of developing pneumonia, so the doctor kept me there for observation. I was thus trapped in bed, alone with my drowsy thoughts. I fell asleep cursing myself for not taking better care of my health.

It was Yuri's voice that woke me up. "Hey! How are you doing?"

"Feeling better, I hope!" said Tetra.

"Ugh," I groaned, fumbling for my glasses on a bedside table. "Did you two come together?"

"Nope! I was here first!" Yuri said.

"We just ran into each other, right at the entrance," added Tetra.

Yuri opened up a folding chair with a clatter. "Talking about ploppers the other day was lots of fun, wasn't it!"

Tetra sat on the other side of my bed. "Ploppers? What's that?"

I finally managed to get my glasses on, and saw they were both wearing surgical masks.

"She's talking about ladder diagrams," I said, wriggling to prop myself up. "Go ahead and show her your graph, Yuri. I'm sure you brought it."

"You need to lay down," Yuri said. She pulled a sheet of paper out of her backpack and spread it on my bed.

Searching for Form

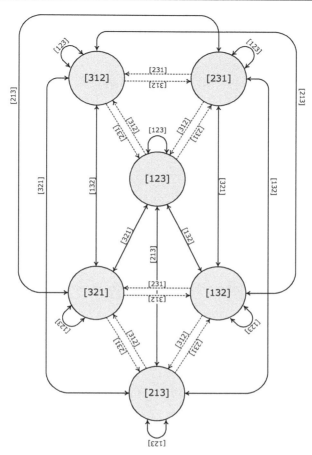

Yuri's diagram for three-line ladder diagrams.

Yuri started explaining her diagram to Tetra, using my bed as a podium. She frequently stumbled at first, but Tetra was an excellent listener, and prompted Yuri's description where needed to keep her going.

"So anyway, I'm going to make this into a research project!" Yuri concluded.

Tetra stared intently at the diagram, thinking deeply. After a time, she said, "Yuri, do you mind if I point out something?" I noticed her voice was different from normal, more big-sisterly.

Yuri made a suspicious face. "I guess not. What is it?"

"I see equilateral triangles in this."

"Equilateral triangles? How's that?"

"First, we can represent the ladder diagram [1 2 3] as an equilateral triangle, with its vertices labeled 1, 2, 3, starting from the top and going counterclockwise. Here, I'll draw one with a line inside it to show the direction."

[1 2 3].

"If we were looking at [2 3 1] instead, it would be rotated like this."

[2 3 1].

"So let's think about what we have to do to change [1 2 3] into [2 3 1]. We can just rotate the triangle 120° clockwise, for example. Doing that of course moves the triangle's vertices, changing them all."

Rotating [1 2 3] gives [2 3 1].

Rotating [1 2 3] twice gives [3 1 2].

"But the only ladder diagrams we can create from [1 2 3] are [2 3 1] and [3 1 2]. No matter how many times we rotate [1 2 3], we'll never get [1 3 2] or [2 1 3] or [3 2 1], right? To make those, we need to flip something. We need one of your 'swappers.' There are three ways we can flip a triangle, along one of its axes of symmetry. Whichever way you choose, two of its vertices will swap places, but the one on the axis of symmetry will stay where it is."

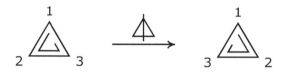

Swapping to change [1 2 3] to [1 3 2].

Swapping to change [1 2 3] to [2 1 3].

Swapping to change [1 2 3] to [3 2 1].

"Oh, and then there's your 'plopper.' That doesn't change any of the triangle's vertices, so everything stays the same."

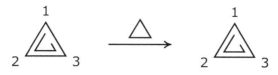

[1 2 3] stays as [1 2 3].

"So anyway, that's the triangles I saw as I listened to your explanation. Do you mind if I add them to your graph?"

Yuri nodded assent, and "big sister" Tetra started drawing.

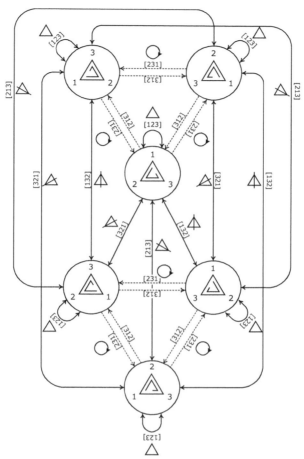

Yuri's graph with Tetra's triangles added.

I eyed Yuri, wary of how she would react to Tetra's commentary on her pet project. She was staring at her updated graph with a stony expression, twiddling her ponytail. After a time, she finally animated.

"Tetra... This is *so cool*! So absolutely totally cool!"

"Thanks," Tetra said. She cocked her head. "I'm not sure if it really means anything, though."

"Still, the triangles fit perfectly!" Yuri started drawing a chart to show correspondences with Tetra's symbols. "One pattern for ploppers, with nothing changing. Three patterns for swappers, with the triangle flipping. Two patterns for scramblers, which match up with rotations. It all fits!"

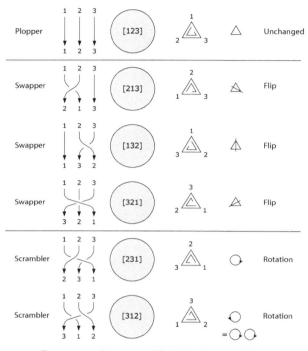

Correspondences with Tetra's notations.

"Sure does," I said. "We can also think of [3 1 2] as a backwards revolution, counterclockwise instead of clockwise."

"Remember how Miruka said symmetry is just one kind of invariance?" Tetra said, speaking slowly. "I think I'm finally starting to see the connection. There are three invariant points with Yuri's plopper, one invariant point with the swappers, and no invariant points with the scramblers."

I sat up in bed, forgetting my fever but feeling my heart beating faster. Tetra and Yuri were on to something. Some relation between symmetry and invariance that seemed just outside of my grasp.

"Anyway," Tetra said, "I just noticed that these 'nothing,' 'flipping,' and 'rotating' operations seem to line up perfectly with all the symmetries of an equilateral triangle, just like Yuri's 'ploppers,' 'swappers,' and 'scramblers' match up with a three-lined ladder diagram, so I figured maybe we could represent those ladder diagrams as triangles too."

Tetra looked at the diagram as if considering some intricate puzzle box.

"Hey, Tetra," Yuri said. "I was thinking about using this for my summer research project for school. This graph is for three-lined ladder diagrams, so there's six patterns in all, but I was going to study four-lined ladders, which would come in $4! = 24$ patterns. Once I'm done, would you mind taking a look, just to make sure I did it right?"

"Who, me? Are you sure? Wouldn't you rather that somebody who's a little better at this kind of thing help you, like—"

"If you mean this guy, he's gonna be too busy studying for exams."

"Well, sure, I wouldn't mind."

They stayed for a while longer, until visiting hours were over.

"Aww, I don't want to go home yet," Yuri said.

"I think we should let him rest," Tetra chided.

"Yeah, I guess. Later, dude!"

Yuri and Tetra both flicked me a 1, 1, 2, 3. I held up five fingers in response, and they departed.

Talking about math was always fun, but I had to admit, I was a little tired.

3.1.2 Fever, Revisited

"39.2 °C," came the far-off voice of the nurse. "Looks like your fever's back up."

My head felt hot and my throat hurt. I couldn't get comfortable in the bed. It felt like I wasn't in my own body. I wanted to fall into a deep sleep, but I was so uncomfortable I was trapped in a weird semi-awake state. Even so, I dreamed.

I was walking in a forest. The trees were entangled in vines, countless vines rising up from the ground to climb their trunks. The vines formed patterns, intricate weavings. I had to untangle them. No, I had to count them. Wait, not the vines, the trees they climbed. No, ignore the trees, it's the forest I need to consider. If I can know the extent of the forest, I'll know how many trees it contains. If I can know the number of trees, I will know the number of patterns they form. I wanted to fly so I could take in the entire forest. A squirrel and a kitten ran by my feet and climbed up a particularly well-formed tree, which spoke to me.

Save the flying for later.

What?

You have a fever. Go back to sleep.

Who's there?

Doesn't matter. Stop talking.

As I slipped back into sleep, I felt something on my lips. Something warm.

3.1.3 Awakening from the Dream

My mother's voice woke me up.

"I've brought you a change of clothes!" She placed a cool hand on my forehead. "Looks like your fever is down."

"I feel much better. Had some crazy weird dreams, though."

My mother looked at the papers on my bed. "You were doing math?"

"Yeah, with Yuri and Tetra. They came to visit." I noticed something different about Yuri's big graph. Ovals that hadn't been there before. "Hand that to me, would you?"

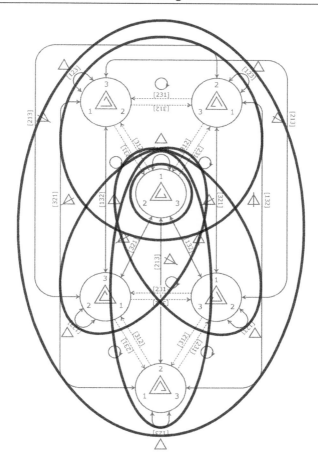

"They came already? I got calls from all three you know. Such sweet friends you have. Oh, I brought some plum kombucha. Want some? And just what is this a picture of? Who drew it?"

"Wait, back up. All *three* called you?"

"Sure. Yuri, Tetra, and Miruka."

3.2 Symmetric Groups

3.2.1 In the Library

"Look! It's Miruka," Yuri said, waving.

I was back in my school library, largely because Yuri had pressed me so hard to take her to my school. When we arrived, Miruka was already there with Tetra.

"How are you feeling?" Tetra asked.

"Much better, thanks." It had taken four days since leaving the hospital to feel like I'd finally truly recovered.

"You were sneezing the other day too, weren't you Miruka?" Tetra asked.

"Was I?" Miruka replied.

"Oh, I remember that uniform!" Tetra said.

Yuri looked down at the uniform she was wearing. "Oh, right. You went to my school. Feels kind of weird to be wearing it, but this guy said it would be weirder to visit his school in plain clothes, so..." She shrugged.

As Tetra and Yuri chattered, I pulled out the paper that had been left behind at the hospital. "Your work, I suppose?" I said to Miruka. "I didn't even know you were there. I must have been asleep."

"You figured it out?" Miruka said, bemused.

"That you'd come? I think I saw you in a dream, but..."

"The graph, I mean."

"Oh, of course. The circles. No, that I didn't figure out."

"Subgroups."

"Subgroups?"

"Hey, Miruka!" Yuri cut in. "You're always doing math talks with these guys, right? Mind taking things down a notch for summer vacation, so I can keep up?" Yuri idolized Miruka, so she had been looking forward to this chance to study math with her.

We moved to a bigger table in the back of the library, sitting with Miruka at the head.

"Okay, Yuri," she began. "Let's use ladder diagrams to talk about groups."

And thus begun a new journey, a summertime trek through group theory.

3.2.2 Group Axioms

"As you found in your investigation of ladder diagrams, you can create a three-lined network in one of six patterns. There are six unique ladders you can create, each fundamentally different according to how you arrange the numbers $1, 2, 3$. Let's call this set S_3."

$$S_3 = \{[1\,2\,3], [1\,3\,2], [2\,1\,3], [2\,3\,1], [3\,1\,2], [3\,2\,1]\}$$

"Nothing hard about that—all I did was line up all the possibilities and use brackets to separate them out. So S_3 is the set of all possible three-lined ladder diagrams. It's important to note that these six elements are not randomly chosen. Each have distinct characteristics, but they're also interrelated.

"I imagine Yuri created this graph because she wanted to show the characteristics of what she calls 'ploppers,' 'swappers,' and 'scramblers,' and how they are related within ladder diagrams. The set S_3 has a structure. A mathematical structure, mind you. We want to get a better grasp on this structure, and figure out how to describe it. To do so, we'll use one of the most fundamental tools of mathematics—groups. So let's start with the definition of groups, and apply their structure to the set S_3."

"One quick question," Tetra said. "Why did you name it S_3? I guess the 3 comes from the three lines in these ladder diagrams, but why 'S'?"

"For 'symmetry,'" Miruka said. "Actually, this group S_3 already has a name in group theory, the 'symmetric group of degree three.'"

"What's that?" Yuri asked. "Is this group theory stuff something I'll be able to follow?"

"It is, but it's easy to confuse its abstractness as difficulty at first, starting with the axioms that define groups."

Definition of groups (the group axioms)

A *group* is any set satisfying the following axioms:

G1 Closure under a group operation $\star$.

G2 Associativity for all elements.

G3 Existence of an identity.

G4 Existence of an inverse for each element.

"Yep, looks hard," Yuri muttered.

"No, because we have your ladder diagram as a concrete example," Miruka said. "A set of ladder diagrams satisfies the conditions of these axioms. It provides a way over the first hurdle for understanding group theory. And that's important because..." Miruka leaned in toward Yuri and whispered, "because this is our inheritance from one of the greatest geniuses of mathematics, Évariste Galois."

"Our inheritance?"

"That's right. Galois raised groups up onto the stage of mathematics as a way to capture the form of equations, and later mathematicians encoded them as these axioms. Today, understanding the group axioms is an important first step toward receiving our inheritance. He's received it, she has too," Miruka said, nodding at me and Tetra in turn. "Think you're ready?"

Yuri's face was solemn. "I am."

Group Axiom G1 (Closure under a group operation)

"First we need to define the group operation $\star$. Let's say when we want to show a ladder diagram x feeding into another ladder y, we'll write this."

$$x \star y$$

"The result will be another ladder. That's all we need for the definition. Using the $\star$ operation, we can write equations that show how we connect one ladder diagram to another. For example, if we want to show connecting [3 2 1] below [3 1 2] to get [2 1 3], we write this."

$$[3\,1\,2] \star [3\,2\,1] = [2\,1\,3]$$

"If x and y are both three-lined networks, their combination $x \star y$ will be a three-lined network too. To say this using the language of group theory, if x and y are both elements of set S_3, then $x \star y$ too is an element of set S_3. Oh, sometimes you'll also hear group elements called 'members.' Same thing. Anyway, let's write this using symbols."

$$\text{If } x \in S_3 \text{ and } y \in S_3, \text{ then } x \star y \in S_3$$

"This is called 'closure.' In this case, it means you'll never join two three-lined ladders and end up with a five-lined ladder, for example.

That's what it means to say that S_3 is closed under the $\star$ operation, and this is the first of the group axioms, G1."

Tetra turned her notebook around to show us. "This would be an operation table for $\star$, right?"

		y					
	⋆	[1 2 3]	[1 3 2]	[2 1 3]	[2 3 1]	[3 1 2]	[3 2 1]
	[1 2 3]	[1 2 3]	[1 3 2]	[2 1 3]	[2 3 1]	[3 1 2]	[3 2 1]
	[1 3 2]	[1 3 2]	[1 2 3]	[3 1 2]	[3 2 1]	[2 1 3]	[2 3 1]
x	[2 1 3]	[2 1 3]	[2 3 1]	[1 2 3]	[1 3 2]	[3 2 1]	[3 1 2]
	[2 3 1]	[2 3 1]	[2 1 3]	[3 2 1]	[3 1 2]	[1 2 3]	[1 3 2]
	[3 1 2]	[3 1 2]	[3 2 1]	[1 3 2]	[1 2 3]	[2 3 1]	[2 1 3]
	[3 2 1]	[3 2 1]	[3 1 2]	[2 3 1]	[2 1 3]	[1 3 2]	[1 2 3]

Operation table for $x \star y$ (gray cells show that $[3\,1\,2] \star [3\,2\,1] = [2\,1\,3]$).

"When did you write all this out?" asked Yuri.

Tetra smiled. "While Miruka was explaining it all."

I was impressed not only with how fast Tetra was, but the fact that she knew what was coming up beforehand.

Yuri looked bothered. "But is this really okay? We can just make up new symbols, like this star?"

"If you give a definition," Miruka said.

"Just a definition?"

"Sure. We defined $x \star y$ as 'an x with a y appended beneath it.' It's clear what that means, that's all that's important. So long as we give a good definition, we can use any symbol to mean anything we want."

"Got it," Yuri said without hesitation.

"Defining an operation that satisfies the group axioms for some set is exactly how we give a set the structure of a group," Miruka said. "Let's test your understanding so far. Can you calculate this?"

$$[2\,3\,1] \star [2\,1\,3] = ?$$

"Hang on." Yuri put on her glasses and began sketching a ladder diagram.

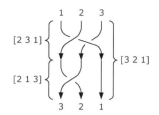

Calculation of $[2\,3\,1] \star [2\,1\,3]$.

"Is this it?"

$$[2\,3\,1] \star [2\,1\,3] = [3\,2\,1]$$

"Well done," Miruka said. "See how you went back to the ladder diagrams to work through it? That's going back to the definition, which is a fine thing to do."

Yuri brightened up. "I guess definitions are more fun than I thought. It's pretty cool to think these things up on your own."

"Next question. Is this true?"

$$[2\,3\,1] \star [2\,1\,3] = [2\,1\,3] \star [2\,3\,1] \quad (?)$$

Yuri thought for a moment, but soon said, "Nope! Because $[2\,3\,1]\star[2\,1\,3] = [3\,2\,1]$, but $[2\,1\,3]\star[2\,3\,1] = [1\,3\,2]$, so they aren't equal."

"Good. This means the operation $\star$ isn't commutative."

"Ah, so $x \star y$ and $y \star x$ are always different," Tetra said.

"No, that can't be true," Yuri said. "They *can* be different, but they don't *have* to be."

Tetra blushed. "Oh, you're right. I guess $x \star [1\,2\,3] = [1\,2\,3] \star x$, for example.

"Also, $x \star y = y \star x$ if $x = y$!" Yuri added. I could practically see sparks flying between the two of them.

"Let's take a look at the second axiom," Miruka said.

Group Axiom G2 (Associativity)

"Associativity says that $(x \star y) \star z = x \star (y \star z)$ holds for any x, y, z. This is clearly true for ladder diagrams, where we would just have ladders x, y, z joined in order on both sides. On the left, we start with $x \star y$, which produces some ladder, then we feed that into z to get $(x \star y) \star z$. On the right, we feed x into $y \star z$, giving us $x \star (y \star z)$. The result will be the same in either case, so Axiom G2 holds."

$$(x \star y) \star z = x \star (y \star z)$$

"Since we can assume associativity, we can just remove the parentheses from $(x \star y) \star z$ or $x \star (y \star z)$, writing $x \star y \star z$ instead."

Group Axiom G3 (existence of an identity)

"The third axiom says a group must have an identity, which is defined like this."

Definition of identity

An element e is called an *identity* (or an *identity element*) for the operation $\star$ if for every element a,

$$a \star e = e \star a = a.$$

"In the case of a ladder diagram, Yuri's 'plopper' [1 2 3] is the identity, because this is always true."

$$a \star [1\,2\,3] = [1\,2\,3] \star a = a$$

"This just says that if we attach a plopper beneath any ladder a, we haven't done anything to change a, and the same is true if we place a plopper on top of a."

"So an identity is like a 0 in addition, or a 1 in multiplication," Tetra said to Yuri.

"How's that?" Yuri asked.

"Because numbers don't change when you add 0 to them, or when you multiply them by 1."

"Oh, right! Just like ploppers never change a ladder they're connected to!"

"On to group axiom G4, then," Miruka said.

Group Axiom G4 (existence of an inverse)

"The fourth group axiom says all elements in a group must have an inverse in the group. An inverse is defined like this."

Definition of inverse

An element b is called the *inverse* of the element a if

$$a \star b = b \star a = e,$$

where e is the identity element.

"For example, the inverse of $[2\,3\,1]$ would be a ladder diagram b that makes this true."

$$[2\,3\,1] \star b = [1\,2\,3]$$

"In other words, when we attach b below $[2\,3\,1]$, the result is the identity element $[1\,2\,3]$. You should be able to see that in this case b would be $[3\,1\,2]$, because $[3\,1\,2]$ is the 'scrambler' that undoes $[2\,3\,1]$. Here it is as an equation."

$$[2\,3\,1] \star [3\,1\,2] = [1\,2\,3]$$

"Let's think of this in terms of actual ladders, where an inverse ladder looks like a mirror reflection of the ladder that it's inverting. It just puts everything back where it was."

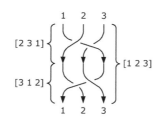

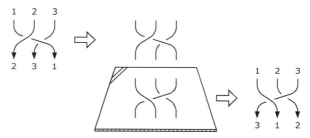

The inverse of [2 3 1] is [3 1 2] (a mirror reflection).

"Each element will have its own inverse—if the element changes, the inverse must too. The inverses for elements in S_3 would be—"

"Like this!" Tetra said, rotating her notebook to show us another table.

Element	[1 2 3]	[1 3 2]	[2 1 3]	[2 3 1]	[3 1 2]	[3 2 1]
Inverse	[1 2 3]	[1 3 2]	[2 1 3]	[3 1 2]	[2 3 1]	[3 2 1]

"Nicely played," Miruka said.
"Wow, you're fast," I said.
"Mmm..." Yuri said.

Tetra blushed. "It's not so hard, you just have to look at the operation table. It tells us which elements are inverses of each other."

★	[1 2 3]	[1 3 2]	[2 1 3]	[2 3 1]	[3 1 2]	[3 2 1]
[1 2 3]	[1 2 3]	[1 3 2]	[2 1 3]	[2 3 1]	[3 1 2]	[3 2 1]
[1 3 2]	[1 3 2]	[1 2 3]	[3 1 2]	[3 2 1]	[2 1 3]	[2 3 1]
[2 1 3]	[2 1 3]	[2 3 1]	[1 2 3]	[1 3 2]	[3 2 1]	[3 1 2]
[2 3 1]	[2 3 1]	[2 1 3]	[3 2 1]	[3 1 2]	[1 2 3]	[1 3 2]
[3 1 2]	[3 1 2]	[3 2 1]	[1 3 2]	[1 2 3]	[2 3 1]	[2 1 3]
[3 2 1]	[3 2 1]	[3 1 2]	[2 3 1]	[2 1 3]	[1 3 2]	[1 2 3]

(with y labeling columns and x labeling rows)

Locations where the operation result is [1 2 3].

Miruka looked at each of us in turn. "So then, we've confirmed that our set S_3 of all three-lined ladder diagrams satisfies the group axioms for the operation ★. And this means—" Miruka opened her arms wide as if about to make a grand announcement. "—the set S_3 of all three-lined ladder diagrams is a group under the operation ★.'"

3.2.3 Axioms and Definitions

"Miruka, this *is* hard," Yuri said. "Why do we have to mess around with the group axioms, anyway?"

"Common perspective," Miruka said.

"Huh?"

"When we look at things as groups, we can consider our findings related to ladder diagrams and to rotations of an equilateral triangle as both being one and the same. Anything satisfying axioms G1 through G4 is a group. That means any theorem we can prove using just those axioms will also be true for all groups. No matter what set we're dealing with, if we can call it a group we can treat it the same as any other."

"Yeah, still not getting it," Yuri said.

"Try this."

Problem 3-1 (Number of identity elements)

Does there exist a group having two identity elements?

"Nope!" Yuri immediately answered.

"Why not?" Miruka asked.

"Well, because there's only one plopper for the ladders."

"That's true, we only found one identity for a three-lined ladder diagram, namely [1 2 3]. But are you sure that's the case for *all* groups? Can you prove that?"

"Well I can't prove anything if you don't tell me what kind of group it is."

"Let's see what he thinks," Miruka said, pointing a thumb at me.

Yeah, kinda expected that.

"Okay, let's prove that no group can have two identity elements. Say we have some group G. From group axiom G3, we can select from G some element having the property of an identity. Say we've selected two, in fact, which we'll call e and f. Now we just need to show that e and f must be the same element.

"Since e is an identity element, the definition of identity says we can apply the operation using it and some element g in G to get this."

$$e \star g = g$$

"Since f is an identity too, we can use it to do the same thing."

$$f \star g = g$$

"Now we know that $e \star g$ and $f \star g$ are both equal to g, so we can say this."

$$e \star g = f \star g$$

"Group axiom G4 assures us of the existence of some inverse for g. Let's call that inverse h, and multiply both sides of this equation by h."

$$(e \star g) \star h = (f \star g) \star h$$

"We know G is associative from group axiom G2, so we can move the parentheses like this."

$$e \star (g \star h) = f \star (g \star h)$$

"But h was the inverse of g, so from the definition of inverse, $g \star h$ is the identity element. That in turn means the $e \star (g \star h)$ on the left equals e and the $f \star (g \star h)$ on the right equals f, so we get this."

$$e = f$$

"And that's it. If you select from group G two elements having the property of an identity, you must have selected the same element twice. It therefore follows that there is no group with two identity elements. QED."

"Wow, proofs are amazing," Yuri said.

Miruka cocked her head. "How's that?"

"Because they save you from having to say 'Well what about this group?' 'What about this one?' It's pretty cool that we can just say one thing about all of them at once. 'If you're a group, you've only got one identity,' things like that."

"That's the power of logic, one of your favorite subjects," I said. "We can make blanket statements like that only because the group axioms define what a group is. It's the axioms that lead us to the definition."

Yuri furrowed her brow. "Definitions from axioms, huh?"

Answer 3-1 (Number of identity elements)

There exists no group with two identity elements.

3.3 CYCLIC GROUPS

3.3.1 Taking a Break

After our discussion of group theory, Tetra proposed we go somewhere to take a break. Stepping outside of the air-conditioned school building, the heat hit me like a wall.

"*Wow*, it's hot!" Yuri shouted. "So Tetra, how were you able to do that?"

"Do what?" Tetra asked.

"Just whip out those tables of group operations and inverses and stuff. It's almost like you had planned the whole thing out with Miruka."

"No, no. I mean, Miruka has told us about the definition of groups before, so I already knew how they're defined. So I just thought about how to apply the group axioms to our set S_3 as she described it. That's how I was able to come up with the tables so quick."

"Huh. So do you study this group theory stuff in high school?"

"You might get the very basics in high school," I cut in. "But to get the full story you have to study on your own, in math books."

3.3.2 Structure

We entered the student lounge, or what our school called an "amenity space." It was in a separate building, and served as a convenient place for students to gather and hang out between classes or after school. Since it was summer vacation, the student store was closed, but thankfully the air conditioner was on.

"So much better!" Yuri said, flapping the neck of her uniform blouse to take in the cool air.

We bought drinks at a vending machine and moved to one of the tables.

Tetra cracked open her juice. "So tell me what you mean by structure, Miruka. When I say structure I'm thinking of buildings, things like that. Like, how a building's shape helps it stay together." Tetra interlaced her fingers to demonstrate a solid structure. "But I don't think that's what you were talking about earlier."

Miruka sipped her iced coffee. "Consider things with structure, and those without. You brought up the example of a building as something that has structure. You could say the same thing about a machine. But clearly gasses and liquids don't have structure. A thing with structure can be broken down into parts, and we can name those parts. We can also compare parts, exchange them, and think about how they're related."

"Okay, sure, I see that. Like the first and second floors in a building."

"Of course, even gasses and liquids have a molecular structure," I said.

Miruka nodded. "Fair enough. Sometimes a change in perspective can reveal hidden structure. The difference between macrostructure and microstructure is simply a question of perspective."

"So we can break sets and groups down into parts too?" Tetra asked.

"Of course," Miruka replied. "Sets into subsets and groups into subgroups."

3.3.3 Subgroups

"When you select part of a set, you create a subset," Miruka said. "For example, we can select only the 'swappers' from S_3 and call the result X. Then X is the set of all swappers, and it's also a subset of S_3."

$$S_3 = \{\ [1\,2\,3],\ [1\,3\,2],\ [2\,1\,3],\ [2\,3\,1],\ [3\,1\,2],\ [3\,2\,1]\}$$
$$X = \{\ [1\,3\,2],\ [2\,1\,3],\ [3\,2\,1]\}$$

"We can show the relation between sets X and S_3 using the 'subset' symbol, $\subset$. You can read $X \subset S_3$ as 'X is a subset of S_3,' or 'X is included in S_3.' Other subsets of S_3 are S_3 itself[1] and the null set, which has no elements at all.

$$X \subset S_3 \qquad \text{X is a subset of } S_3$$
$$S_3 \subset S_3 \qquad S_3 \text{ itself is a subset of } S_3$$
$$\{\} \subset S_3 \qquad \text{the null set is a subset of } S_3$$

"I remember this," Tetra said, nodding. I noticed Yuri was nodding along with her.

"Just like we can talk about sets and subsets, we can also talk about groups and subgroups."

"That would be when we consider a subset as a group?"

"Right. We have to be a little bit more careful, though. Tell me, if you select some number of elements from a group, is the result a subgroup?"

"No," Yuri said. "At least, not always."

"Why not?"

"Because it doesn't satisfy."

---

[1] Some authors prefer to write $A \subseteq B$ for "A is a subset of B, possibly equal to B," and $A \subset B$ for "A is a subset of B, but not equal to B." In that case, $A \subseteq A$ is true, but $A \subset A$ is false.

"A bit more precision," Miruka shot back. "What doesn't satisfy what?"

"Because the part you selected, no, wait..." Yuri paused to choose her words carefully. "Because a subset might not satisfy the group axioms."

Miruka placed a hand on Yuri's head. "Very good. I'm glad to see you understand this. You will always get a subset when you select elements from a set, but that's not necessarily the case for subgroups. Take this set X of all swappers we just created. That won't be a subgroup of S_3. In other words, X isn't a group. Why not, Tetra?"

$$X = \{[1\,3\,2], [2\,1\,3], [3\,2\,1]\} \quad \text{not a group}$$

"Um, let's see. Well, X doesn't have an identity, so it doesn't satisfy group axiom G3. That's enough to say X it isn't a group."

"Now I get it!" I said. "Those circles you drew on Yuri's chart. You were indicating subgroups, weren't you. That's why every circle included the identity element $[1\,2\,3]$."

"Well spotted. Did you notice there are six circles?" Miruka picked up the pencil. "Here are the six subgroups of S_3."

Galois Theory

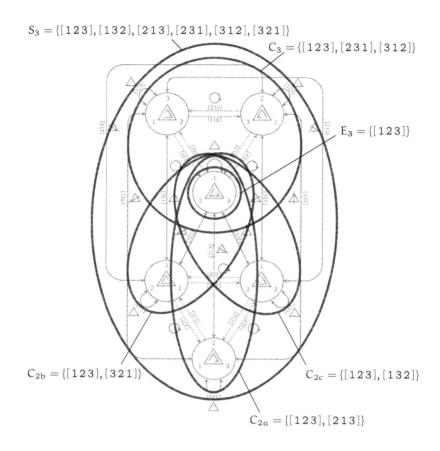

$S_3 = \{[1\,2\,3], [1\,3\,2], [2\,1\,3], [2\,3\,1], [3\,1\,2], [3\,2\,1]\}$
$C_3 = \{[1\,2\,3], [2\,3\,1], [3\,1\,2]\}$
$E_3 = \{[1\,2\,3]\}$
$C_{2b} = \{[1\,2\,3], [3\,2\,1]\}$
$C_{2c} = \{[1\,2\,3], [1\,3\,2]\}$
$C_{2a} = \{[1\,2\,3], [2\,1\,3]\}$

$$S_3 = \{[1\,2\,3], [1\,3\,2], [2\,1\,3], [2\,3\,1], [3\,1\,2], [3\,2\,1]\}$$
$$C_3 = \{[1\,2\,3], [2\,3\,1], [3\,1\,2]\}$$
$$C_{2a} = \{[1\,2\,3], [2\,1\,3]\}$$
$$C_{2b} = \{[1\,2\,3], [3\,2\,1]\}$$
$$C_{2c} = \{[1\,2\,3], [1\,3\,2]\}$$
$$E_3 = \{[1\,2\,3]\}$$

Subgroups of the symmetric group S_3.

"S_3 includes itself as a subgroup, since every group is a subset of itself. We create C_3 using the scramblers to produce another subgroup of S_3. This corresponds to Tetra's rotations of equilateral triangles. If we rotate an equilateral triangle by 120° three times, we end up where we started. The inverse of a 120° rotation is a rotation by −120°. A group like this is called a cyclic group of order 3.

"We create C_{2a}, C_{2b}, and C_{2c} using the swappers. These too are subgroups of S_3, with each corresponding to flipping an equilateral triangle along a different axis of symmetry. Two flips and you're back to where you started. These are cyclic groups of order 2.

"E_3 is a group comprising just one element—our identity, the plopper—but it's still a subgroup of S_3. This is called the identity group, or the trivial group.

"Any questions, Yuri?"

"One. Saying S_3 is a subset of itself sounds an awful lot like saying S_3 is a subgroup of itself."

"It does. And?"

I looked at Miruka. She clearly knew what Yuri was trying to say, but wanted to be sure she said it herself.

"It's just..." Yuri thought for a moment before continuing. "It seems like if the null set is a subset of S_3, it should also be a subgroup of S_3."

"But Yuri," Tetra said, "the null set—*mrph*"

"You hush." With her hand still over Tetra's mouth, Miruka said, almost singing, "Yuri, Yuri, my lovely logical Yuri, can the null set really be a subgroup of S_3?"

"Well, I mean—" Yuri froze, and after a moment something clicked. "Oh, wait. The null set has no elements, which means it can't have an identity element, which means it can't be a group!"

Miruka nodded. "That's right."

Tetra nodded too. "*Nrmph.*"

"To review, we are looking for structure in the symmetric group S_3. In all, we found six subgroups. This is one characteristic of S_3."

3.3.4 Order

Miruka removed her hand from Tetra's mouth.

"Whew," Tetra said. "I thought you were going to smother me!"
"Yuri had to answer that question on her own," Miruka said.
"Group theory is fun!" Yuri said.
"Then let me introduce you to a new vocabulary word that will be useful in our exploration of the symmetric group S_3. The number of elements in a group is called its order. So the order of S_3 is 6."

"So it's like the size of the group, right?" Tetra said. "The bigger the order the bigger the group, and vice versa."

"You could say that," Miruka said. "You could also say that knowing the 'size' of something in some sense is fundamental to exploring its structure."

Yuri grinned at me. "It's only natural to wonder how many things there are, right cuz?"

"So is S_3 the only group with order 6?" Tetra asked.

"A natural question to ask," Miruka said. "Find a concept, define your terms, and wonder what else there is to learn. That's how you discover problems to solve. Problems like, 'Are there any other groups with order 6?' That feeling of wanting to know more is the fuel of mathematics."

"Uh, right. So... are there?"

"Now what fun would it be if I just went and told you?"

Problem 3-2 (Groups of order 6)

Does there exist any group other than S_3 having order 6?

3.3.5 Cyclic Groups

We took a quick break to clean up our drinks, but decided to stay in the lounge rather than lose valuable math time trekking back to the library. It wasn't like we needed the books there, after all.

Yuri kicked us back into gear. "Before, you called C_3 a cyclic group, but what's that?"

$$C_3 = \{[1\,2\,3], [2\,3\,1], [3\,1\,2]\}$$

"A group that cycles," Tetra said. "Isn't that what it sounds like? Something spinning around and around and around?"

"But do you know what that means mathematically?" Miruka asked.

"Er..."

"Okay then, let's measure your understanding of cyclic groups. Define one."

Definitions reconfirm our understanding, I thought. *A fine followup to my usual motto, "Examples are the key to understanding." We already have an example of a cyclic group, so we understand what one is at that level. Now it's time for the next step, creating our own definition.*

"Well, it's a cyclic group," Tetra said, "so I guess its elements go around... No, wait. I shouldn't say 'around,' right? In my head I keep picturing a rotating triangle, but that doesn't make for a mathematical definition. So... I'm sorry, I guess I don't understand this after all."

"How about you?" Miruka said, looking at me.

"A definition for cyclic groups? Well, I understand why we might want to imagine something rotating, but in the case of a group, I think it might be closer to repeatedly following in our own footsteps."

"Meaning what?" Yuri said.

"I think it's about repeatedly applying the operation to the same element. Take [2 3 1], for example. If we keep applying the operation to it, we end up back where we started."

$$
\begin{aligned}
[2\,3\,1] &= [2\,3\,1] \\
[2\,3\,1] \star [2\,3\,1] &= [3\,1\,2] \\
[2\,3\,1] \star [2\,3\,1] \star [2\,3\,1] &= [1\,2\,3] \\
[2\,3\,1] \star [2\,3\,1] \star [2\,3\,1] \star [2\,3\,1] &= [2\,3\,1] \quad \leftarrow \text{back!} \\
[2\,3\,1] \star [2\,3\,1] \star [2\,3\,1] \star [2\,3\,1] \star [2\,3\,1] &= [3\,1\,2] \\
[2\,3\,1] \star [2\,3\,1] \star [2\,3\,1] \star [2\,3\,1] \star [2\,3\,1] \star [2\,3\,1] &= [1\,2\,3] \\
[2\,3\,1] \star [2\,3\,1] \star [2\,3\,1] \star [2\,3\,1] \star [2\,3\,1] \star [2\,3\,1] \star [2\,3\,1] &= [2\,3\,1] \quad \leftarrow \text{back!} \\
&\vdots
\end{aligned}
$$

"Neat!" Yuri said.

"When we did this we only got three elements, [2 3 1], [3 1 2], and [1 2 3], and these are exactly the elements that make up C_3, which Miruka called a cyclic group of order 3.

$$C_3 = \{[1\,2\,3],\ [2\,3\,1],\ [3\,1\,2]\}$$

"Correct, but long-winded," was Miruka's harsh evaluation. "All you needed to say is that a cyclic group is a group generated from a single element."

"Uh... Really?"

"Back to C_3. Take one element, [2 3 1] for example, and apply the operation to it to get what we'll call a product. We'll also call repeated application of the operation 'exponentiation.' So what you found is powers of [2 3 1], its square and its cube and so on."

"Hey, this feels a lot like when we were 'squaring' ladders, doesn't it." Yuri said.

"It does indeed," I said.

"One thing, though," Miruka said. "You focused on getting back to where you started at the fourth and seventh powers, which is fine, but you should also notice how the identity element is appearing at the third and sixth powers."

$$[2\,3\,1]^1 = [2\,3\,1]$$
$$[2\,3\,1]^2 = [3\,1\,2]$$
$$[2\,3\,1]^3 = [1\,2\,3] \quad \leftarrow \text{identity!}$$
$$[2\,3\,1]^4 = [2\,3\,1] \quad \leftarrow \text{back!}$$
$$[2\,3\,1]^5 = [3\,1\,2]$$
$$[2\,3\,1]^6 = [1\,2\,3] \quad \leftarrow \text{identity!}$$
$$[2\,3\,1]^7 = [2\,3\,1] \quad \leftarrow \text{back!}$$
$$\vdots$$

"Interesting," Tetra said. "It's like the identity element shows up and resets everything, so we start over from the beginning. Okay, now I think I really am getting this cyclic group thing."

"So now you see how we can also write C_3 like this."

$$C_3 = \{[2\,3\,1]^1,\ [2\,3\,1]^2,\ [2\,3\,1]^3\}$$

"And now it makes much more sense when you say the group is generated from just one element, [2 3 1] in this case. By the way, [2 3 1] here is called a generator of C_3."

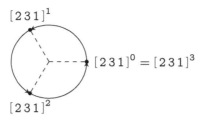

"If you want to generalize," Miruka continued, "you can write a cyclic group of order n with a generator a like this, with a^n being the identity element."

$$\{a^1, a^2, a^3, \ldots, a^{n-1}, a^n\}$$

"Couldn't we also write the identity element as a^0?" Tetra asked.

"You could. $a^0 = a^n =$ identity, so you could also describe a cyclic group like this."

$$\{a^0, a^1, a^2, a^3, \ldots, a^{n-1}\}$$

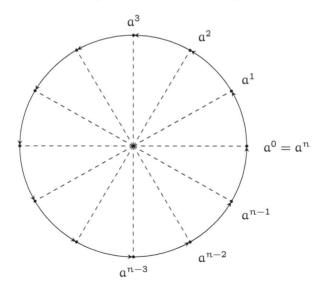

3.3.6 Abelian Groups

"Another question for you," Miruka said. "Are all cyclic groups abelian?"

"No idea," Yuri said.

"No!" Miruka slapped the table, causing Yuri to shrink back. "That's not what you say. You should be asking what the definition of an abelian group is.

"Oops, sorry. So... what's the definition of an abelian group?"

"Better. An abelian group—named after Niels Henrik Abel, by the way—is a group for which the commutative law holds. For that reason they're also sometimes called commutative groups. The commutative law doesn't necessarily hold for all groups. In other words, you can't assume $x \star y = y \star x$. To investigate whether all cyclic groups are abelian, we need to determine whether the commutative law does indeed hold for them. But anyway, just remember that this unfamiliar term 'abelian group' is just another way of saying 'a group that satisfies the commutative law,' something you should understand well." Miruka's tone suddenly softened. "So, Yuri."

"Yes?!"

"Let's find out if $x \star y = y \star x$ for an arbitrary x, y. Say the generator for our cyclic group is a, and x and y are arbitrary elements. Then we can write $x = a^j, y = a^k$ for some integers j and k, so we can write this."

$$x \star y = a^j \star a^k$$
$$= \underbrace{(a \star a \star \cdots \star a)}_{j \text{ of these}} \star \underbrace{(a \star a \star \cdots \star a)}_{k \text{ of these}}$$

"Now we use associativity of the group..."

$$= \underbrace{a \star a \star \cdots \star a}_{j + k \text{ of these}}$$

"... and associativity again to regroup the k and j items."

$$= \underbrace{(a \star a \star \cdots \star a)}_{k \text{ of these}} \star \underbrace{(a \star a \star \cdots \star a)}_{j \text{ of these}}$$
$$= a^k \star a^j$$
$$= y \star x$$

"And so we've shown that $x \star y = y \star x$. One restriction, though—this proof assumes a finite order. If we're dealing with a cyclic group of infinite order, we need to worry about negative powers, in other words exponents for inverse elements. Anyway, we've shown that all cyclic groups are abelian, but we can't say the reverse. Some abelian groups are not cyclic."

"I still can't get over my image of something spinning around," Tetra said. "I guess I'll have to get used to the idea of a group generated from a single element before I can do these proofs, but... Hey, wait a minute..."

"What's up?" I asked.

"I think I've found one! Another group of order 6!"

"Oh, that problem Miruka gave us a while back?"[2]

"Right! To make a group of order 6, we just have to create a cyclic group. Specifically, we just have to define a group with a generating element whose sixth power is the identity element!"

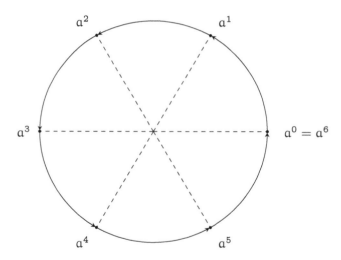

"I think I even know how to create a specific one," Tetra continued. "We just create a group that's like a rotating hexagon, with a no-backtracking rule. Then the generating element will go through $\frac{1}{6}$ of a rotation, and should create a group of order 6!"

---

[2] Problem 3-2, pg. 86

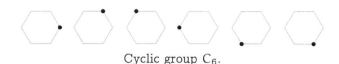

Cyclic group C_6.

"That's C_6, the cyclic group of order six," Miruka said. "However, my problem asked for a group of order 6 *other than* the symmetric group S_3. To complete the problem, you have to show that C_6 is a fundamentally different group from S_3. In other words, you have to show that the two aren't isomorphic."[3]

"Makes sense," Tetra said.

"Not that that's difficult—the symmetric group S_3 isn't cyclic, so there's no way it can be isomorphic to the cyclic group C_6."

"Ah, of course. Because S_3 isn't cyclic."

"How?"

"How?" Tetra repeated.

"How can we say that S_3 isn't cyclic?"

"Oh, uh... Because, it can't spin around?"

"Wrong," Miruka shot back. "You're relying on that mental image too much."

Yuri spoke up. "It's because it doesn't have a generating element! No matter what power you raise any of the elements in S_3 to, you can't make the whole thing!"

"You're on the right track, but your wording needs work," Miruka replied. "Instead of 'it doesn't have a generating element,' say 'it can't be generated using any single element.' While there's no single element that can generate all of S_3, you can generate it using two elements, $[2\,1\,3]$ and $[2\,3\,1]$. But as you were trying to say, you can try finding powers of the six elements in the symmetric group yourself to see that none of them generate S_3. In contrast, there is an element that can generate the cyclic group C_6, so it follows that the two are not isomorphic despite both having order 6."

---

[3] For a description of isomorphism, see *Math Girls 2: Fermat's Last Theorem*, section 6.2.10.

> **Answer 3-2 (Groups of order 6)**
>
> A group of order 6 other than the symmetric group S_3 does exist. (The cyclic group C_6 has order 6, but is not isomorphic to the symmetric group S_3.)

"Mmm..." Tetra groaned. "I'm still stuck with the spinning-around image for cyclic groups. I really need to get comfortable with this 'generated from a single element' definition."

Yuri let out a deep sigh of satisfaction. "Wow, math is fun. I just love taking squishy ideas and using definitions to hammer them into shape."

Miruka looked at both and smiled. I found myself looking at her beautiful black hair, thinking about how much I loved her... lectures. We spoke with each other, we taught each other, we thought and searched and asked questions together. This time together was... well, it was irreplaceable.

I shook my head to clear it. To switch gears, I mentally ran through everything we had talked about.

- The definition of groups and the group axioms (operation, closure, associativity, identity, inverse)

- Subgroups

- Cyclic groups and generating elements

- Cyclic groups as abelian groups

- Group order

We had treated numbers not just as numbers, forms not just as forms, operations not just as operations. Instead, we had used groups to handle them all in one package. We used numbers not for calculations, but as a tool to find hidden structures. Through groups, we could make a unified presentation of numbers and forms and operations. They were all connected.

Using mathematical expressions, we can write and calculate rotations and symmetries of figures. We can consider movements of figures as products, and repeated operations as exponentiation. Further, if an operation on a set takes us back to where we started when we repeat it enough times, we can consider that set as a cyclic group.

Through groups, all this mathematical knowledge that I had learned separately was being brought together and bound together. Such a good feeling. Even better, I knew there was still more to learn beyond all this. Much more. What would happen if we had an infinite set? What significance might the group order have? Were there any more interesting relations between groups and subgroups? I was sure I would encounter questions like this if I studied groups further.

Miruka must have read my mind. "Don't worry, eventually we'll fly through the skies to view the forest of groups from above," she said, confirming that our journey had only just begun. She smiled. "Glad to see your fever's down, by the way."

> Starting with the work of Lagrange, continuing with that by Ruffini and Abel, and finally completed through the genius of Galois, we now know that the basis for the theory of equations is symmetries among permuted roots hidden in the methods for solving equations, and when we use the concept of groups to drag the matter out into the light, we find that the principle lies in extending the invariance of groups to fields.
>
> SHIGA KOJI [14]

CHAPTER 4

In a Yoke with You

> No matter how simple the emotion, no matter how simple the poem, no matter how much they seem to have spontaneously appeared, those things did not simply arise from nature. They result from a desire to find form in the formless, stability in that which is unstable.
>
> HIDEO KOBAYASHI
> *Words*

4.1 IN THE LIBRARY

4.1.1 Tetra

"Hey! Over here!" Tetra yelled, waving.

"Oh, you're here too?" I said.

During the morning I had attended class at my summer seminar, and come to my school's library after it was done. There were several other students studying there, seniors like me who had college entrance exams coming up. Tetra was sitting among them, in front of an open notebook.

"I am. Miruka's over there." She pointed toward the window, where Miruka was busy writing something.

"Looks like you're hard at work too," I said. "Studying math?"

"Sort of. I was thinking about a research topic from Mr. Muraki." She showed me a card, blushing slightly.

$$x^{12} - 1$$

Mr. Muraki was one of the math teachers at our school. After learning that we were studying math on our own, he started giving us interesting problems to work on. The card Tetra showed me was one of the index cards he often used for that purpose. Some were easy, some were very hard, and sometimes they came as what we called "research topics," simple equations with no explanation, not even what the problem was supposed to be about. "Think about this however you want," he seemed to be saying, so we would use whatever equation he'd given us to create our own problem, solve it, and write up a report to return to him. He would read whatever we wrote, but these problems weren't for extra credit, and they didn't come with a due date. He didn't even ask us to write up the reports; we just created them out of our own interest. We took them quite seriously, despite the fact that we did them for the pure fun of it.

"Interesting," I said. "Did you come up with a problem?"

"I think so!"

4.1.2 Factorization

"When I saw the $x^{12} - 1$ on this card, I thought it might be fun to try factoring it. The twelfth power of x makes for a pretty big exponent, so I figured I should try factoring it into a product of smaller-degree expressions. So here's my problem."

Problem 4-1 (Factorization)

Factor $x^{12} - 1$.

"Factorization means writing things as a product, something like this."

$$x^{12} - 1 = (x - \alpha_1)(x - \alpha_2)(x - \alpha_3) \cdots (x - \alpha_{12})$$

"There will be twelve alphas here, from α_1 to α_{12}, so my goal is to find twelve specific numbers that fit. So first, the reason there has to be twelve of them is...uh...because $x^{12} - 1$ is a twelfth-degree polynomial in x. If we're going to break this twelfth-degree polynomial down into a product of first-degree expressions like $(x - \alpha_k)$, we have to gather up twelve of them. Without that, we wouldn't get the x^{12} term. As to what these twelve numbers have to be, well, what we want to factor is this polynomial $x^{12} - 1$, so if we set that equal to zero as the equation $x^{12} - 1 = 0$, then we should get solutions like $x = \alpha_1, \alpha_2, \alpha_3, \ldots, \alpha_{12}$. That's true because $(x - \alpha_1)(x - \alpha_2)(x - \alpha_3) \cdots (x - \alpha_{12}) = 0$ implies that x must be one of $\alpha_1, \alpha_2, \alpha_3, \ldots, \alpha_{12}$."

$x^{12} - 1$ the polynomial we want to factor

$x^{12} - 1 = 0$ an equation having solutions $x = \alpha_1, \alpha_2, \alpha_3, \ldots, \alpha_{12}$

"In other words, the problem 'factor the polynomial $x^{12} - 1$' is essentially the same as the problem 'find the solutions for $x^{12} - 1 = 0$.'"

4.1.3 Scope of Numbers

Tetra paused to catch her breath. "Good so far?"

"All good," I said. "A very easy-to-follow explanation. One thing about the wording, though. You should say that the problem 'find the solutions for $x^{12} - 1 = 0$' is essentially the same as the problem 'find the roots of the polynomial $x^{12} - 1$.' We talk about the *solutions* to equations, but the *roots* of polynomials. I have seen some math books that use 'root' when talking about equations, though."

"Solutions and roots," Tetra said. "Okay, I'll remember that."

"But anyway, you're definitely thinking about this in the right way. You're given a polynomial, you want to factor it into first-degree polynomials, so you look for solutions—or the roots—when you set the polynomial equal to zero. That's a natural way to think about things. No problem up to that point. There's one thing you haven't addressed, though. Are you trying to do this factorization in the scope of the real numbers, or the complex numbers?"

"Uh...I hadn't thought about that."

"I mean, you want to change $x^{12}-1$ into a product of degree one expressions, right? In other words, you want to look for degree one factors."

"That's right. I'm looking for twelve factors in the form $(x-\alpha_k)$. I want to find the α_k's that give a solution."

"So are these α_k values real numbers? Or are they complex numbers? Are you thinking about *where* you're looking for these numbers?"

"Um, not really. So is something like this wrong?"

Tetra showed me something she'd written in her notebook.

if $x=1$	$x^{12}-1 = 1^{12}-1$	$= 1-1 = 0$
if $x=-1$	$x^{12}-1 = (-1)^{12}-1$	$= 1-1 = 0$
if $x=i$	$x^{12}-1 = i^{12}-1$	$= 1-1 = 0$
if $x=-i$	$x^{12}-1 = (-i)^{12}-1$	$= 1-1 = 0$

"No, no. That's not wrong. You've found four of the twelve solutions, namely $x=1,-1,i,-i$. But you still need to be conscious of what set of numbers you're thinking about. If you're exploring within the scope of complex numbers, you can definitely factor the polynomial into first-degree expressions. If you're sticking to the reals or the rationals though, well, things won't necessarily work out that way."

"Well, two of the solutions I found were i and $-i$, which aren't real numbers, so I guess I'm searching within the complex numbers. Not that I was consciously doing so."

"Great. And you've found four factors so far, right?"

"Right, $1, -1, i,$ and $-i$. That's how far I've gotten."

$$x^{12}-1 = (x-1)(x+1)(x-i)(x+i)(\cdots\cdots)$$

(factoring with complex coefficients)

"Do you know what happens if you factor only in the realm of real numbers?" I asked.

"Well, I guess I couldn't use i in that case, so... what does that mean?"

"Well, if we expand the $(x-i)(x+i)$ part we'll stay within the scope of the real numbers."

$$(x-i)(x+i) = x^2 + xi - ix - i^2 = x^2 + 1$$

"Here's what we get."

$$x^{12} - 1 = (x-1)(x+1)\underline{(x^2+1)}(\cdots\cdots)$$

(factoring with real coefficients)

"I see."
"So now you're in the middle of a voyage in search of what to put in place of the dots."
"Right. Actually, it's all well and good that I've replaced the factorization problem with solving an equation, but that doesn't really make the problem any easier. Even if I'm thinking in terms of the complex numbers..."

$$x^{12} - 1 = (x-1)(x+1)(x-i)(x+i)(\underbrace{\cdots\cdots})$$

"...I'm still not sure what to do with the underlined part."
"I see. You have no idea what kind of expression will replace that?"
"What do you mean?"
"For example, what will be the degree of the replacing expression?"
"Eight, I guess?"
"Sure, since the degree of a product is the sum of the degrees."

$$\underbrace{x^{12}-1}_{\text{degree 12}} = \underbrace{(x-1)}_{\text{degree 1}}\underbrace{(x+1)}_{\text{degree 1}}\underbrace{(x-i)}_{\text{degree 1}}\underbrace{(x+i)}_{\text{degree 1}}\underbrace{(\cdots\cdots)}_{\text{degree 8}}$$

$$12 = 1 + 1 + 1 + 1 + 8$$

4.1.4 Polynomial Division

Before resuming, I paused to reflect on how comfortable I was in this library, and how much fun it was to work through this problem that Tetra had created.

"We can make some more progress using polynomial division," I said. "Like this."

$$x^{12} - 1 = (x-1)(x+1)(x-i)(x+i)(\underbrace{\cdots\cdots})$$

"We can divide both sides of this by $(x-1)(x+1)(x-i)(x+i)$ to isolate the dotted part."

$$\frac{x^{12}-1}{(x-1)(x+1)(x-i)(x+i)} = (\ldots\ldots)$$

"See? Now if we expand the denominator we get $(x-1)(x+1)(x-i)(x+i) = (x^2-1)(x^2+1) = x^4-1$, so we want to divide $x^{12}-1$ by x^4-1."

$$\frac{x^{12}-1}{x^4-1} = (\ldots\ldots)$$

"You learned polynomial division in school, right?"
Tetra nodded silently, and started writing.

$$\begin{array}{r}
x^8 + x^4 + 1 \\
x^4-1 \overline{\smash{\big)}\ x^{12} -1} \\
\underline{-x^{12}+x^8 } \\
x^8 \\
\underline{-x^8+x^4 } \\
x^4-1 \\
\underline{-x^4+1} \\
0
\end{array}$$

"So it looks like we can factor $x^{12}-1$ like this," I said.

$$x^{12}-1 = (x-1)(x+1)(x-i)(x+i)\underbrace{(x^8+x^4+1)}$$

"And $x^8 + x^4 + 1$ is an eighth-degree polynomial!"
"Sure is. So now what we want to do is factor it."
"Which means we want to find the solution to $x^8 + x^4 + 1 = 0$, right?"
"Right. So letting $y = x^4 \ldots$" I paused to do some calculations in my head. "Hmm, looks like the root of ω might be a problem$\ldots$"

I sat there, worrying over how to proceed. I glanced over at Miruka. Our black-haired genius was sitting in the same place, still writing.

"Want me to call her over?" Tetra asked.

"Why?"

"Oh, uh, nothing. Just..." she looked down. "Well, we often turn to Miruka when we get stuck. That seems to be the pattern or whatever, but... I mean, I know it's important that we think on our own, but she's our go-to when we need a hint, or want to know the deeper meaning behind some bit of mathematics."

"Yeah, I guess. But we aren't there quite yet. We just need a change of perspective."

"Meaning?"

"Let's go back to the equation $x^{12} - 1 = 0$."

4.1.5 The Twelfth Root of 1

"We start by moving the constant part of $x^{12} - 1 = 0$ to the right, giving us this."

$$x^{12} - 1 = 0 \qquad \text{the equation}$$
$$x^{12} = 1 \qquad \text{move constant to the right}$$

"In other words, the problem of finding the solutions for $x^{12} - 1 = 0$ is fundamentally the same as finding numbers whose twelfth power equals 1, which we call the 12th roots of unity. That means the $\alpha_1, \alpha_2, \alpha_3, \ldots, \alpha_{12}$ we're looking for are these roots of unity. There are twelve of them, and we want to find them all. You've already found four of them, $1, -1, i,$ and $-i$, so there are eight left. But rather than jump right at it, let's start small. We'll let $n = 1, 2, 3, 4$ and so on, and look for these nth roots of unity among the complex numbers to see just what kind of numbers they are."

1st root of unity

"The 1st root of unity is the number that equals 1 when it's raised to the first power. In other words, the solution to the first-degree equation $x^1 = 1$. The only solution is $x = 1$, so 1 is the only 1st root of unity."

$$1 \qquad \text{1st root of unity}$$

2nd roots of unity

"Similarly, the 2nd roots of unity would be the solutions to the second-degree equation $x^2 = 1$, in other words the square root of 1. There are two of those, 1 and -1."

$$1, -1 \qquad \text{2nd roots of unity (the square root)}$$

3rd roots of unity

"Continuing on, 3rd roots of unity are the solutions to the third-degree equation $x^3 = 1$, so we're talking about cube roots. There will be three, one of which is easy to find, 1. So we're looking for something like this."

$$x^3 - 1 = (x-1)(\ldots \ldots \ldots)$$

"We can factor $x^3 - 1$ by dividing it by $x - 1 \ldots$"

$$x^3 - 1 = (x-1)(x^2 + x + 1) \qquad \text{factor } x^3 - 1$$

"... and solving $x^2 + x + 1 = 0$ to find the other two roots of unity."

$$x = \frac{-1 \pm \sqrt{-3}}{2} \qquad \text{from the quadratic equation}$$

$$= \frac{-1 \pm \sqrt{3}i}{2} \qquad \text{rewrite } \sqrt{-3} \text{ as } \sqrt{3}i$$

"Let's write $\frac{-1+\sqrt{3}i}{2}$ as omega, so $\frac{-1-\sqrt{3}i}{2}$ would be ω^2."

$$\omega = \frac{-1 + \sqrt{3}i}{2}, \quad \omega^2 = \frac{-1 - \sqrt{3}i}{2}$$

"Now we have the 3rd roots of unity, 1, ω, and ω^2."

$$1, \omega, \omega^2 \qquad \text{3rd roots of unity (cube roots)}$$

4th roots of unity

"Next is the 4th roots of unity, the solutions to the fourth-degree equation $x^4 = 1$, but we've already talked about those."

$$1, -1, i, -i \qquad \text{4th roots of unity}$$

"Okay, so we have the 1st through 4th roots of unity. Let's write these as points in the complex plane."

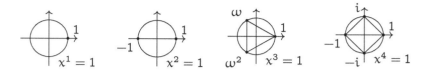

"So my question to you is... what do you think the 12th roots of 1 will look like?"

4.1.6 Regular n-gons

"This is neat!" Tetra said as she sketched in her notebook. "Setting aside the first and second ones, the third roots of unity form an equilateral triangle, and the fourth roots form a square!"

- 1st root of unity $\longrightarrow$ Single point in the complex plane

- 2nd root of unity $\longrightarrow$ Two points in the complex plane

- 3rd root of unity $\longrightarrow$ Three points in the complex plane (equilateral triangle)

- 4th root of unity $\longrightarrow$ Four points in the complex plane (square)

"It is interesting," I said. "So what do you think will happen in general for a nth root of unity?"

"I think I remember you telling me this before[1]. You get a regular polygon, right?"

"That's right. Specifically, when you plot the nth roots of unity in the complex plane, you get the vertices of a regular n-gon centered at the origin, inscribed in a circle with radius 1."

"So the 12th roots of 1 would make, um... a dodecagon!"

---
[1] See *Math Girls 2: Fermat's Last Theorem*, Ch. 6.

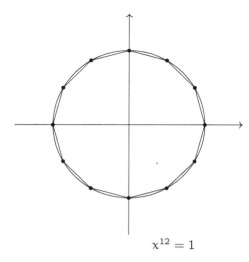

$x^{12} = 1$

12th roots of unity.

4.1.7 Trigonometric Functions

"Kinda strange, isn't it?" Tetra said. "That each of these twelve vertices on a dodecahedron in the complex plane become 1 if you raise them to the twelfth power?"

"Not only that, these are the *only* numbers that equal 1 when you raise them to the twelfth power."

"That *is* strange... I mean, I can see how that works for $1, -1, i$, and $-i$. Like, of course 1 to the twelfth power is 1—it's 1 for every power. And -1, since that's 1 when you square it, its twelfth power will be 1 too. The cube of ω is 1, so it makes sense that its twelfth power is 1, and the fourth power of $\pm i$ is 1, so same thing there. So I guess all the numbers work that way."

"Yeah, but rather than think of the twelve numbers separately, we can use trigonometric functions to represent them all at once."

"With trigonometry?"

"Sure, like this. We can represent a point on the unit circle as $(\cos \theta, \sin \theta)$, where the thetas are the argument."

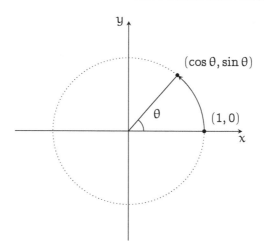

A point on the unit circle.

"The 'argument' is the angle from the x-axis to the point, going counterclockwise, right?" Tetra asked.

"That's right. So when we want to use complex numbers to represent points on the unit circle, we can multiply the $\sin\theta$ used for the y-coordinate by i, giving $\cos\theta + i\sin\theta$."

$$\begin{array}{ccc} \text{Point} & \longleftrightarrow & \text{complex number} \\ (\cos\theta, \sin\theta) & \longleftrightarrow & \cos\theta + i\sin\theta \end{array}$$

"Sure, I've seen that."

"The argument for the real number 1 is 0. Let's start from there and create a 12-gon."

"A dodecagon."

"A dodecagon, right. All we have to do is advance in increments of one-twelfth of 2π, and we should get back to 0 after twelve steps. To make things simpler, let's use this to represent a one-twelfth step."

$$\theta_{12} = \frac{2\pi}{12}$$

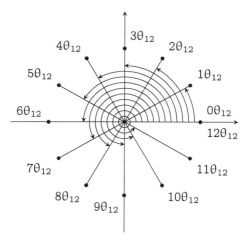

Vertices and arguments for a dodecagon.

"Okay, I'm good so far."

"The arguments are $0\theta_{12}, 1\theta_{12}, 2\theta_{12}, 3\theta_{12}, \ldots, 11\theta_{12}$. When you get to $12\theta_{12}$ that's 360°, so you're back at $0\theta_{12}$. We've set $k\theta_{12}$ as the argument, so we can use complex numbers to represent the dodecagon's vertices as $\cos k\theta_{12} + i \sin k\theta_{12}$, where k is some integer. Do you remember de Moivre's formula?"

"Sounds familiar, but..."

"Here, I'll write it out."

De Moivre's formula (for trigonometric functions)

$$\underbrace{\cos n\theta + i \sin n\theta}_{n\text{th multiple of argument}} = \underbrace{(\cos\theta + i \sin\theta)^n}_{n\text{th power of the complex number}}$$

"De Moivre's formula shows us that if we raise a complex number on the unit circle to the twelfth power, we get a complex number with its argument multiplied by 12. Let's try it and see."

$(\cos k\theta_{12} + i \sin k\theta_{12})^{12}$ 12th power of a dodecagon vertex
$= \cos 12k\theta_{12} + i \sin 12k\theta_{12}$ from de Moivre's formula
$= \cos 2\pi k + i \sin 2\pi k$ because $12 \cdot \theta_{12} = 12 \cdot \dfrac{2\pi}{12} = 2\pi$
$= 1$ because $\cos 2\pi k = 1, \sin 2\pi k = 0$

"See? We raised a complex number corresponding to an arbitrary vertex on a dodecagon to the twelfth power, and got 1. Looking at it another way, every vertex on a dodecagon is a solution to the equation $x^{12} = 1$."

Tetra let out an odd groan. "So hang on. Doesn't this mean that when you say the solutions to $x^{12} = 1$ are vertices on a dodecagon, that's the same as saying when you multiply a fraction $\frac{\text{integer}}{12}$ by 12, you get an integer?"

"I guess it does, in a sense. Playing around with the complex plane is fun, isn't it?"

4.1.8 Onward

"All of math is fun," Tetra said, flipping through her notebook. "I love how Mr. Muraki's cards are so, what... broad? It's like he's saying to just go have fun." She continued, speaking almost dreamily. "I remember when I first told him you were helping me with math, quite a while back. He told me to come see him from time to time, that he'd give me these cards. That what I got out of them was up to me." She nodded to herself. "That's right about the time I decided I definitely want to go to college. Well, not quite. It's not that I want to go to college so much as I want to keep learning. You only live once, after all, so it feels like I should learn what I can, appreciate how far humanity has progressed, and help that progress along as much as I can, even if it's just the smallest of nudges forward."

I wasn't sure what to add to that, so I just remained silent.

"I don't think I can accomplish much during just the four short years of college," Tetra continued. "But I can do my best to learn as much as possible. And I guess that's pretty much my plans, for now at least."

"Plans, right," I said.

Tetra was looking at me with those huge eyes of hers. I was impressed with how she was thinking about her future, despite still only being a second year student. Only vaguely, perhaps, but still. Maybe she was still a bit flighty at times, but even so there was a constancy about her, a flexible but strong will supporting her curiosity, her love of learning.

4.2 Cyclic Groups

4.2.1 Miruka

"Revolutions?"

I looked up to see Miruka standing behind us. Apparently she had finished whatever she had been writing. She sat down next to me. She gave her hair a shake, creating a citrus-scented breeze.

"Of a sort," I said.

I realized that despite it being summer vacation we had fallen into our usual school-days pattern—gathering in the library to do math. Posing and solving problems, reading and critiquing each other's solutions... same as every day. Math was always at the center of our little group, like the origin on a coordinate plane. It was the reference point from which we measured where we were.

"We were looking for the twelve solutions to $x^{12} = 1$, the 12th roots of unity," Tetra said. "We used trigonometric functions and de Moivre's formula to show how raising the roots to the twelfth power equals 1!"

"Calculating, as usual," Miruka said, looking straight at me.

"Calculating is fun," I countered, slightly miffed. "Besides, we did graph a 12-gon, so there."

"I do love dodecagons," she said, cocking her head to look at Tetra's notebook. "It looks like you were creating n-gons from nth roots of unity, but considering individual values for n is no fun. Let's play with all of them at once."

Miruka reached out and placed her hand on mine.

So warm.

Unfortunately, she was just taking the pencil from my hand.

4.2.2 Twelve Complex Numbers

"You used θ_{12} to represent an argument of $\frac{2\pi}{12}$," Miruka began. "Let's try naming the vertices of our dodecagon instead, using zetas instead of thetas. We'll let ζ_{12} represent a point on the unit circle centered at the origin with an argument of $\frac{2\pi}{12}$."

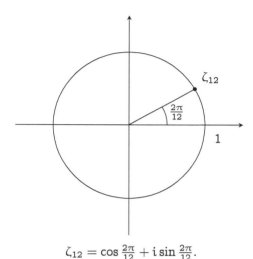

$$\zeta_{12} = \cos\frac{2\pi}{12} + i\sin\frac{2\pi}{12}.$$

"By the way, I've named this zeta, but it doesn't have anything to do with the zeta function[2], it's just another Greek letter here. Of course, the '12' subscript comes from the fact that we're considering a 12-gon. As we can immediately see from de Moivre's formula, the vertices of a 12-gon each correspond to a power of ζ_{12}. Specifically, the vertices correspond to these twelve complex numbers."

$$\zeta_{12}^1,\ \zeta_{12}^2,\ \zeta_{12}^3,\ \zeta_{12}^4,\ \zeta_{12}^5,\ \zeta_{12}^6,\ \zeta_{12}^7,\ \zeta_{12}^8,\ \zeta_{12}^9,\ \zeta_{12}^{10},\ \zeta_{12}^{11},\ \zeta_{12}^{12}$$

"There's a loop here, namely $\zeta_{12}^{12} = \zeta_{12}^0 = 1$. Here's a graph."

---
[2] See *Math Girls*, Ch. 8

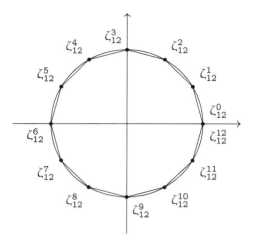

Vertices of a 12-gon as powers of ζ_{12}.

Tetra raised a hand. "I noticed something." She turned to me. "In your version, you changed the arguments as $1\theta_{12}, 2\theta_{12}, 3\theta_{12}, \ldots$, finally multiplying θ_{12} by 12 to go all the way around."

I nodded. "That's right."

"But it looks like Miruka is changing the complex number as $\zeta_{12}^1, \zeta_{12}^2, \zeta_{12}^3, \ldots$, so we make it back to the start with the twelfth power of ζ_{12}. The two approaches seem so similar, and yet so different..."

"That's de Moivre's formula showing its face," I said. "Like it says, a multiple of the argument is the same as a power of the complex number as a whole. Look at it with the n's circled."

$$\underbrace{\cos(\widehat{n})\theta + i\sin(\widehat{n})\theta}_{\widehat{n}\text{th multiple of argument}} = \underbrace{(\cos\theta + i\sin\theta)^{\widehat{n}}}_{\widehat{n}\text{th power of the complex number}}$$

"Oh, I see. I guess I didn't read it closely enough the first time."

"If you want a good look at the relation between multiples and powers of n," Miruka said, "in other words, the relation between multiplication and exponentiation, it's easier to consider exponential functions than trigonometric functions. This is Euler's equation."

$$\cos\theta + i\sin\theta = e^{i\theta}$$

"We can use this to rewrite de Moivre's formula."

In a Yoke with You

> **de Moivre's formula (for exponential functions)**
>
> $$\underbrace{e^{in\theta}}_{n\text{th multiple of argument}} = \underbrace{\left(e^{i\theta}\right)^n}_{n\text{th power of the complex number}}$$

"Interesting," I said. "This lets us look at de Moivre's formula in the form of an exponential law. No clearer picture of the relation between multiplication and exponents than $a^{mn} = (a^m)^n$."

4.2.3 Making a Table

Miruka pushed her metal-framed glasses up her nose. "The square of ζ_{12} is one of the 6th roots of unity, since raising ζ_{12} squared to the sixth power equals 1. I guess that's easier to see written out. Here."

$$\left(\zeta_{12}^2\right)^6 = \zeta_{12}^{2\times 6} = \zeta_{12}^{12} = 1$$

"In the same way, if we let $\zeta_n = \cos\frac{2\pi}{n} + i\sin\frac{2\pi}{n}$, we get this."

$$\text{sixth power of } \zeta_{12} = \text{third power of } \zeta_6$$
$$= \text{second power of } \zeta_4$$
$$= \text{first power of } \zeta_2$$

"In other words..."

$$\zeta_{12}^6 = \zeta_6^3 = \zeta_4^2 = \zeta_2^1$$

"Almost like reducing a fraction, right?"

$$\frac{6}{12} = \frac{3}{6} = \frac{2}{4} = \frac{1}{2}$$

"Let's use this to create a table of kth powers of ζ_n."

											ζ_1^1
					ζ_2^1						ζ_2^2
			ζ_3^1				ζ_3^2				ζ_3^3
		ζ_4^1			ζ_4^2			ζ_4^3			ζ_4^4
	ζ_6^1		ζ_6^2		ζ_6^3		ζ_6^4		ζ_6^5		ζ_6^6
ζ_{12}^1	ζ_{12}^2	ζ_{12}^3	ζ_{12}^4	ζ_{12}^5	ζ_{12}^6	ζ_{12}^7	ζ_{12}^8	ζ_{12}^9	ζ_{12}^{10}	ζ_{12}^{11}	ζ_{12}^{12}

$$k\text{th powers of } \zeta_n = \cos\frac{2\pi}{n} + i\sin\frac{2\pi}{n}.$$

"There's a pattern here, isn't there?" Tetra said.

"And you've arranged these so that all the numbers in each column have the same value, right?" I added.

4.2.4 Polygons with Shared Vertices

Miruka turned to a new page in her notebook—actually, *my* notebook—and continued her talk.

"Next, let's think about polygons that share vertices with the 12-gon we've been playing with."

"Sharing vertices? In what sense?" Tetra asked.

"Well, let's start with a 1-gon and a 2-gon and see."

"You mean polygons with one and two sides? What on earth would those look like..."

"Use your imagination," Miruka said as she sketched.

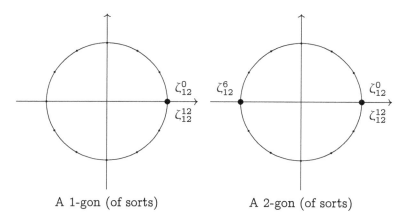

A 1-gon (of sorts)   A 2-gon (of sorts)

"Oh, I get it!" Tetra said. "We're just looking at the number of vertices there are."

"Now let's look at a 3-gon and a 4-gon."

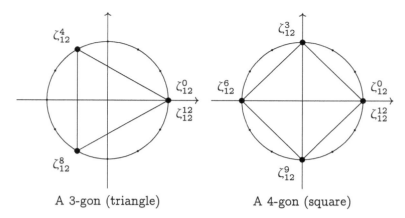

A 3-gon (triangle) A 4-gon (square)

"We can't use the vertices of a 12-gon to create a 5-gon, so the 6-gon comes next. We similarly skip everything from 7 to 11, so we end with the 12-gon itself."

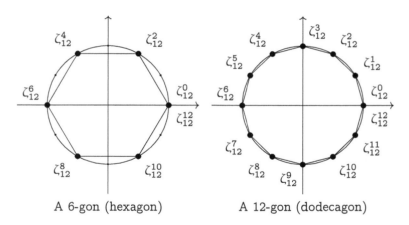

A 6-gon (hexagon) A 12-gon (dodecagon)

"You're right," Tetra said. "A pentagon wouldn't fit in there neatly, would it."

"Count the vertices," Miruka said with a grin. "The only regular n-gons we can create using the vertices of a 12-gon are 1-, 2-, 3-, 4-, 6-, and 12-gons."

"Oh, of course!" I said. "The divisors of 12!"

$$\{1, 2, 3, 4, 6, 12\}$$

"Indeed."

"But I guess that makes sense if you think about the form of a regular polygon."

Tetra raised a hand. "Um... Something I noticed. Is thinking about shared vertices in a polygon like this an example of looking at structure?"

Miruka narrowed her eyes. "How so?"

"It just seems like we're pulling something apart and looking at how the parts fit together."

"Hmph." She waved a finger to trace out a Φ. "I like it."

4.2.5 Primitive 12th Roots of Unity

Miruka abruptly stood and went around the table to stand behind Tetra.

"So we've seen that we can represent the 12th roots of unity as the vertices of a 12-gon. Let's study these roots of unity from a different perspective. For example, we know we can raise the imaginary unit i to the fourth power and get 1. That happens long before we get to the twelfth power, doesn't it?"

Tetra spun in her chair to face Miruka. "Right! I had noticed that! You get to 1 just by raising i or $-i$ to the fourth power, or even just by squaring -1."

Miruka raised an eyebrow.

"Er, sorry for interrupting," Tetra said.

"Not at all, you've saved me some time." Miruka sat next to Tetra. "We called a number that equals 1 when raised to the nth power an 'nth root of unity.' But let's tighten our conditions a bit. Specifically, let's take a look at the *smallest* positive power of our root needed to produce a 1. If the smallest power is n, then we call that root a '*primitive* nth root of unity.'"

nth root of unity: a number whose nth power equals 1

primitive nth root of unity: a number whose nth power is the smallest positive power that equals 1

"So it has a name and everything!" Tetra said.

"Let's see if you understand what it means. Tell me, what's the primitive 1st root of unity?"

"That's easy! The number that first becomes 1 when you raise it to the first power is 1! So 1 itself must be the first primitive root of unity!"

"Well done. Next—"

Tetra held up both hands in a halting gesture. "Hold up, Miruka! 'Examples are the key to understanding,' right? I think I understand the definition of nth roots of unity, so I'd like to try making some examples."

Good for you, Tetra.

"The primitive 2nd root of unity is -1, because that's the only one you actually have to square to get 1. And the primitive 3rd roots of unity would be, uh... Oh, right. We ignore the 1 in $1, \omega, \omega^2$, so ω and ω^2 are both primitive roots. Okay, I'm getting this. To find the primitive nth roots of unity, we just think about kth roots of unity with the k's lined up in increasing order, $k = 1, 2, 3, 4, \ldots$, and just use the nth roots of unity that don't show up for smaller k. Right?"

"You'll never go wrong if you consider polygons," Miruka said. "Here, let me graph them in order, leaving circles for numbers we've seen so far."

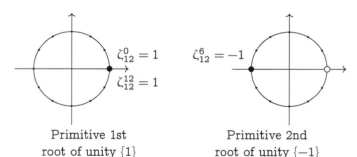

Primitive 1st
root of unity $\{1\}$

Primitive 2nd
root of unity $\{-1\}$

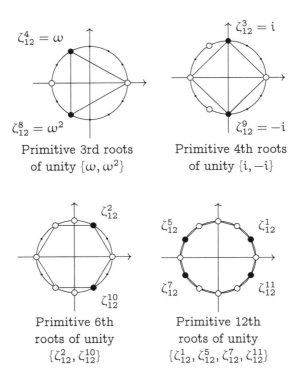

Primitive 3rd roots of unity $\{\omega, \omega^2\}$

Primitive 4th roots of unity $\{i, -i\}$

Primitive 6th roots of unity $\{\zeta_{12}^2, \zeta_{12}^{10}\}$

Primitive 12th roots of unity $\{\zeta_{12}^1, \zeta_{12}^5, \zeta_{12}^7, \zeta_{12}^{11}\}$

4.2.6 Cyclotomic Polynomials

"I think that pretty much covers primitive nth roots of unity," Miruka said. "Where to next?"

Where to indeed. Today's talk had already become a whirlwind tour.

"Well, we've already created graphs..." Tetra said.

I had a sudden flash of insight. "How about polynomials with roots that are primitive nth roots of unity?"

"Ah, very interesting," Miruka said. "Okay, then, let's start by assuming any coefficients are rational numbers."

"Polynomials with primitive nth roots of unity as roots..." Tetra mumbled, deep in thought. "Well, now I know about roots of polynomials, so I guess it should be pretty easy to create polynomials like that. We just have to expand some first-degree products of linear expressions, right?"

"Sounds like it," I said. "It's pretty easy to create polynomials with primitive 1st and 2nd roots of unity, like this."

$$x - \zeta_{12}^0 = x - 1 \qquad \text{polynomial with primitive 1st root of unity as roots}$$

$$x - \zeta_{12}^6 = x - (-1)$$
$$= x + 1 \qquad \text{polynomial with primitive 2nd root of unity as roots}$$

"Hey, save the easy stuff for me!" Tetra complained. "Let's see, the primitive 3rd roots of unity are ω and ω^2, so..."

$$\begin{aligned}(x - \zeta_{12}^4)(x - \zeta_{12}^8) &= (x - \omega)(x - \omega^2) \\ &= x^2 - (\omega + \omega^2)x + \omega^3 \\ &= x^2 - (\omega + \omega^2)x + 1 \qquad \text{because } \omega^3 = 1 \\ &= \text{uh}\ldots\end{aligned}$$

"Oops, I think I'm stuck."

"Well, $\omega^2 + \omega + 1 = 0$, so you can use $\omega^2 + \omega = -1$," I suggested.

$$\begin{aligned}(x - \zeta_{12}^4)(x - \zeta_{12}^8) &= x^2 - (\omega + \omega^2)x + 1 \\ &= x^2 - (-1)x + 1 \qquad \text{because } \omega^2 + \omega = -1 \\ &= x^2 + x + 1 \qquad \text{polynomial with primitive 3rd roots of unity as roots}\end{aligned}$$

"I see," Tetra said, nodding.

"Next is primitive 4th roots of unity, which are i and $-i$, so..."

$$\begin{aligned}(x - \zeta_{12}^3)(x - \zeta_{12}^9) &= (x - i)(x - (-i)) \\ &= (x - i)(x + i) \\ &= x^2 + 1 \qquad \text{polynomial with primitive 4th roots of unity as roots}\end{aligned}$$

"Let me try the primitive 6th roots of unity," Tetra said.

$$
\begin{aligned}
(x - \zeta_{12}^2)(x - \zeta_{12}^{10}) &= x^2 - (\zeta_{12}^2 + \zeta_{12}^{10})x + \zeta_{12}^2 \zeta_{12}^{10} \\
&= x^2 - (\zeta_{12}^2 + \zeta_{12}^{10})x + \zeta_{12}^{2+10} \\
&= x^2 - (\zeta_{12}^2 + \zeta_{12}^{10})x + \zeta_{12}^{12} \\
&= x^2 - (\zeta_{12}^2 + \zeta_{12}^{10})x + 1 \\
&= \text{erm}\ldots
\end{aligned}
$$

"What can I do with this $\zeta_{12}^2 + \zeta_{12}^{10}$?" Tetra asked.

"Think of vector addition, and you'll see it equals 1," Miruka said.

"Vector addition? How?"

"Like this."

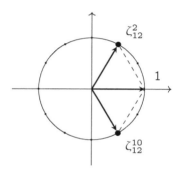

"Oh, okay."

$$
\begin{aligned}
&(x - \zeta_{12}^2)(x - \zeta_{12}^{10}) \\
&= x^2 - (\zeta_{12}^2 + \zeta_{12}^{10})x + 1 \\
&= x^2 - 1x + 1 \quad \text{use } \zeta_{12}^2 + \zeta_{12}^{10} = 1 \\
&= x^2 - x + 1 \quad \begin{array}{l}\text{polynomial with primitive} \\ \text{6th roots of unity as roots}\end{array}
\end{aligned}
$$

"I'll give the primitive 12th roots of unity a shot, then."

$$
\begin{aligned}
&(x - \zeta_{12}^1)(x - \zeta_{12}^5)(x - \zeta_{12}^7)(x - \zeta_{12}^{11}) \\
&= (x^2 - (\zeta_{12}^1 + \zeta_{12}^5)x + \zeta_{12}^1 \zeta_{12}^5)(x^2 - (\zeta_{12}^7 + \zeta_{12}^{11})x + \zeta_{12}^7 \zeta_{12}^{11}) \\
&= \text{ugh}\ldots
\end{aligned}
$$

"No, wait! I think I can do this using vector addition and the exponent rule!"

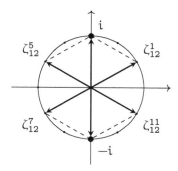

$$(x - \zeta_{12}^1)(x - \zeta_{12}^5)(x - \zeta_{12}^7)(x - \zeta_{12}^{11})$$
$$= (x^2 - \underbrace{(\zeta_{12}^1 + \zeta_{12}^5)}_{i}x + \underbrace{\zeta_{12}^1 \zeta_{12}^5}_{\zeta_{12}^{1+5} = -1})(x^2 - \underbrace{(\zeta_{12}^7 + \zeta_{12}^{11})}_{-i}x + \underbrace{\zeta_{12}^7 \zeta_{12}^{11}}_{\zeta_{12}^{7+11} = -1})$$
$$= (x^2 - ix - 1)(x^2 + ix - 1)$$
$$= x^4 + \cancel{ix^3} - x^2 - \cancel{ix^3} + \cancel{x^2} + \cancel{ix} - x^2 - \cancel{ix} + 1$$
$$= x^4 - x^2 + 1 \qquad \text{polynomial with primitive 12th roots of unity as roots}$$

"And that's it, right?" I said.

Miruka nodded. "Let's use phi to designate polynomials like this, letting $\Phi_k(x)$ represent a polynomial having primitive kth roots of unity as its roots. To determine them uniquely, we'll insist that the coefficient of the the highest-degree term be 1, in other words, a 'monic' polynomial."

$$\Phi_1(x) = x - 1$$
$$\Phi_2(x) = x + 1$$
$$\Phi_3(x) = x^2 + x + 1$$
$$\Phi_4(x) = x^2 + 1$$
$$\Phi_6(x) = x^2 - x + 1$$
$$\Phi_{12}(x) = x^4 - x^2 + 1$$

"Polynomials like this are called cyclotomic polynomials. Okay, time to see if you're keeping up. What happens if you multiply all these cyclotomic polynomials together?"

$$\Phi_1(x)\Phi_2(x)\Phi_3(x)\Phi_4(x)\Phi_6(x)\Phi_{12}(x) = ?$$

"Give me a minute!" Tetra said, turning a page in her notebook.

"Whoa, whoa, whoa," I said, waving a hand to stop Tetra.

"Hey, you!" Miruka said, reaching out to cover my mouth but unable to reach from across the table. She settled for kicking my shin instead.

"Uh, I guess I shouldn't be working this out?" Tetra said.

"You can find the answer without calculations," Miruka said. "Think about what happens if you plot the roots of these multiplied cyclotomic polynomials in the complex plane. You would end up plotting every point at one-twelfth increments around the unit circle, with no dupes and no leaks. In other words, this product of cyclotomic polynomials will equal $x^{12} - 1$."

$$\Phi_1(x)\Phi_2(x)\Phi_3(x)\Phi_4(x)\Phi_6(x)\Phi_{12}(x) = x^{12} - 1$$

"Put the opposite way, $x^{12} - 1$ can be factored in the scope of the complex numbers like this."

$$x^{12} - 1 = \underbrace{(x - \zeta_1^1)}_{\Phi_1(x)} \underbrace{(x - \zeta_2^1)}_{\Phi_2(x)} \underbrace{(x - \zeta_3^1)(x - \zeta_3^2)}_{\Phi_3(x)}$$
$$\underbrace{(x - \zeta_4^1)(x - \zeta_4^3)}_{\Phi_4(x)} \underbrace{(x - \zeta_6^1)(x - \zeta_6^5)}_{\Phi_6(x)}$$
$$\underbrace{(x - \zeta_{12}^1)(x - \zeta_{12}^5)(x - \zeta_{12}^7)(x - \zeta_{12}^{11})}_{\Phi_{12}(x)}$$

"See how cyclotomic polynomials act like primes for $x^{12} - 1$? We can use relatively prime, or 'coprime' integers n and k to write a generalized cyclotomic polynomial like this."

$$\Phi_n(x) = \prod_{n \perp k}(x - \zeta_n^k) \qquad n \perp k \text{ means "n and k are coprime"}$$

"Wow!" Tetra said, raising her voice. She was waving her hands about, clearly excited. "It's like, all the individual parts seem so obvious, but when they all get connected like this they, I don't know, resonate! I mean, look at everything we've brought together! Circles and polygons and primitive nth roots of unity and integers and polynomial factorization and solutions to equations and powers of complex numbers and trigonometric functions— They're all connected!"

"Pretty cool, I agree," I said.

"We can add one more thing to the web," Miruka said. "My dear Euler's totient function $\varphi(n)$ gives the number of primitive nth roots of unity. In the range $1 \leqslant k < n$, this $\varphi(n)$ tells us how many natural numbers are relatively prime to n, and furthermore how many generators for cyclic groups there are." Miruka traced out a φ in the air. "Too bad Yuri isn't here. She loves relatively prime numbers." Miruka stared at me coldly. "Why didn't you bring her?"

4.2.7 Cyclotomic Equations

"Wow, it's amazing how much one of Mr. Muraki's cards can expand your world," Tetra said.

"I have to agree," I said, nodding.

"What kind of equations did you say these are?" Tetra asked.

"Cyclotomic," Miruka said.

"Cyclotomic... I guess the 'cyclo' part has something to do with cycles? Since we're going around and around on a circle. The '-tomic' part, though... Hmmm. Maybe that's like 'atomic,' something you can't break apart? So a cyclotomic equation is one that breaks a circle up into its atomic parts?"

Miruka nodded. "No doubt. The solutions to an nth-degree cyclotomic equation break the circumference of the unit circle up into n equal parts. An equation in the form $x^n - 1 = 0$ is a cyclotomic equation of degree n, so for example $x^{12} - 1 = 0$ would be a 12th-degree cyclotomic equation."

$$x^{12} - 1 = 0 \quad \text{a 12th-degree cyclotomic equation}$$

I groaned. I knew we could use de Moivre's formula to derive the 12th roots of unity from $x^{12} = 1$. I had also heard of Euler's totient function before, somewhere. But these cyclotomic equations were new to me. How beautiful, this merging of so many disparate ideas. And it all arose from something as simple as $x^{12} - 1$. Just staring at that expression would never reveal the treasures hidden within; we had to break it up into its more primitive elements. Who would think that doing so could be so fascinating? Decomposition into primitive elements and recombining them into something new... Analysis and synthesis... Who knew what hidden structures they might reveal?

Miruka was looking out the window. "Doesn't this remind you of something? Some variation on the ω waltz, perhaps?"

"It does," I said.

We had glimpsed $\left\{ \zeta_{12}^0, \zeta_{12}^4, \zeta_{12}^8 \right\}$ hiding within a 12-gon. A small, triangular structure. That in turn was linked to the ω waltz we had shared.[3]

"The words we use, or maybe I should say the way we write things, is important, isn't it?" Tetra said. "Like how when we write $\cos\theta + i\sin\theta$, we can immediately see the argument θ and even the coordinates in the complex plane. When we use the notation ζ_{12}, then $\zeta_{12}^6 = \zeta_2^1$ looks like fractions, which feels so neat and tidy. And $e^{i\theta}$ is such a nice form for the exponent laws to make sense. When we write things differently, they have different nuances. It makes me feel like I'm starting to understand the feelings of who wrote these things, the message behind the math."

Answer 4-1a (Factorization)

$x^{12} - 1$ can be factored as

$$x^{12} - 1$$
$$= \varphi_1(x)\varphi_2(x)\varphi_3(x)\varphi_4(x)\varphi_6(x)\varphi_{12}(x)$$
$$= \underbrace{(x-1)}_{\varphi_1(x)} \underbrace{(x+1)}_{\varphi_2(x)} \underbrace{(x^2+x+1)}_{\varphi_3(x)} \underbrace{(x^2+1)}_{\varphi_4(x)} \underbrace{(x^2-x+1)}_{\varphi_6(x)} \underbrace{(x^4-x^2+1)}_{\varphi_{12}(x)}$$

(assuming rational coefficients).

---

[3] See *Math Girls*, Chapter 3.

Answer 4-1b (Factorization)

$x^{12} - 1$ can be factored as

$$x^{12} - 1$$
$$= \underbrace{(x - \zeta_1^1)}_{\varphi_1(x)} \underbrace{(x - \zeta_2^1)}_{\varphi_2(x)} \underbrace{(x - \zeta_3^1)(x - \zeta_3^2)}_{\varphi_3(x)}$$
$$\underbrace{(x - \zeta_4^1)(x - \zeta_4^3)}_{\varphi_4(x)} \underbrace{(x - \zeta_6^1)(x - \zeta_6^5)}_{\varphi_6(x)}$$
$$\underbrace{(x - \zeta_{12}^1)(x - \zeta_{12}^5)(x - \zeta_{12}^7)(x - \zeta_{12}^{11})}_{\varphi_{12}(x)}$$

(assuming complex coefficients), where $\zeta_n = \cos\frac{2\pi}{n} + i\sin\frac{2\pi}{n} = e^{\frac{2\pi i}{n}}$.

"Here's another fun way to look at it," Miruka said.

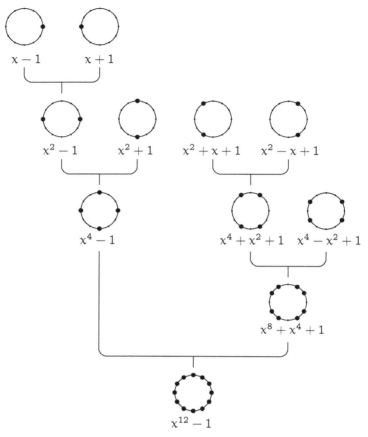

Factorization of $x^{12} - 1$ (for rational coefficients).

4.2.8 The Yoke's on You

"Something interesting about this graph," Tetra said. "All the dots are symmetric up and down, like stars reflected on water."

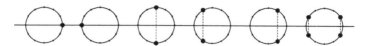

"Complex conjugates," Miruka said. "Pairs of complex numbers where only the sign of the imaginary part is different, like $a + bi$ and $a - bi$. They don't have to be on the unit circle."

"Complex number solutions to quadratic equations with real coefficients are always complex conjugates," I said.

"The name comes from the Latin *conjugat*, which means 'in a yoke together.'"

"Like, the yellow part of an egg?"

"Not a yolk!" Tetra said. "A yoke! That thing farmers put on oxen pulling a plow, to make them move in the same direction."

"Another vocabulary word for my list, I guess."

"Think of complex conjugates as being yoked by the equation they're in," Miruka said. "If one moves, the other has to follow along, bound together by this yoke they share."

"Something like seeing yourself in a mirror," I said. "Our image in a mirror can only follow our movements. The mirror binds us to our reflected image."

"Or like a married couple?" Tetra suggested. "Moving through life together, bound by a promise."

"That's an interesting take on it."

"But isn't that kind of what marriage is like? A promise to share a yoke." Tetra nodded deeply in self-affirmation.

"In any case," Miruka said, "conjugate roots are bound by the yoke of their equation. There's a deep mystery in there, one that was revealed by the genius of Galois."

4.2.9 Cyclic Groups and Generating Elements

"I feel like I'm much better friends with dodecagons now!" Tetra said.

"Then let's take a look at ζ_{12} from a different perspective," Miruka said. "First, let me define $\langle a \rangle$ like this."

$$\langle a \rangle = \text{the set of all numbers obtained by raising a number } a \text{ to the } n\text{th power } (n = 1, 2, 3, \ldots)$$

"Okay."

"So if we let $\zeta_{12} = \cos \frac{2\pi}{12} + i \sin \frac{2\pi}{12}$, we get this."

$$\langle \zeta_{12} \rangle = \{\zeta_{12}^1, \zeta_{12}^2, \zeta_{12}^3, \zeta_{12}^4, \zeta_{12}^5, \zeta_{12}^6, \zeta_{12}^7, \zeta_{12}^8, \zeta_{12}^9, \zeta_{12}^{10}, \zeta_{12}^{11}, \zeta_{12}^{12}\}$$

"Sure, make sense."

"So tell me, why does the set $\langle\,\zeta_{12}\,\rangle$ only have twelve elements, even though we said there are infinitely many n's, $n = 1, 2, 3, \ldots$?"

"Because you're going around and around. If you say ζ_{12}^{13}, that's the same as ζ_{12}^{1}. You can't get past these twelve elements, no matter how high a power you raise them to."

"Good enough. Okay, next problem."

Problem 4-2 (Number of generators)

How many values for integer k in the range $1 \leqslant k < 12$ satisfy the following relation?

$$\langle\,\zeta_{12}\,\rangle = \langle\,\zeta_{12}^{k}\,\rangle$$

"Huh?" Tetra looked startled at first, but chewed on a fingernail as she thought deeply about Miruka's problem.

Intrigued, I thought about how I might reword the problem.

> Repeated exponentiation of ζ_{12} generates each of the 12th roots of unity. How many numbers among the 12th roots of unity similarly generate the full set $\langle\,\zeta_{12}\,\rangle$ through repeated exponentiation?

Wordy. Messy. But writing it out this way showed me how Miruka's definition of $\langle\,a\,\rangle$ as "the set of all numbers obtained by raising a number a to the nth power" made the problem so much simpler and clearer. Of course, failing to truly understand the meaning of that definition could only make the problem harder to understand.

Tetra was still silently reviewing her notes. Miruka was staring at Tetra. Naturally, I was staring at Miruka.

"I think I got it," Tetra said. "There's four."

"Which four?" Miruka asked.

"1, 5, 7, and 11?"

"Very good." Tetra breathed a sigh of relief. I knew what was coming next, though. "And what kind of numbers are those?" *Nailed it.*

"What kind? Um, they're numbers that can't be evenly divided by 2, 3, 4, 6, or 12, so... Oh, I get it. They're numbers other than 1 that can't be evenly divided by the divisors of 12."

"Not wrong. What would you say?" Miruka asked, turning to me.

"I'd call $1, 5, 7$, and 11 numbers whose greatest common factor with 12 is 1. In other words, numbers that are relatively prime with 12."

Miruka nodded.

Tetra reddened. "Relatively prime, of course. How could I forget."

"Each of ζ_{12}^1, ζ_{12}^5, ζ_{12}^7, and ζ_{12}^{11} are primitive 12th roots of unity. Repeated exponentiation of any of them will generate each of the roots of $x^{12} - 1$. Also, they generate a group under multiplication of complex numbers. That group is generated from a single number, so it's a cyclic group. In other words, any of the primitive 12th roots of unity generate the cyclic group $\langle \zeta_{12} \rangle$."

$$\langle \zeta_{12}^1 \rangle = \langle \zeta_{12}^5 \rangle = \langle \zeta_{12}^7 \rangle = \langle \zeta_{12}^{11} \rangle$$
$$= \{\zeta_{12}^1, \zeta_{12}^2, \zeta_{12}^3, \zeta_{12}^4, \zeta_{12}^5, \zeta_{12}^6, \zeta_{12}^7, \zeta_{12}^8, \zeta_{12}^9, \zeta_{12}^{10}, \zeta_{12}^{11}, \zeta_{12}^{12}\}$$

Answer 4-2 (Number of generators)

There are four integers k in the range $1 \leqslant k < 12$ that satisfy

$$\langle \zeta_{12} \rangle = \langle \zeta_{12}^k \rangle.$$

4.3 Mock Exams

4.3.1 At the Testing Site

I closed my eyes as I listened to the test monitor's perfunctory instructions.

Only your registration certificate, writing implements, and a watch are allowed on your desks. Do not touch your exam booklet until instructed to do so. If you have any questions, please silently raise your hand. Be sure to...

I was at a high school one town over, the site of a mock college entrance exam being conducted by the prep school I was attending. There was a palpable tension in the air. The air conditioner wasn't doing its job. A different school, different smells. I figured my actual exam would be something like this too, so getting used to taking a test in a strange environment was part of my preparations. As the monitor droned on, I reflected on everything Miruka, Tetra, and I had talked about the other day.

Mr. Muraki's deceptively simple $x^{12} - 1$ had led us through a wide variety of topics.

- Polynomial factorization and solutions to equations

- Regular n-gons

- nth roots of unity and primitive nth roots of unity

- Cyclotomic polynomials and cyclotomic equations

- Relatively prime integers

- Cyclic groups and generating elements

- Conjugate roots, yoked together by an expression

I loved this interconnectedness within mathematics, just as much as I loved Miruka's lectures and Tetra's unique perspectives. Both of them had an attractiveness I had never known before. A deepness. I was left doubting how well I understood them, just as I was insecure in my understanding of math. Of understanding myself, even.

The prep books I'd read to get ready for my exams often used the phrase "easy to understand." That was fine for something as simple as a test problem, but the truly important things in life are rarely easy to understand. Things like—

"You may begin," came the monitor's harsh call.

I opened my eyes. *Let's do this. For the important things in life.*

While the sixteen-year-old Galois was dissatisfied with his life at school, he had his mathematics to focus on, so he was not unhappy.

KOICHIRO HARADA [23]

CHAPTER 5

Trisected Angles

> There grew up, all round about the park, such a vast number of trees, great and small, bushes and brambles, twining one within another, that neither man nor beast could pass thro'; so that nothing could be seen but the very top of the towers of the palace...
>
> CHARLES PERRAULT,
> TRANS. BY A.E. JOHNSON
> *The Sleeping Beauty in the Wood*

5.1 THE WORLD OF GRAPHS

5.1.1 Yuri

"Get up, lazybones! You have exams to study for!" Yuri shouted.

"Ugh. You have your own exams to worry about," I said.

Yuri replied with her own "ugh."

We were in my room. It was already past noon. Now well into summer vacation, Yuri had been visiting even more frequently than usual.

I had gotten back the results from my mock exam. I hadn't done horribly, but not great either. No truly stupid blunders, but the answers I had gotten wrong weren't just careless mistakes. I had spent the time since comparing my answers with those on the answer sheet, taking detailed notes and making sure I knew exactly where I'd gone wrong so I could improve my score. It was tedious work, making it hard to stay motivated, but necessary nonetheless.

"I'm actually kinda' busy today," I said.

"Have you decided where you're going to apply?"

"Maybe."

Actually I *had* decided where I wanted to go to college. More on my mind, however, was what I wanted to do there. The *why* part of going to college. That wasn't something I really wanted to have a heart-to-heart with Yuri about, though.

"That means yes," she said.

"I won't get in if I don't study, though, so..."

"Yeah, whatever."

"No whatevers!"

"But I came here to ask you a favor!"

I sighed. "What kind of favor?" I started making flashcards for the English words I had misspelled. *redundant... mischievous...*

"I need an escort. I can't go alone." Yuri crossed her arms and cast me an upward glance in her best damsel-in-distress impression.

"An escort? Where are we going, a ball?"

"No, silly! To the Narabikura Library!"

5.1.2 Trisecting Angles

"Here you go," I said, handing Yuri a drink I had bought at the station.

"Thanks!" she said. "Wow, it's hot."

Yuri had of course won out. We were in the train, on the way to the Narabikura Library.

"Want a sip?" Yuri asked.

"Not as much as I want to know why we're going to the library."

"Have you ever heard of trisecting angles?"

"Sure I have. That's a classic problem in mathematics. Like, ancient Greece classic."

> **Problem 5-1 (Trisecting angles)**
>
> Is it possible to trisect a given angle using only a straightedge and a compass?

"That's the one! The answer is 'no,' right?"

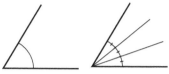

A trisected angle

"That's right, there's already a proof for that. Specifically, that you can't *always* trisect a given angle."

"But hang on, you can use a ruler and compass to make a triangle, right?" She pulled out a notebook with care.

Well that was quick.

Procedure for creating an equilateral triangle using straightedge and compass

1. Use the straightedge to draw a line between two points A and B.

2. Use the compass to draw a circle with its center at point A and passing through point B.

   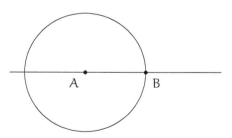

3. Use the compass to draw a circle with its center at point B and passing through point A, and let one of the intersections between the two circles be point C.

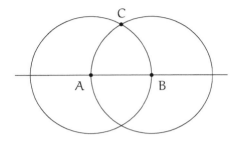

4. Use the straightedge to draw a line (segment) through points A and C.

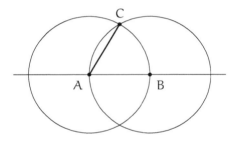

5. Use the straightedge to draw a line (segment) through points B and C.

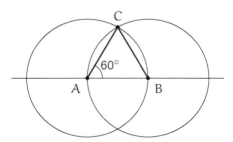

"We also know that each of the interior angles in an equilateral triangle is 60°, right? So isn't that like breaking a 180° angle up into three parts, since $180° \div 3 = 60°$? So you can't say it's impossible to use a ruler and compass to divide an angle up into three equal parts."

"Yeah, but Yuri, that's not what I'm saying. I'm saying you can't necessarily trisect *any given* angle. In other words, it's possible for some angles, but not for others. 180° just happens to be one of the angles that you *can* trisect."

"Huh. Okay. So what kind of angle is it possible for?"

"Well, besides 180° we can create 90° angles, so we can trisect 270°. Since we can create 90° and 60° angles, we can create $90° - 60° = 30°$ angles, which means we can trisect 90° angles."

"Then there's a whole bunch of them. So what kind of angle is impossible to trisect?"

"To be honest, I don't remember. It's been a while since I read about this."

"Is 60° one of the impossible ones?" Yuri asked in a small voice.

"That sounds familiar. Yeah, I think it might be."

"Which means it's impossible to use a ruler and compass to create a $60° \div 3 = 20°$ angle?"

Problem 5-2 (Constructing a 20° angle)

Is it possible to use a straightedge and compass to construct a 20° angle?

"I guess that's what it implies, yeah. If we can't trisect a 60° angle, that means we can't construct 20° angles."

"Doesn't seem like it should be all that hard, though. In geometry problems there's always some tricky way of extending lines or whatever to find a solution. Maybe we just haven't found the trick."

"No, that's not the problem. Do you want to hear why?"

And with that we fell deep into a discussion of angle trisection. Math is fun, even in a train.

5.1.3 Misunderstandings about Angle Trisection

"Okay," I said, "from what I remember reading about trisecting angles, the problem simply asks whether it's possible to trisect an arbitrary angle using only a straightedge and a compass. I can see how you'd think it shouldn't be all that hard, but there are a few things about that problem that are easy to misunderstand.

"The first is misunderstanding what 'impossible' means in a mathematical sense. If we say something is mathematically impossible, we have a proof backing that up. It doesn't mean we haven't found something because we haven't tried hard enough. So if we have a mathematical proof of the impossibility of a construction, that means the construction is impossible, no matter how hard you try. It isn't a matter of just needing to try something new. Extending lines might be a fine idea, for example, but to extend a line you need points to guide you, and you can't just add points anywhere you like.

"Another trap you can fall into is misunderstanding 'you can't necessarily trisect an angle' as meaning 'you can't trisect *any* angle.' Another way of saying 'you can't necessarily trisect an angle' would be 'there is at least one angle that you can't trisect.' So if we can find even a single angle that can't be trisected, that would be enough to establish a proof.

"Oh, and another common misunderstanding is not quite getting the preconditions for the problem. For example, you're only allowed to use two tools—a straightedge and a compass—and you're also only allowed to use them a finite number of times to perform your construction. You can't invent some new tool that would make things easier for you.

"Finally, you have to be careful about confusing 'existence' with 'constructibility.' Of course any angle can be split into three equal angles. But just because one-third of an angle exists, that doesn't mean you can construct it. Remember, the problem only allows you the two tools, and finite uses of them, and that may not be enough for the construction."

Yuri drained the last of her drink. "Yeah, but in essence the problem is just asking if you can create a diagram, right? There's so many ways to do that. I can see how you could prove a diagram *can* be created, but proving one *can't* be created? It's just so... broad."

"Meaning that even if we have a proof, we might have missed something?"

"Something like that, I guess."

"Actually, no. Tell you what, let's think about this problem together. I'm not sure I can make it to the end of a proof, but let's see how close we can get."

Yuri popped on her glasses. "Math time!"

5.1.4 Straightedges and Compasses

"Okay, let's go slow," I said. "The first thing we have to talk about are the straightedge and the compass, the only two tools you're allowed to use to perform the trisection. If you were allowed to use other tools, like a protractor, the problem would be trivial.

"First off, the straightedge. You can use a ruler for this, but regardless of what you use, you're only allowed to do one thing—draw lines between two points. We can assume that our straightedge can draw a line between those points no matter how far apart or how close together they are. You can't draw a line through two points that are on top of each other, though. One other thing you can't do is use markings on the straightedge, like the lines on a ruler to measure things. So you can't use the fact that a line is some number of centimeters long, for example. Lines between two points—that's it.

The straightedge can (only) be used to draw a line between two points

"Next, the compass. You've used a compass to draw circles before, right?"

"Of course."

"Then you know how to do the only thing you're allowed to use it for in this problem—drawing a circle centered on one point and passing through another."

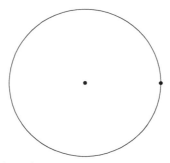

The compass can (only) be used to draw a circle centered on one point and passing through another

"And that's it. That's all the rules allow us to do when we're trying to find some way to trisect our angle. We can do these things as many times as we want, as long as it's a finite number of operations, so we can combine these operations in various ways to make lots of different figures, like you just did to create an equilateral triangle."

5.1.5 Constructibility

I paused to take a breath. "Good so far?" I asked Yuri, who had been listening silently.

"I'm good. Straightedge and compass. Lines and circles between two points. Nothing hard about that. One thing, though."

"What's that?"

"How do you get started? We won't have any points at the beginning!"

"Well spotted. You're right, both of our operations require two points, so we can assume we'll be starting the problem with two points available to us. Beyond those, the only points we can use are ones that we've constructed as intersections between lines and circles, lines and lines, or circles and circles. We'll also say that after we've used the compass to draw a circle, we can move its needle to another point and draw another circle with the same radius there[1]."

---

[1] While this use of the compass to "measure a radius" is not explicitly allowed by the rules of straightedge construction, doing so is possible by applying the compass equivalence theorem (*Elements*, Book I, Proposition II). This theorem and its proof are omitted here for space and simplicity of presentation.

"Okay, I get all that, but I'm still not seeing how a proof of not being able to create something works in practice. You said yourself that we can construct various diagrams, but to me it seems like we can construct pretty much anything. When you're dealing with numbers I can see how you can show that something is impossible, like 'it's impossible for $\sqrt{2}$ to be a rational number,' or 'it's impossible for 3 to be an even number.' But I still don't see how you can say 'you cannot construct this kind of figure.'"

"I see." I stopped to think about what Yuri was saying. She seemed to be becoming more stubborn in wanting to be sure she understood problems. She was also becoming better at describing what she wasn't understanding.

"Well?" she said.

"Right. I guess we need to think more about constructibility. How about we take a little trip, taking a problem from the world of figures to the world of numbers."

"A trip?"

"A trip."

Even as the train continued advancing toward the library, my head was already plunging into the world of numbers.

5.2 THE WORLD OF NUMBERS

5.2.1 A Concrete Example

"We want to take a look at the kind of figure we can construct using only a straightedge and a compass," I said. "As we've said, we need points to work with, so let's start by thinking about what kind of points we can use, namely constructible points."

"Sounds like a plan," Yuri said.

"We can use (x, y) coordinates to represent points on a Euclidean plane, so we'll be looking into the numbers used as x- and y-coordinates for our constructible points.

"So we can think of those as constructible numbers?"

"Constructible numbers, right. You've heard of them before?"

"Somewhere, yeah," Yuri mumbled, playing with her hair. "Actually, there's a meeting today, for the festival preparation committee."

"A festival?"

"Yeah, they're meeting at the Narabikura Library, Miruka and... and them."

So that's where we're going. If Miruka's coming, it must be some math thing.

"Is this festival why you're so interested in angle trisections all of a sudden?" I asked.

"Something like that. Just wanted to brush up."

Same pattern as always, I figured. A bunch of math people were getting together, where we would talk about math and solve problems. Miruka would no doubt provide some insightful analysis. That's what we were headed to.

"Back to math," Yuri said. "Constructible numbers."

"Right. So let's focus on constructible numbers, namely, the numbers that appear as coordinates for points that are constructible using a straightedge and compass."

"Gotcha."

"Let's say our bootstrap points are $(0,0)$ and $(1,0)$. That means we can treat 0 and 1 as constructible numbers, and we can use these points to start creating lines and circles. Now if we can create an intersection between two lines, two circles, or a line and a circle, we can add the x- and y-coordinates of that point to our list of constructible numbers."

"Gimme an example."

"Well, we can start by drawing a line through $(0,0)$ and $(1,0)$ and calling that the x-axis."

"Bang, there it is."

"Then if we draw a circle centered at $(0,0)$ with radius 1, we'll get intersections with the x-axis at $(1,0)$ and $(-1,0)$. That shows us that -1 is also a constructible number. Then we can move the compass needle to $(1,0)$ and draw a circle passing through $(0,0)$, which will also intersect the x-axis at $(2,0)$, showing that 2 is a constructible number. By repeating this, we can add $\ldots, -3, -2, -1, 0, 1, 2, 3, \ldots$, in other words all integers, to our list of constructible numbers."

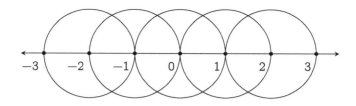

"Aha!" Yuri said.

"Since we can use a straightedge and compass to draw a line that's orthogonal to another line, the y-axis is constructible too."

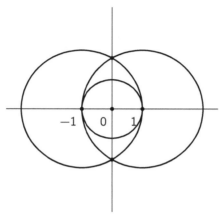

A straightedge and compass can be used to create orthogonal lines

"Since we can do this, it's pretty straightforward to extend that to a lattice of points."

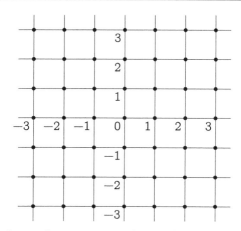

A straightedge and compass can be used to create a lattice of points

" 'Constructible numbers' is kinda long, so how about if we give these a name, say, D. That's what we'll call the set of all constructible numbers. Then if a is a constructible number, we can show that by writing $a \in D$."

$$a \text{ is a constructible number} \iff a \in D$$

"So here's what we have so far."

$0 \in D$ 0 is a constructible number (given)
$1 \in D$ 1 is a constructible number (given)
$\ldots, -2, -1, 0, 1, 2, \ldots \in D$ all integers are constructible numbers

"Letting the set of all integers be $\mathbb{Z} = \{\ldots, -3, -2, -1, 0, 1, 2, 3, \ldots\}$, we can use the subset symbol $\subset$ to write this."

$$\{\ldots, -3, -2, -1, 0, 1, 2, 3, \ldots\} \subset D$$
$$\mathbb{Z} \subset D$$

"We're trying to figure out more about this set D, right?" Yuri asked.

"We are, but actually instead of a set, let's call D a field."

"A field? What's that?"

"Right. Um, simply put, think of it as a set that you can add, subtract, multiply, and divide in."

"Wait, are we talking about geometry or arithmetic here?"

"Both, since you can use a straightedge and compass to do arithmetic. For example..."

5.2.2 Arithmetic by Construction

"For example, here's how we can do addition using a straightedge and compass."

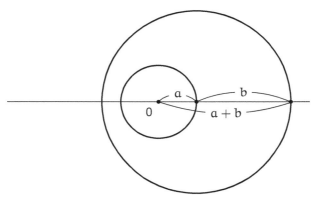

Constructing $a + b$ from a and b

"We can also use a straightedge and compass to perform subtraction in a similar way. Just assign positives or negatives according to whether the point is to the right or the left of 0."

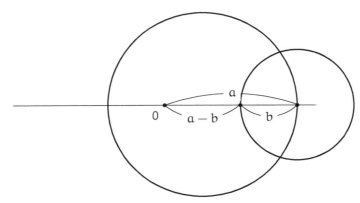

Constructing $a - b$ from a and b

"You can even use a straightedge and compass for multiplication. Just use proportional triangles."

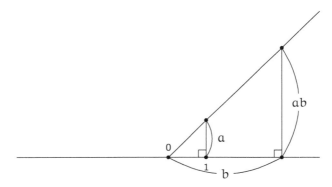

Constructing ab from a and b

"Reverse that, and you can perform division."

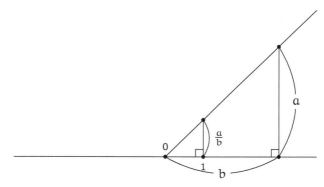

Constructing $\frac{a}{b}$ from a and b

"Okay, that's pretty impressive," Yuri said. "Arithmetic with a straightedge and compass. Who'd have thought."

"Glad you see how it works. This means not only the integers are constructible, but also all the rational numbers, numbers that can be written as one integer divided by another nonzero integer."

"Because you can create rationals by dividing integers?"

"Right. So $\mathbb{Z} \subset D$, and since D is closed under arithmetic, we can say that $\mathbb{Q} \subset D$."

$$\mathbb{Q} \subset D$$

"Hmm."

"And since it's closed under arithmetic and satisfies all the other requirements, we can say the set of all constructible numbers is a field[2]."

$$a, b \in D \Rightarrow a + b \in D$$
$$a, b \in D \Rightarrow a - b \in D$$
$$a, b \in D \Rightarrow a \times b \in D$$
$$a, b \in D \Rightarrow a \div b \in D \quad (b \neq 0)$$

---

[2] See *Math Girls 2: Fermat's Last Theorem*, Section 7.4 for a definition of fields.

5.2.3 Extracting Square Roots by Construction

"Okay, so numbers you can create through arithmetic are important in construction problems, right?" Yuri said.

I paused for a think. "Except that it feels like I'm missing something. Seems like there was something else, something other than arithmetic... Ah, right! Finding square roots. I'm pretty sure there was some way to do that."

"I know! You can find $\sqrt{2}$ from the diagonal of a square!"

"Yeah, but I don't think it was just 2. I'm pretty sure there was a way to find the root of any number a greater than zero as a construction. Some way to find $\sqrt{a}$ using just a straightedge and compass."

"Think you can figure it out?"

I spent some time trying various things in my notebook before I finally remembered how to do it.

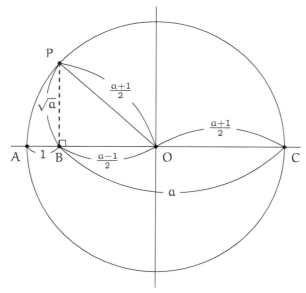

Finding a square root with straightedge and compass ($\sqrt{a}$ from a)

Trisected Angles 147

Method for Finding a Root Using a Straightedge and Compass

1. Find a point B exactly 1 to the right of a point A (1 is a constructible number).

2. Find a point C exactly a to the right of point B (where a is a given constructible number).

3. Find a point O exactly $\frac{a+1}{2}$ to the left of point C (the addition and division required to find $\frac{a+1}{2}$ can be performed using a straightedge and compass).

4. Draw a circle passing through C with point O at its center.

5. Draw a line perpendicular to $\overline{AC}$ from point B to a point P on the circle.

6. The distance between points B and P is $\sqrt{a}$.

"We can use the Pythagorean theorem to confirm that the distance between points B and P is $\sqrt{a}$."

$$\overline{BP}^2 + \overline{BO}^2 = \overline{OP}^2 \qquad \text{from the Pythagorean theorem}$$

$$\overline{BP}^2 + \left(\frac{a-1}{2}\right)^2 = \left(\frac{a+1}{2}\right)^2 \qquad \text{length in terms of } a$$

$$4\overline{BP}^2 + a^2 - 2a + 1 = a^2 + 2a + 1 \qquad \text{expand and cancel denominators}$$

$$\overline{BP}^2 = a \qquad \text{calculate}$$

$$\overline{BP} = \sqrt{a} \qquad \text{positive square root is } \overline{BP}$$

"It works!" Yuri said.

"So we've shown that we can do calculations using a straightedge and compass, and we saw how all rational numbers are constructible. We also saw that if a is a positive rational, then $\sqrt{a}$ is constructible too. So numbers like $\sqrt{2}$, $\sqrt{3}$, and $\sqrt{0.5}$ are all constructible numbers."

$$\sqrt{2} \in D$$
$$\sqrt{3} \in D$$
$$\sqrt{0.5} \in D$$
$$\sqrt{a} \in D \qquad (a \in \mathbb{Q}, a > 0)$$

"Right."
"We can also find repeated roots. This is true..."

$$a \in D \Rightarrow \sqrt{a} \in D \quad (a > 0)$$

"...so these are true too."

$$\sqrt{a} \in D$$
$$\sqrt{\sqrt{a}} \in D$$
$$\sqrt{\sqrt{\sqrt{a}}} \in D$$
$$\sqrt{\sqrt{\sqrt{\sqrt{a}}}} \in D$$

"Roots of roots of roots!" Yuri said.

"Also, we can perform arithmetic on the numbers we constructed that way, so if p, q, and r are all positive rational numbers, we can do things like this."

$$\sqrt{\sqrt{p} + \sqrt{q}} \in D$$
$$\sqrt{\sqrt{p} + \sqrt{\sqrt{q}} + \sqrt{r}} \in D$$
$$\sqrt{\sqrt{\sqrt{p} + \sqrt{\sqrt{q}}} + \sqrt{r}} \in D$$

"Well if we can do all that stuff, can't we just construct *any* number?" Yuri asked.

"No, not quite. Not with just arithmetic and square roots. For example, $\sqrt{\sqrt{a}} = \sqrt[4]{a}$, and $\sqrt{\sqrt{\sqrt{a}}} = \sqrt[8]{a}$, so the only numbers we can find simply by repeated square roots are 2^n-th roots."

"Oh. Well why are we limited to square roots, then?"

"We can write straight lines as linear equations, and circles as quadratic equations. We can find their intersection as a system of those equations. So the only numbers we can construct are those that are solutions to linear or quadratic equations."

"And?"

"Well, we can solve linear equations using only arithmetic. And remember the quadratic formula? It only uses arithmetic and square roots to give the solutions to a quadratic equation."

"So?"

"In other words, the only constructible numbers are those that we can derive through repeated arithmetic and taking square roots. You'll see that better if you set up some systems of equations for finding the coordinates of intersections between lines and circles. Circles can be tricky, though."

Just as I was starting to write the equations for a couple of circles, the train stopped.

"We're here!" Yuri said.

"We'll leave the circles as homework. We have a long climb to the Narabikura Library."

"Ugh, I'd forgotten about that."

"The journey of a thousand miles, and all that."

"Yeah, yeah. Next you're going to say something about there being only a finite number of steps to our destination, right?"

5.3 THE WORLD OF TRIGONOMETRIC FUNCTIONS

5.3.1 At the Narabikura Library

The Narabikura Library sat atop a seaside hill. The white dome atop its third floor came into view as we trudged uphill from the station. A little further, and the beautifully symmetric form of the library itself appeared. Turning away from the library toward the sea, you could see a lighthouse standing on the end of a cape. When the weather was clear you could see beyond that all the way to the horizon. It was a truly beautiful view.

It was a privately owned library, founded by Dr. Narabikura as a facility for promoting the sciences. It had an extensive collection of math and science books, and even hosted small conferences. Dr. Narabikura was the director of a mathematical research laboratory in the U.S., and was also Miruka's aunt. I had never met her in person.

We had come here several times for conferences and such. We had also used the conference rooms for our own explorations of math. It was a wonderful place for people who loved studying on their own, and for meeting other such people.

But today...

"Uh, Yuri?" I said, pointing at a sign on the door that said "Closed."

"That's...strange."

"You sure you got the day right?"

"I thought so, but... Wait, it's open!"

Yuri opened the door, and we stepped into the empty lobby. The air was chilly and smelled of brine and books. I could see the upper floors through windows surrounding a high atrium. Those floors appeared empty as well.

We heard somebody whistling.

"What was that?!" Yuri said.

"Shh!"

I looked around. I saw no one, and heard only the oddly familiar whistled melody. We walked toward the tune, and I peeked around a bookshelf. Seated there on a sofa was a redheaded girl. She had a notebook computer on her lap, and was rapidly typing as she whistled.

"Lisa?" I said.

She turned toward me, and Lisa Narabikura, first-year high school student and Dr. Narabikura's daughter, answered in a husky voice.

"Hey."

5.3.2 Lisa

Yuri and I sat on the sofa next to Lisa in the silent lobby.

"I heard there was a prep committee meeting today," Yuri said.

"There was," an expressionless Lisa replied. "Missed it."

Lisa's red hair looked as if it had been roughly chopped at with a pair of dull scissors. It was a wild look, but Lisa herself was cool and quiet. I had never heard her speak at length. In fact, she only rarely strung more than two words together. She seemed to enjoy communicating with computers more than with her fellow humans.

"*Whaaat?* He said he would meet me here!"

He?

"He's gone." Lisa said.

"The meeting starts at three o'clock, right?"

"Ten. Until lunch."

"You got the time wrong?" I asked.

"I guess? And after all that work I did to get ready!"

"They talked about posters," Lisa said, still expressionless.

"Posters? What posters?"

"In prep," she added, then fell silent, leaving me just as confused.

"So this...festival, was it? It's a conference with something to do about trisecting angles?"

"Partly," then silence again.

Partly?

"You're on the staff, I guess?"

Lisa nodded.

"And today's meeting—for the preparation committee, was it?—is already over."

Again a nod, a 1-bit packet of information that didn't advance the conversation by much.

"Was Miruka here?"

Lisa furrowed her eyebrows. "She was. Long gone." She let out a small cough.

"Did *he* say anything?" Yuri asked.

"Arithmetic and roots using straightedge and compass," Lisa said.

"I mean, like, a message or anything?" Yuri mumbled.

Despite the disjointed communications, I figured that this "festival" they were talking about was something like a small-scale conference, a way for math aficionados to use the summer break to get together and watch or even give some presentations. I could see how Miruka couldn't keep herself away from something like that, a perfect way for her to engage not only with high school students, but university students and even faculty.

And this mystery boy that Yuri's so interested in is involved too. Even Lisa is on the staff. Makes sense, I guess. Brusque as she can be, she's excellent in supporting roles.

"Did he use 60°?" Yuri asked Lisa.

"$\frac{\pi}{3}$." Lisa said.

Yuri turned to me. "What's a 'pi-over-three'?"

"An angle," I said, trying to clarify Lisa's curtness. "She's just using radian units, where π radians is 180°. That means π divided by 3 would be 60°. So I guess they were talking about trisecting a 60° angle."

"Huh. But I want to know what they said!"

"Anything left of what they talked about?" I asked Lisa. "A whiteboard or something?"

"No," she said, not even looking up from her computer.

"Ah well. Let's head back, Yuri. We can figure it out ourselves."

"I guess," Yuri said, but her dissatisfaction was evident.

"cos $\frac{\pi}{9}$," Lisa said.

"What's that?" Yuri asked.

"Same as cos 20°," I said.

"Oh, I get it!"

"You get what?"

"That if cos 20° is a constructible number we can construct 20° angles, and vice versa—if we can construct 20° angles then cos 20° is a constructible number. So we just have to look into cos 20°."

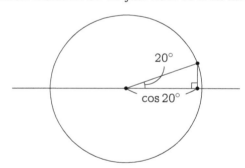

After that Yuri "conversed" with Lisa for a time, making sure she had the schedule for future meetings.

"Okay, I guess we're headed back home then," I said to Lisa, who nodded.

Yuri and I left the library, stepping out into the oppressive heat. Turning back, I saw Lisa watching us from the doorway.

"I get that this festival is something like a conference," I said to her, "but what are the posters all about?"

"Research summaries," Lisa said.

"Ah, of course. Can't expect people to stop and read through your notebooks, I guess."

Lisa nodded.

"Still, I'm not sure how that works. Mathematics is kind of abstract, right? A pathway that you follow. I don't know how posters can show all that."

Another nod.

"Well, anyway, see ya later."

"Building a tower," Lisa said, expressionless as usual.

"A tower?"

5.3.3 Parting Words

Yuri was silent for most of the train ride home. It was only when we were almost at our station when she finally opened her mouth.

"I'm so stupid."

"Why would you think that?" I asked. It had been a while since I'd seen her in self-deprecation mode.

"I mean, I know we can't always be together, but still..."

I started to ask with who, but stopped. *Him*, of course. Yuri's boyfriend, the one she'd missed today. Her math buddy who had transferred to a different school.

"Just give him a call," I said, being as casual as possible. "After what happened today I'd think you have something to talk about."

"I dunno. Maybe. Hey, sorry but I'm going to bail on you today."

"Huh?"

"Sorry to have dragged you away from your studies. We didn't even get to properly talk about trisecting angles."

"No sweat. We can do it some other time."

"And I have the mysteries of $\cos 20°$ as homework."

"Right. If you can show that $\cos 20°$ is a member of the set D of constructible numbers we were talking about, your proof will be complete. Not to imply that's necessarily an easy thing. You need to find a way to create $\cos 20°$ using only arithmetic and square roots."

"Give me some kind of secret weapon. Derivatives or integrals or so-and-so's equation or something."

I smiled and rubbed Yuri's head. "There are no shortcuts."

Normally she would have protested my touching her hair, but today she just sighed in resignation.

5.4 The World of Equations

5.4.1 Seeing Structures

That night I was alone in my room, studying for entrance exams. Trying to, at least; $\cos 20°$ kept fluttering about in my brain, distracting me. I tapped it out on my calculator.

$$\cos 20° = 0.9396926207859083840541092773 2473\cdots$$

Not what I'm looking for.

What I really wanted was to know whether $\cos 20°$ was a constructible number. Specifically, could I create it from the rational numbers, using only arithmetic and square roots? Actually, knowing the answer, I wanted to come up with a proof that I couldn't. Just staring at the number wouldn't help, though. I needed to truly perceive $\cos 20°$, to grasp its characteristics. I needed to see its structure in my mind's eye.

Problem 5-3 (Constructibility of $\cos 20°$)

Is $\cos 20°$ a constructible number?

I felt sorry for Yuri, who'd had such a bad day, missing her boyfriend and all. I recalled her parting words: *Derivatives or integrals or so-and-so's equation or something.*

Then I'd rubbed her head and—

Wait, equations?

I wanted to know the characteristics of $\cos 20°$. So maybe one way of doing that would be to find equations to which $\cos 20°$ was a solution to?

This just might work!

I stood and started pacing around my room. I rapped my knuckles on my bookcase, and tried to keep my excitement under control. I was sure this was a discovery—a problem that would lead me to the solution I was after. I wrote on a new page in my notebook:

To what equation is $\cos 20°$ a solution?

For example, the irrational number $\sqrt{2}$ is one solution to the equation $x^2 - 2 = 0$, and the imaginary unit i is one solution to the equation $x^2 + 1 = 0$. So what did that tell me about $\cos 20°$?

Think. Think...

I dredged up everything I knew about the cos function. Trigonometric function. x-coordinate on the unit circle. $\cos^2 \theta + \sin^2 \theta = 1$. Values from -1 to 1. Calculations of inner products. The law of cosines. Angles. Angles? Angles! Trisected angles!

$20°$ is one-third of $60°$. I didn't know much about $\cos 20°$, but I knew plenty about $\cos 60°$—that was just $\frac{1}{2}$. I could use that to tie together $\cos 20°$ and $\cos 60°$, using a multiple-angle identity!

Specifically, I could derive the triple-angle identity by letting a matrix for rotation by 3θ be equal to the cube of a matrix for rotation by θ. That would give me an equation for which $\cos 20°$ is a solution.

matrix for rotation by 3θ = cube of matrix for rotation by θ

$$\begin{pmatrix} \cos 3\theta & -\sin 3\theta \\ \sin 3\theta & \cos 3\theta \end{pmatrix} = \begin{pmatrix} \cos \theta & -\sin \theta \\ \sin \theta & \cos \theta \end{pmatrix}^3$$

I could calculate out the right side.

$$\begin{pmatrix} \cos \theta & -\sin \theta \\ \sin \theta & \cos \theta \end{pmatrix}^3$$

$$= \begin{pmatrix} \cos \theta & -\sin \theta \\ \sin \theta & \cos \theta \end{pmatrix}^2 \begin{pmatrix} \cos \theta & -\sin \theta \\ \sin \theta & \cos \theta \end{pmatrix}$$

$$= \begin{pmatrix} \cos^2 \theta - \sin^2 \theta & -\cos \theta \sin \theta - \sin \theta \cos \theta \\ \sin \theta \cos \theta + \cos \theta \sin \theta & -\sin^2 \theta + \cos^2 \theta \end{pmatrix} \begin{pmatrix} \cos \theta & -\sin \theta \\ \sin \theta & \cos \theta \end{pmatrix}$$

$$= \begin{pmatrix} \cos^3 \theta - 3 \cos \theta \sin^2 \theta & \sin^3 \theta - 3 \cos^2 \theta \sin \theta \\ -\sin^3 \theta + 3 \cos^2 \theta \sin \theta & \cos^3 \theta - 3 \cos \theta \sin^2 \theta \end{pmatrix}$$

So now I had this, with a matrix for rotation by 3θ on the left and the cube of a matrix for rotation by θ on the right.

$$\begin{pmatrix} \cos 3\theta & -\sin 3\theta \\ \sin 3\theta & \cos 3\theta \end{pmatrix} = \begin{pmatrix} \cos^3\theta - 3\cos\theta\sin^2\theta & \sin^3\theta - 3\cos^2\theta\sin\theta \\ -\sin^3\theta + 3\cos^2\theta\sin\theta & \cos^3\theta - 3\cos\theta\sin^2\theta \end{pmatrix}$$

Corresponding entries in the matrices would be equal, allowing me to derive the triple-angle identity.

$\cos 3\theta = \cos^3\theta - 3\cos\theta\sin^2\theta$ from comparison of entries

$\cos 3\theta = \cos^3\theta - 3\cos\theta(1 - \cos^2\theta)$ because $\sin^2\theta = 1 - \cos^2\theta$

$\cos 3\theta = 4\cos^3\theta - 3\cos\theta$ calculate

The triple-angle identity

$$\cos 3\theta = 4\cos^3\theta - 3\cos\theta$$

Next I could insert $\theta = 20°$ into this identity.

$$\cos 60° = 4\cos^3 20° - 3\cos 20°$$

Since $\cos 60° = \frac{1}{2}$,

$$\frac{1}{2} = 4\cos^3 20° - 3\cos 20°,$$

and multiplying both sides by 2 to clean up, I got this equation with $\cos 20°$ as a solution.

$$8\cos^3 20° - 6\cos 20° - 1 = 0$$

What I wanted to know was the value of $\cos 20°$. Letting X be the $\cos 20°$ part, I could create a third-degree equation in X having $X = \cos 20°$ as a solution.

$8X^3 - 6X - 1 = 0$ equation satisfied by $X = \cos 20°$

Trisected Angles

I could make this even simpler by letting $x = 2X$.

$8X^3 - 6X - 1 = 0$ equation satisfied by $X = \cos 20°$

$(2X)^3 - 3(2X) - 1 = 0$ factor out $2X$

$x^3 - 3x - 1 = 0$ let $x = 2X$

I'd set $x = 2X$, so $x = 2\cos 20°$ is a solution for $x^3 - 3x - 1 = 0$.

This is good. No need for a straightedge or compass any more.

So my focus of interest was now this third-degree equation.

$$x^3 - 3x - 1 = 0$$

This equation would have three solutions, one of which is $2\cos 20°$. If I could prove that this equation had no constructible numbers as solutions, I could assuredly say that $2\cos 20°$ cannot be constructed using a straightedge and compass. If $2\cos 20°$ cannot be constructed, then that number divided in half, $\cos 20°$, can't either. If $\cos 20°$ isn't constructible, then neither is $20°$. And if $20°$ isn't constructible, then $60°$ can't be trisected using a straightedge and compass.

Exactly what I want to prove!

The key to whether it is possible to trisect a $60°$ angle was somewhere within this equation $x^3 - 3x - 1 = 0$. This was something like a trisecting equation for $60°$!

Problem 5-4 (Trisecting equation for $60°$)

Does

$$x^3 - 3x - 1 = 0$$

have a constructible solution?

Whoops, gotta be careful here.

If $x^3 - 3x - 1 = 0$ doesn't have a single constructible solution, then I can certainly say that $2\cos 20°$ cannot be constructed using a straightedge and compass. It's possible that some constructible solution other than $2\cos 20°$ exists. If that's the case...

Well, no point in worrying about that just yet. For now, I need to focus on whether there are any constructible solutions to $x^3 - 3x - 1 = 0$.

5.4.2 Warming Up with Rational Numbers

I stared at this single equation I'd written in my notebook.

$$x^3 - 3x - 1 = 0 \qquad \text{trisecting equation for } 60°$$

I wanted to prove that this equation didn't have a single constructible solution. Since I wanted to prove the nonexistence of something, I was pretty sure I'd be using proof by contradiction. I already had a sense of how the proof would go.

First I would hypothesize that the equation did indeed have a constructible solution, which I could call α. If I did the right calculations using that α, I should be able to come across a contradiction. That constructible number α might be a pretty complicated number, though. It could have lots of nested roots, for example, representing many repeated uses of the compass. It could be something like this.

$$2 + \sqrt{3 + \sqrt{\sqrt{5} + \sqrt{\sqrt{7}} + 11\sqrt{13} + \sqrt{17}}}$$

Not sure how I could pin down a number like that. Maybe I need to pull back and regroup...

I still didn't have a good enough grasp on the constructible numbers. Maybe I should take the long way around and warm up using just the rational numbers $\mathbb{Q}$?

Okay, then, let's set aside the constructible numbers $\mathbb{D}$ *for now and focus on* $\mathbb{Q}$.

Problem 5-5 (Trisecting equation for $60°$ and rational numbers)

Does

$$x^3 - 3x - 1 = 0$$

have any *rational* solutions?

Since $\mathbb{Q} \subset \mathbb{D}$, I could hypothesize that $x^3 - 3x - 1 = 0$ didn't have a solution in $\mathbb{Q}$, either.

What I want to prove: $x^3 - 3x - 1 = 0$ has no rational number solution.

I could use proof by contradiction to show this, too, using the negation of what I wanted to prove.

Negated proposition: $x^3 - 3x - 1 = 0$ has a rational number solution.

I could write this presumed solution as $\frac{A}{B}$, where A and B are integers and $B \neq 0$. I could also assume that A and B are relatively prime without loss of generality. That would mean the greatest common factor between A and B is 1. Put another way, I could assume the fraction $\frac{A}{B}$ is in reduced form.

$x^3 - 3x - 1 = 0$ trisecting equation for $60°$

$\left(\dfrac{A}{B}\right)^3 - 3\left(\dfrac{A}{B}\right) - 1 = 0$ substitute $x = \dfrac{A}{B}$

$A^3 - 3AB^2 - B^3 = 0$ multiply both sides by B^3

$A^3 = 3AB^2 + B^3$ move $-3AB^2 - B^3$ to the right

$A^3 = (3A + B)B^2$ factor out B^2

So I now had the equation in the form $A^3 = (3A+B)B^2$.
Making progress...

Integer problems tend to be easier to deal with when they come in the form of a product, because that lets us explore them using prime factors. A and B are both integers. Prime factors show the structure of integers, so I decided to focus on A's prime factors, numbers that would divide A with no remainder. I could select one of those prime factors of integer A and call it p.

Hang on, is it possible that no such p exists?

I stared at $A^3 = (3A+B)B^2$, thinking about cases where A equaled 0, 1, and -1.

If $A = 0$, the left side is $A^3 = 0$. The right side would be $(3A + B)B^2 = B^3$, meaning $B = 0$, so this contradicts our assumption that $B \neq 0$. So I know that $A \neq 0$.

If $A = 1$, the left side is $A^3 = 1$. Then the right side would be $(3A + B)B^2 = (B + 3)B^2$. This will never equal 1, so $A \neq 1$.

If $A = -1$, the left side is $A^3 = -1$. Then the right side would be $(3A+B)B^2 = (B-3)B^2$. Again, this will never equal -1, so $A \neq -1$.

From this, A wouldn't be any of 0, 1, or -1, so I could select a prime factor p.

I could also assume that $A > 0$ without loss of generality, because if A was negative I could just flip the sign of B to make A positive without changing the value of $\frac{A}{B}$. So from here on out, I would just work under that assumption.

So letting p be one prime factor of A, since A would be divisible by p, the A^3 on the left side of $A^3 = (3A + B)B^2$ should be divisible by p too. However, the $(3A+B)B^2$ on the right would *not* be divisible by p. I knew that because $3A + B$ wouldn't be divisible by p, which in turn I knew because p does evenly divide A, and thus also $3A$, meaning that the remainder after dividing $3A + B$ by p would be equal to the remainder after dividing B by p. I had set up A and B to be relatively prime, so the remainder when dividing B by p couldn't possibly be 0.

Further, B^2 wasn't evenly divisible by p, because if it was, since p is a prime number B would be divisible by p too. That would mean both A and B would be divisible by p, a contradiction since they're supposed to be relatively prime—in other words, their greatest common factor was 1. So no, the $(3A + B)B^2$ on the right is not divisible by p.

Summing up what I had found regarding $A^3 = (3A + B)B^2 \ldots$

- I *could* evenly divide the left side using a prime factor p of A.

- I *could not* evenly divide the right side using that prime factor p of A.

A contradiction!

So using proof by contradiction, I had shown that $x^3-3x-1=0$ has no rational solutions. Done.

> **Answer 5-5 (Trisecting equation for 60° and rational numbers)**
>
> The equation
> $$x^3 - 3x - 1 = 0$$
> has no rational number solution.

This was all well and good, but it just meant I was done with my "warm-up" exercise, showing that the equation had no *rational* solutions. This didn't mean it had no *constructible* solutions, numbers that could be created by some series of arithmetic and square root operations on rational numbers. A finite number of operations, yes, but I still wasn't sure how to deal with the possibility of repeated operations.

5.4.3 One Step at a Time

I sat there in my room, unwilling to give up the battle just yet.

I had shown that the trisecting equation for 60° had no rational solutions, but what I really wanted to know was whether it had any *constructible* solutions.

> **Problem 5-4 (Trisecting equation for 60°)**
>
> Does
> $$x^3 - 3x - 1 = 0$$
> have a constructible solution?

Of particular concern was how to handle repeated square root operations.

Well, regardless of how many operations there are, they'll come one step at a time. Might as well start with the first step.

I could represent the number resulting from a single square root operation as $p + q\sqrt{r}$. So now the question I wanted to answer was whether $x^3 - 3x - 1 = 0$ had a solution in the form $p + q\sqrt{r}$.

I sat back and closed my eyes, wondering what kind of numbers p, q, and r might be.

5.4.4 The Next Step?

My eyes flew open, and I realized I had fallen asleep with my desk as a pillow. I looked at my clock: 1:30 AM.

Where was I... ? Ah, right.

I wanted to know if $x^3 - 3x - 1 = 0$ had a solution in the form $p + q\sqrt{r}$.

I had an idea: I would reduce "many operations" to "a sequence of separate operations." In other words, I had a new proposition to to prove.

> No solution to $x^3 - 3x - 1 = 0$ can be represented in the form $p + q\sqrt{r}$ for $p, q, r \in \mathbb{Q}$.

A proof of this proposition would show that the solution could not be obtained as a sequence of square root operations.

Hold up now, calm down. Let's think this out in turn.

First off, let $K = \mathbb{Q}$. Now I have this as my proposition.

> No solution to $x^3 - 3x - 1 = 0$ can be represented in the form $p + q\sqrt{r}$ for $p, q, r \in K$ (I hope).

Next, I'll fix $r \in K$, and let K' be the set of all numbers in the form $p + q\sqrt{r}$, with $p, q \in K$. Now I've got this.

> No solution to $x^3 - 3x - 1 = 0$ can be represented in the form $p' + q'\sqrt{r'}$ for $p', q', r' \in K'$ (I hope).

I could keep doing this, saying...

> No solution to $x^3 - 3x - 1 = 0$ can be represented in the form $p'' + q''\sqrt{r''}$ for $p'', q'', r'' \in K''$ (I hope).

> No solution to $x^3 - 3x - 1 = 0$ can be represented in the form $p''' + q'''\sqrt{r'''}$ for $p''', q''', r''' \in K'''$ (I hope).

Trisected Angles

…and so on. Then, I consider the sets of numbers $\mathbb{Q} = K, K', K'', K''', \ldots$ in turn.

Sets of numbers? Hmm, this is a field. Not only that, this is a field extension, a field with an element adjoined, like Miruka was talking about the other day! I was dropping a new number, $\sqrt{r}$, into the set of numbers resulting from arithmetic, and that single drop diffused throughout the entire thing. Okay, let's think carefully here. Letting K be a field, $r \in K$, and $\sqrt{r} \notin K$, I'm defining K' like this.

$$K' = \{p + q\sqrt{r} \mid p, q \in K\}$$

This K' is K with $\sqrt{r}$ adjoined. In other words,

$$K' = K(\sqrt{r})$$

I could explicitly perform arithmetic operations to show that K' was $K(\sqrt{r})$.

So now what I wanted to show was that if the equation $x^3 - 3x - 1 = 0$ has no solution in the field K, then it has no solution in the field $K(\sqrt{r})$. If I could show this one step, I could repeat the process to show that the solution could not be created through a finite number of square root operations.

I wrote this down to be sure I knew where I was:

> Precondition: $x^3 - 3x - 1 = 0$ has no solution in field K.
>
> What I want to prove: $x^3 - 3x - 1 = 0$ has no solution in field $K(\sqrt{r})$.

I could use proof by contradiction.

> Negated proposition: $x^3 - 3x - 1 = 0$ has a solution in field $K(\sqrt{r})$.

My goal was to use this to derive a contradiction. Assuming that a solution did exist, I could write it in the form $p + q\sqrt{r} \in K(\sqrt{r})$. I knew that $q \neq 0$, because if it did I would have $p + q\sqrt{r} = p \in K$, contradicting the proposition that $x^3 - 3x - 1 = 0$ has no solution in the field K. I could similarly say that $r \neq 0$. Since $p + q\sqrt{r}$ is the solution, I could substitute $x = p + q\sqrt{r}$ into $x^3 - 3x - 1$, and know that the result would be 0.

$$x^3 - 3x - 1 = (p + q\sqrt{r})^3 - 3(p + q\sqrt{r}) - 1$$
$$= (p^3 + 3p^2 q\sqrt{r} + 3pq^2\sqrt{r}^2 + q^3\sqrt{r}^3) - 3p - 3q\sqrt{r} - 1$$

I wasn't considering imaginary solutions, so I decided to use $\sqrt{r}^2 = r$ and $\sqrt{r}^3 = r\sqrt{r}$, with $r > 0$.

$$= (p^3 - 3p - 1 + 3pq^2 r) + 3p^2 q\sqrt{r} + q^3 r\sqrt{r} - 3q\sqrt{r}$$

Factoring out a $q\sqrt{r}$ ...

$$= (p^3 - 3p - 1 + 3pq^2 r) + (3p^2 + q^2 r - 3)q\sqrt{r}$$
$$= 0$$

Therefore,

$$\underbrace{(p^3 - 3p - 1 + 3pq^2 r)}_{\text{in K}} + \underbrace{(3p^2 + q^2 r - 3)q}_{\text{in K}} \sqrt{r} = 0.$$

In other words, letting $P = p^3 - 3p - 1 + 3pq^2 r$ and $Q = (3p^2 + q^2 r - 3)q$, I had $P \in K$ and $Q \in K$, giving $P + Q\sqrt{r} = 0$.

Okay, so what does that mean?

I was in the process of showing there was no solution in the field $K(\sqrt{r})$ under the condition of no solution existing in the field K, so... so...

I closed my eyes. To think.

5.4.5 Discovery?

My eyes opened. Desk pillow. Clock. 3:30 AM.

Wow, it's late. Now where was I?

Given that there is no solution to $x^3 - 3x - 1 = 0$ in field K, assuming there exists in the field $K(\sqrt{r})$ a solution in the form $p + q\sqrt{r}$ will lead to a contradiction. I hoped.

I could say that $P + Q\sqrt{r} = 0$ by letting $P = p^3 - 3p - 1 + 3pq^2 r$ and $Q = (3p^2 + q^2 r - 3)q$. The only question was ... how could I use this to derive a contradiction? This, I did not know.

I felt wide awake, oddly enough, but I was very thirsty. I went to the kitchen and drank a glass of water. The night air felt heavy, somehow.

I recalled having a dream while I slept. A dream about noticing something, something about...a bond? *A bond like...a relationship? What did I notice?* A relationship...like a family relationship? The mother–daughter relationship between Lisa and her mother, Dr. Narabikura? The aunt–niece relationship between Dr. Narabikura and Miruka? I paused for a moment, realizing that this made Miruka and Lisa cousins. Speaking of cousins, the relationship between Yuri and me? No, nothing like that, something even closer, more binding. Something like...

Like being yoked together!

I recalled what Tetra had said: "Moving through life together, bound by a promise."

Also, what Miruka had said: "Conjugate roots are bound by the yoke of their equation."

That's it!

If $a + bi$ is one solution to a quadratic equation, then $a - bi$ is the other solution. In other words, $a + bi$ and $a - bi$ are complex conjugates. Writing i as $\sqrt{-1}$, these would be

$$a + b\sqrt{-1} \quad \text{and} \quad a - b\sqrt{-1}.$$

By analogy, if $p + q\sqrt{r}$ is a solution to $x^3 - 3x - 1 = 0$, wouldn't that equation have $p - q\sqrt{r}$ as a solution too? Wouldn't these two solutions be bound together by the yoke $x^3 - 3x - 1 = 0$?

I need to check this out! I hurried back to my room.

I could check whether $x = p - q\sqrt{r}$ was a solution to $x^3 - 3x - 1 = 0$ easily enough, by substitution.

$$\begin{aligned}
x^3 - 3x - 1 &= (p - q\sqrt{r})^3 - 3(p - q\sqrt{r}) - 1 \quad \text{substitute } x = p - q\sqrt{r} \\
&= (p^3 - 3p^2 q\sqrt{r} + 3pq^2\sqrt{r}^2 - q^3\sqrt{r}^3) - 3p + 3q\sqrt{r} - 1 \\
&= p^3 - 3p^2 q\sqrt{r} + 3pq^2 r - q^3 r\sqrt{r} - 3p + 3q\sqrt{r} - 1 \\
&= (p^3 - 3p - 1 + 3pq^2 r) - (3p^2 + q^2 r - 3)q\sqrt{r} \\
&= P - Q\sqrt{r}
\end{aligned}$$

Here I had let $P = p^3 - 3p - 1 + 3pq^2r$ and $Q = (3p^2 + q^2r - 3)q$. And these were the same P and Q as before!

Very interesting! The results of substituting $p + q\sqrt{r}$ and $p - q\sqrt{r}$ into the x in $x^3 - 3x - 1$ are respectively $P + Q\sqrt{r}$ and $P - Q\sqrt{r}$!

But hang on, something's not quite right.

What I wanted to be able to say was that $P - Q\sqrt{r} = 0$, because I wanted to be able to say that $p - q\sqrt{r}$ was a solution to $x^3 - 3x - 1 = 0$. But was that really the case?

Problem 5-6 (Something not quite right)

Letting K be an extension field of the rational numbers, does
$$P + Q\sqrt{r} = 0 \implies P - Q\sqrt{r} = 0$$
hold for $P, Q, r \in K$, $\sqrt{r} \notin K$?

It *would* hold. I was sure of it. I was also sure I could prove it, because I already knew a similar problem.

I can do this. My head is clear. Let's do it.

5.4.6 Predictions and Theorems

I felt like I was leaving angle trisection far behind, but whatever. I could piece things together once the dust settled. For now, I wanted to put my efforts into proving that if $P + Q\sqrt{r} = 0$, then $P - Q\sqrt{r} = 0$. I had one lead on that—I knew a similar problem.

In the practice problems I'd been studying for my entrance exams, I sometimes used numbers like $p + q\sqrt{2}$ for $p, q \in \mathbb{Q}$. Most of the time, that resulted in using this theorem:

Theorem: $p + q\sqrt{2} = 0 \iff p = q = 0$ $(p, q, 2 \in \mathbb{Q}, \sqrt{2} \notin \mathbb{Q})$.

My suspicion, my prediction, was that something similar would hold in field K.

Prediction: $p + q\sqrt{r} = 0 \iff p = q = 0$ $(p, q, r \in K, \sqrt{r} \notin K)$

Of course, a prediction is not a proof, and I needed a proof to turn my prediction into a theorem.

I needed to think about this field K I'd created by extending $\mathbb{Q}$ and the field $K(\sqrt{r})$ I'd created by adjoining $\sqrt{r}$ to K. Of course, I would assume $r \in K$, $\sqrt{r} \notin K$.

Proof that $p + q\sqrt{r} = 0 \impliedby p = q = 0$:

The proof in the $\impliedby$ direction is simple enough; if $p = q = 0$, then clearly $p + q\sqrt{r} = 0$.

Proof that $p + q\sqrt{r} = 0 \implies p = q = 0$:

So what about in the $\implies$ direction? This wasn't too bad either. Assume that $p + q\sqrt{r} = 0$. Then if $q = 0$, certainly $p = 0$. On the other hand, if $q \neq 0$ ...

$$p + q\sqrt{r} = 0$$
$$q\sqrt{r} = -p$$
$$\sqrt{r} = -\frac{p}{q} \qquad \text{because } q \neq 0$$

Here, the $\sqrt{r}$ on the left side wouldn't belong to K, but the $-\frac{p}{q}$ on the right side would. That would be a contradiction, so it follows that q must equal 0. We can therefore say that $p + q\sqrt{r} = 0 \implies p = q = 0$. End of proof.

And thus prediction becomes theorem.

Theorem: $p + q\sqrt{r} = 0 \iff p = q = 0 \quad (p, q, r \in K, \sqrt{r} \notin K)$

This resolved my feeling that something wasn't quite right— now from $P + Q\sqrt{r} = 0$ I knew that $P = Q = 0$, and thus that $P - Q\sqrt{r} = 0$.

Answer 5-6 (Something not quite right)

Letting K be an extension field of the rational numbers,

$$P + Q\sqrt{r} = 0 \implies P - Q\sqrt{r} = 0$$

holds for $P, Q, r \in K$, $\sqrt{r} \notin K$.

From this, I knew that if $p+q\sqrt{r}$ was a solution for the trisecting equation $x^3 - 3x - 1 = 0$, then $p - q\sqrt{r}$ would be a solution too.

In other words, $p + q\sqrt{r}$ and $p - q\sqrt{r}$ are yoked together! Good so far! But where to next...?

5.4.7 Where to Next

To decide where to head next, I needed to figure out exactly where I was.

I had found two solutions to the equation $x^3 - 3x - 1 = 0$. But this was a third-degree equation, so there should be three solutions. Naming those alpha, beta, and gamma, I could write the solutions like this:

$$\begin{cases} \alpha = p + q\sqrt{r} \\ \beta = p - q\sqrt{r} \\ \gamma = ??? \end{cases}$$

I also knew there was no solution for $x^3 - 3x - 1 = 0$ in K, so I was safe in assuming $q \neq 0$. Also, $\beta = p - q\sqrt{r} \notin K$.

Okay, then, what about γ? Can I say that $\gamma \in K$? Or that $\gamma \notin K$, for that matter?

I was working my way through a proof by contradiction, so what I really wanted was to derive a contradiction. If I could show that $\gamma \in K$, my proof would be complete—I had assumed that there was no solution in the field K, so that would be my contradiction. Unfortunately, however, I had no idea what kind of number γ was.

Mmm... The equation is $x^3 - 3x - 1 = 0$... Its solutions are α, β, γ, and I know two of those: $\alpha = p + q\sqrt{r}$ and $\beta = p - q\sqrt{r}$. I want to know something—anything!—about γ. What to do... What to do...

I have no idea.

I recalled something that Tetra had said, about wishing she had a math book with a finger that would point at what she was missing. Now I knew exactly how she felt. What were we talking about then? Something about sums and products of solutions...

I had $\alpha = p + q\sqrt{r}$ and $\beta = p - q\sqrt{r}$, so their sum would be $\alpha + \beta = (p + q\sqrt{r}) + (p - q\sqrt{r}) = 2p$. And 2p belongs to K, so...

Trisected Angles

That's it! The relation between roots and coefficients! I can use that relation for third-degree equations!

Using the identity

$$(x-\alpha)(x-\beta)(x-\gamma) = x^3 - (\alpha+\beta+\gamma)x^2 + (\alpha\beta+\beta\gamma+\gamma\alpha)x - \alpha\beta\gamma,$$

I could write the polynomial $x^3 - 3x - 1$ as

$$x^3 - 3x - 1 = x^3 - (\alpha+\beta+\gamma)x^2 + (\alpha\beta+\beta\gamma+\gamma\alpha)x - \alpha\beta\gamma.$$

From the coefficient on x^2 on the right, and the lack of a x^2 term on the left, I got

$$\alpha + \beta + \gamma = 0,$$

giving me a relation between the roots and coefficients of $x^3 - 3x - 1$. So ...

$\alpha + \beta + \gamma = 0$	relation between roots and coefficients
$(p + q\sqrt{r}) + (p - q\sqrt{r}) + \gamma = 0$	use $\alpha = p + q\sqrt{r}, \beta = p - q\sqrt{r}$
$2p + \gamma = 0$	calculate
$\gamma = -2p$	move 2p to the right

There it is!

I had found that $\gamma = -2p$. From $p \in K$, I knew that $-2p \in K$, and thus $\gamma \in K$.

> What I had assumed: $x^3 - 3x - 1 = 0$ has no solution in field K.
>
> What I had found: $x^3 - 3x - 1 = 0$ has a solution in field K.

There was my contradiction. This negated the assumption I had set up for my proof by contradiction,[3] proving this:

> Assuming that $x^3 - 3x - 1 = 0$ has no solution in field K, $x^3 - 3x - 1 = 0$ has no solution in field $K(\sqrt{r})$.

---
[3] pg. 163

To give a clean proof, I guess I should use mathematical induction. Okey-dokey, here we go.

Letting n be a nonnegative integer, I'll inductively define the fields $K_0, K_1, \ldots, K_n$ by

- $K_0 = \mathbb{Q}$

- Given $r_k \in K_k$ with $\sqrt{r_k} \notin K_k$, let

$$K_{k+1} = K_k(\sqrt{r_k}) = \{p + q\sqrt{r_k} \mid p, q \in K_k\},$$

where $k = 0, 1, 2, \ldots n - 1$.

I would define proposition $P(n)$ like this:

Proposition $P(n)$: The equation $x^3 - 3x - 1 = 0$ has no solution in the field K_n.

Now I wanted to use mathematical induction to show that $P(n)$ holds for all values of n.

Step (a) I had already proved the $P(0)$ case in my warmup,[4] where I showed that the equation $x^3 - 3x - 1 = 0$ has no solutions in $\mathbb{Q}$, in other words in the field K_0.

Step (b) I had just proved the proposition for the $P(k) \Rightarrow P(k+1)$ case. Namely, assuming that $x^3 - 3x - 1 = 0$ has no solution in the field K_k, then it has no solution in the field $K_{k+1} = K_k(\sqrt{r_k})$.

Since I had already completed both of these steps, by mathematical induction $P(n)$ was true for all nonnegative integers n. Specifically, the equation $x^3 - 3x - 1 = 0$ has no solution in the field K_n. In other words, this equation has no solution that can be derived by starting with a rational number and performing a finite number of arithmetic and square root operations on it. In other other words, it has no constructible solution. *Done!*

---

[4]pg. 161

> **Answer 5-4 (Trisecting equation for $60°$)**
>
> The equation
> $$x^3 - 3x - 1 = 0$$
> has no constructible solutions.

I had found that $x^3 - 3x - 1 = 0$ has no solution in any of the fields $K_0 (= \mathbb{Q}), K_1, K_2, K_3, \ldots$ This meant that the solutions for that equation could not be constructed using a straightedge and compass some finite number of times. Specifically, $\cos 20°$ is not a constructible number.

> **Answer 5-3 (Constructibility of $\cos 20°$)**
>
> $\cos 20°$ is not a constructible number.

It therefore followed that neither $2\cos 20°$ nor $20°$ are constructible using straightedge and compass.

> **Answer 5-2 (Constructing a $20°$ angle)**
>
> A $20°$ angle cannot be created using a straightedge and compass.

Since it's impossible to construct a $20°$ angle, it's impossible to trisect a $60°$ angle using a straightedge and compass.

Finally!

I had finally arrived at my original goal—an answer to the angle trisection problem.

> **Answer 5-1 (Trisecting angles)**
>
> It is not always possible to use a straightedge and compass to trisect a given angle. We can use $60°$ as a counterexample.

I had first learned about this problem in some introductory mathematics book I'd read long ago. The problem had been around since ancient Greece, but it wasn't solved until the nineteenth century.

> It is not always possible to use a straightedge and compass to trisect a given angle.

A nugget of human wisdom, so simple that a single reading was enough to remember it forever. But sitting down and working out a proof on my own was a thrill like none other. The proof I'd worked out had existed for nearly two centuries, so all my work that night wasn't of any academic value. But it was invaluable to me. It was truly a joy to think through things myself, even clumsily, putting pencil to paper until I arrived at a solution.

My thoughts turned to the upcoming math festival at the Narabikura Library. I wondered what would happen there, and realized I actually wanted to go. I knew I probably shouldn't be spending time on things like that, though. I *really* needed to study for entrance exams.

Might as well just see when it is, though. I took a look at the note that Lisa had given me. *End of summer vacation, huh.*

Beneath the date, Lisa had written the name for the event.

The "Galois Festival"?

What we are asserting here is that there exists no method for precisely trisecting any given angle through a finite number of operations using a straightedge and compass. Thus, if we can find even a single angle for which such a procedure is impossible, we can consider our assertion to have been proved.

KENTARO YANO [9]
Angle Trisections

CHAPTER 6

Pillars of the Heavens

> There is nothing wrong with abstraction and generality—they are still cornerstones of the mathematical enterprise. But... general ideas should be abstracted *from* something, not conjured from thin air.
>
> IAN STEWART
> *Galois Theory, 3rd Ed.*

6.1 Dimensions

6.1.1 Neighborhood Festival

"I'm here!" called a voice from my front door, late the next evening.

There I found Yuri, wearing a bright orange summer kimono. She spun around to show it off. She was holding a *kinchaku* drawstring purse and had an elaborate *kanzashi* hairpin in her ponytail. Pale wooden *geta* sandals were on her feet.

"Cute, right?"

"I can't remember the last time I saw you in Japanese clothes. Looks good on you."

"Of course it does!"

My mother joined us. "Well hello, Yuri! Are you going to the festival?"

"I am! Mind if I take him along?"

"Not at all! He can be your bodyguard."
Yuri turned to me. "You heard her. Let's go!"
"Wait, what festival?"

6.1.2 Fourth Dimension

There were more people than I would have expected for a local summer festival. The streets teemed with couples, families, and packs of small kids.

Yuri and I walked down a row of vendor's stalls and game booths. Everything you'd expect was there: fried noodles and other street foods, crepes and cotton candy, goldfish catching and target shooting galleries. It wasn't long before Yuri started complaining, though, about how her kimono made it hard to move and her geta were hurting her feet. I told her she should spend less time in jeans.

After touring what was on display, we bought some fried noodles and sat on a bench. Yuri took off one of her geta and started rubbing her foot. "I think I'm getting a blister."

"You're getting manicures now?" I said, watching her.

"For feet it's called a pedicure, doofus."

"Yeah, yeah."

Yuri slipped her clog back onto her foot. "Hey, do you know *Fourth Dimension?*"

"The time-travel TV show?"

"That's the one!" She started singing the theme song. *"Inward! Outward! Upward! Beyond time! We journey through the fourth dime—nsion! Ba-dum!"*

"Right, right," I said, waving her into silence. "But a mathematical nth dimension just means you need n numbers to specify the location of a point in that space, not—"

"I know, I know. Pickled ginger, yuck." She started transferring scarlet strands of ginger from her tray of noodles to mine. "So the four dimensions are length, width, height, and time, right?"

"I guess, in the sense that you can use those four values to specify a single point. It works in that show, where their time-machine spaceship lets them travel not just in space, but in time as well. The idea is that they can travel not just forward–backward, left–right, up–down, but forward and backward in time, too."

"So time is the fourth dimension, right?"

"Well, you've gotta be careful there. It's not that time is *the* fourth dimension, it's that it can be *a* fourth dimension, depending on context. In a mathematical sense, time is just one example of a fourth dimension. In math you can also use five dimensions, or six, or generalize things out to n dimensions. But none of those will necessarily have anything to do with time."

"Huh." Yuri tossed her empty tray into a trashcan.

"Okay, here's a problem that might clear things up."

"Shoot."

"A line is a one-dimensional object, but its nature changes depending on whether it's located within a two-dimensional or a three-dimensional space."

"How so?"

"Well, can you imagine two lines in a two-dimensional space?"

"Sure. Just draw two lines on a piece of paper, right?"

"Yeah, but you can draw them in different ways, right? They can be right on top of each other, or they can intersect at a point, or they can be parallel. And they'll have to be in one of those three positional relationships."

"Oh, we've talked about this before. When we talked about systems of equations."[1]

"I'm glad you remember. Okay, so here's the problem. Say your two lines are in a three-dimensional space, instead of on a plane. What other positional relationship is possible, other than aligning, intersecting, or running parallel?"

"Answering that question requires octopus balls for fuel."

"Seriously? You're still hungry?"

"I'm a growing girl!"

---

[1] *Math Girls 4: Randomized Algorithms*, Chapter 7

> **Problem 6-1 (Positional relation between two lines)**
>
> When drawing a two lines in a 3-dimensional space, what positional relation other than
>
> · superposition,
>
> · intersection, and
>
> · parallelism
>
> is possible?

6.1.3 Octopus Balls

"Hot! Hot!" Yuri huffed around the octopus ball in her mouth.

"Now that you're fueled up, back to dimensions," I said. "Let's say you want to specify a single point in a plane. Besides just pointing and saying 'here,' you can also say something like 'starting from the origin, go this far in the horizontal direction, and so far in the vertical direction.' Maps do this, for example, using number pairs—longitude and latitude."

"I've never thought of longitude and latitude as a number pair, but now that you mention it..."

"A space like that, one where you can specify a point using a pair of numbers, is called a two-dimensional space. If instead you need three numbers, that's a three-dimensional space. And you can continue on in that manner—a four-dimensional space is one where you need four numbers to specify a point. That's hard to draw on paper, though."

"I'd think so."

"Like we were talking about, we can think of a space where you have depth, width, height, and time as a four-dimensional space. But don't forget, time here is just one possible fourth dimension."

"Got it," Yuri said, twirling her purse. "Oh, hey! I've already figured out the answer to your problem. Easy! Like this!"

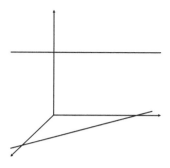

"Very good. That's called 'skew.'"
"Skew, got it."

Answer 6-1 (Positional relation between two lines)

When drawing a two lines in a 3-dimensional space, in addition to

- superposition,

- intersection, and

- parallelism,

the lines can be in a

- skew

positional relation.

6.1.4 Supports

"So what exactly is a dimension?" Yuri asked.

"What do you mean?" I said. "It's exactly what we've been talking about. In an n-dimensional space—"

"No, not that. Like, a line is a one-dimensional object, right? But you can put one in a three-dimensional space?"

"That's right."

Yuri sat thinking about what she wanted to say, reminding me of Tetra. Years apart and attending different schools, they didn't have many opportunities to meet, but we did get together to study math sometimes. I wondered if Tetra's dedication to learning was rubbing off.

"So what's bugging me is that, like, any point in three-dimensional space has three numbers associated with it, but then you've got this one-dimensional line that only needs one number, so... Grrr, I don't know how to put it in words."

I imagined what she might be thinking, and tried drawing the rest out.

"Maybe what you're trying to ask is something like this. It takes one number to specify a point on a one-dimensional line, but it takes three numbers to specify a point in three-dimensional space. So if you have a line in a three-dimensional space, how many numbers does it take to specify a point on that line?"

"Exactly! So how many?"

"An excellent question," I said in well-deserved praise. "It still only requires one number, since you've added the condition that the point must be on the line. It's easy to see if we use vectors."

I drew a simple graph in my notebook.

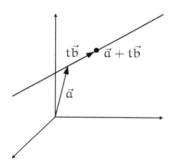

A line in a 3-dimensional space.

"In this figure, you can represent any point on the line as a sum of vectors $\vec{a}$ and $t\vec{b}$. Just go from the origin to the end of $\vec{a}$, then follow t times $\vec{b}$."

"Not following. How is that one dimension?"

"Well, look at the t variable here. Vectors $\vec{a}$ and $\vec{b}$ define the line, so you can specify any point on the line just by setting a value for t. Only one value is needed, so the line is still one-dimensional."

"I don't know much about vectors, but you're saying we just put some number into this t? And when that number changes, the point on the line will move around?"

"That's exactly right, that's what $\vec{a}+t\vec{b}$ is saying. We describe a line using vectors $\vec{a}$ and $\vec{b}$, and this one variable t determines where on that line the point is. Just by changing t, we can move the point anywhere in the direction of $\vec{b}$. That feels very line-like, right?"

"Still a bit math-nerdy for me, but you somehow make it sound like fun."

Yuri took my arm. She smelled like soap.

"It's getting late," I said. "Maybe we should head back."

The night became quieter as we worked our way home, the noise of the festival replaced by the clonking of Yuri's wooden sandals. I looked up and saw that the sky was full of stars.

"Pretty, isn't it?" Yuri said, following my gaze.

"It is."

"It's funny to think how space is so huge, but we can still use just three numbers to describe any place in it."

"Some cosmologists might disagree. There are theories that describe the world as having even more than four dimensions."

"Seriously?"

"I don't really know all the details, but I think the idea is that we can better understand the nature of space and time by using even more numbers to represent the condition of a given point in space."

"So you're saying there's dimensions other than depth, width, height, and time? I can't even imagine how."

"Neither can I. I don't even know if it's possible to represent in some way that we can see. I guess researchers today mainly look at space through their equations."

Yuri spread the sleeves of her kimono wide and shouted, "Space is huuu—ge!"

"Yuri, please."

"I'm still hungry," she said.

"I wonder how many dimensions are in your belly."

6.2 Linear Space

6.2.1 In the Library

"Four-dimensional space, huh?" Tetra said when I told her about my conversation with Yuri.

We were in the library, where I had again gone for an afternoon study session. I was taking a break, talking with Tetra to pass the time. Miruka was there too, scribbling away at something.

"Looks like Yuri's into that TV show everyone's watching," I said.

"That science fiction show with the fourth dimension, right? But I think that show uses 'fourth dimension' to describe, like, a parallel world or something, like so many science fiction stories do."

"You're probably right."

" 'Fourth dimension' is just one of those cool words." Tetra rolled her eyes. "So is 'space,' I guess. It just sounds so... *big*. There are spaces in math too, right Miruka?"

"There are," Miruka said, not looking up from what she was writing. "But a space in math means pretty much the same thing as a set. Usually, a space is a set with some sort of structure. There are sample spaces, probability spaces, vector spaces..."

Tetra raised a hand. "We've talked about sample spaces and probability spaces before,[2] but what's a vector space?"

"The most space-like space," Miruka said, looking up. "They're also called linear spaces."

"Sounds familiar," I said.

"Do you know the vector space axioms?"

"I do not."

Miruka narrowed her eyes. "Hmph."

"I have seen something about linear spaces in a book somewhere, though," I hurried to add. "Something about vectors and scalars, right?"

"Hang on, guys, I'm getting confused" Tetra said. "First off, vector spaces and linear spaces are the same thing, right?"

"Exactly the same," Miruka said.

"Okay. Also, I think of vectors as being something like arrows, but what's a scalar? Like, something that scales?"

---

[2] *Math Girls 4: Randomized Algorithms*, Chapter 4

"Scales?" I said.

"You know, as in 'scale up' and 'scale down.'"

"That works," Miruka said. "A linear space has vectors and scalars. Scalars can be used to stretch or shrink vectors. So sure, scaling up and scaling down, in a sense."

"Yeah, I guess so," I said, secretly relieved to be reminded what a scalar is.

"We also say that linearly independent vectors 'span' the vector space. Count up the maximum number of linearly independent vectors you can fit into the space, and that's the space's dimension."

"Of course!" I said, feeling something click in my head. Tetra apparently hadn't clicked yet.

"Um... So many new words..."

"Let's talk about vector spaces," Miruka said. And thus her lecture began.

6.2.2 Coordinate Planes

"We'll start with the coordinate plane," Miruka said. "We use two numbers to represent a point in the plane. For example, we call this point $(3, 2)$. The 3 is the x-coordinate, and the 2 is the y-coordinate."

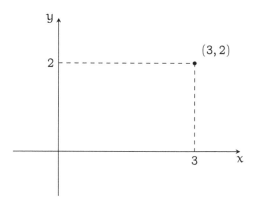

A coordinate plane.

"Got it," Tetra said.

"So tell me, Tetra. What are coordinates?" Miruka asked.

"Well, if you think of the coordinate plane as something like graph paper, I guess coordinates would be, like, the lines?"

"Sure. And when you make those lines, you have to decide how far apart to place them, right? So let's consider a situation like this."

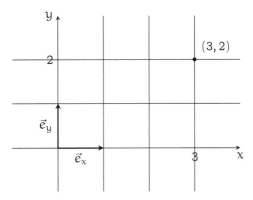

Figure 6.1: Point $(3,2)$ and vectors $\vec{e}_x, \vec{e}_y$.

"I've drawn two vectors here, $\vec{e}_x$ and $\vec{e}_y$. Vector $\vec{e}_x$ marks off one unit along the x-axis, and vector $\vec{e}_y$ does the same along the y-axis."

"I see that," Tetra said.

"So you see how an x-coordinate is a coefficient of $\vec{e}_x$ and a y-coordinate is some coefficient of $\vec{e}_y$?"

"Sure, that's the first thing you learn about the coordinate plane, right?"

"More or less. We're starting out with a review of what we already know, after all. So we can represent a point (a_x, a_y) as a sum of $\vec{e}_x$ multiplied by a_x and $\vec{e}_y$ multiplied by a_y. For example, point $(3,2)$ is the sum of $\vec{e}_x$ times 3 and $\vec{e}_y$ times 2, and we can write that like this."

$$3\vec{e}_x + 2\vec{e}_y$$

Tetra nodded. "So this is a correspondence for point $(3, 2)$, right?"

$$\text{Point } (3,2) \quad \longleftrightarrow \quad \underbrace{3}_{\text{x coord.}} \vec{e}_x + \underbrace{2}_{\text{y coord.}} \vec{e}_y$$

"Very good. And we can use a set to represent the entire plane made up from real-number coordinate values, like this."

Set of coordinate points $= \{a_x \vec{e}_x + a_y \vec{e}_y \mid a_x \in \mathbb{R}, a_y \in \mathbb{R}\}$

"Nothing difficult here."

"Hang on, though," Tetra said. "Maybe it isn't difficult, but I have no idea where you're going with all this."

Miruka twirled a finger in the air. "We're starting here because the coordinate plane is such a familiar mathematical object. But now it's time to get a little bit more abstract. Let's take a closer look at $a_x \vec{e}_x + a_y \vec{e}_y$."

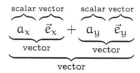

"See the structure there? Specifically..."

· A vector multiplied by a scalar is a vector.

· The sum of a vector and a vector is also a vector.

"See that? This is the fundamental aspect of a vector space."

"Hey, I've seen this before..." Tetra said, an odd tone in her voice. "Oh! I know! Multiply–multiply–add! Just like matrices!"

"Except that in this case we're going to call the form $a_x \vec{e}_x + a_y \vec{e}_y$ a *linear combination* of vectors $\vec{e}_x$ and $\vec{e}_y$."

6.2.3 Vector Spaces

"Okay, enough warmup," Miruka said. "Let's get into a more general discussion of vector spaces. First off, we need to think of a set of scalars S and a set of vectors V, and add to those two operations, addition and multiplication."

"So we need to think about what those operations do," I said.

"Right." Miruka's speech increased in tempo. "When we define a specific vector space we also have to define these two operations. The first will be scalar multiplication of vectors, so we're defining

$sv \in V$ for elements $s \in S$ and $v \in V$. Multiplication of a scalar and a vector, in other words. In terms of the $a_x \vec{e}_x$ we were just talking about, S is the set of real numbers $\mathbb{R}$, and V is the set of all vectors in the two-dimensional plane. Similarly, $a_x \vec{e}_x \in V$ is defined for $a_x \in \mathbb{R}$ and $\vec{e}_x \in V$."

"Okay, good with that," Tetra said.

"The next operation is for addition of two vectors, so we're defining $v + w \in V$ for $v \in V$ and $w \in V$. Before, we were talking about $a_x \vec{e}_x + a_y \vec{e}_y$, the sum of $a_x \vec{e}_x$ and $a_y \vec{e}_y$, right?"

"We were," I said.

"When V and S satisfy some rules, we call V a 'vector space over S.' Here's everything laid out nice and neat as axioms."

Vector space axioms

Given elements v, w of an abelian group V and elements s, t of a field S, V is called *a vector space over* S if it satisfies the following axioms:

VS1 sv is an element of V. (Scalar multiplication of vectors)

VS2 $s(v + w) = sv + sw$ holds. (Distributivity across vector sums)

VS3 $(s+t)v = sv + tv$ holds. (Distributivity across scalar sums) (The + operator on the left is a scalar sum, that on the right is a vector sum.)

VS4 $(st)v = s(tv)$ holds. (Associativity of scalar multiplication)

VS5 $1v = v$ holds. (Scalar multiplication identity)

"If V is a vector space over S, an element of the abelian group V is called a vector, and an element of the field S is called a scalar."

"So that's where vectors and scalars come from," Tetra said.

"A problem for you," Miruka said. "Let's consider the coordinate plane we normally use as a vector space over the set of real numbers $\mathbb{R}$. That would mean $\mathbb{R}$ is equal to the set of scalars S, right?"

"Sure, I see that," Tetra said after some hesitation.

"Okay, so if we're considering the coordinate plane to be a vector space over $\mathbb{R}$, then what kind of set is V, the set of all vectors?"

"Hmm, let's see..."

"Oh!" I said. "V is—"

"I'm not asking you," Miruka snapped.

"I'm sorry, I'm not seeing anything," Tetra said.

"V is the overall coordinate plane, so an element of V has to be a point in the coordinate plane."

"Oh, okay. I guess."

"The vector set V in the vector space is a set in which we've defined addition, right? So V is an abelian group, in other words, a group in which the commutative law holds."

Tetra held up her hands. "Hang on for a second, let me make sure I've got this all straight. We're considering the coordinate plane to be a vector space over the real numbers. That means the vector set V is equal to the coordinate plane. Vectors are points in the coordinate plane, and V is an abelian group. Which means, uh, that we've introduced an addition operation for points in the coordinate plane?"

"Exactly. You know how to add points, right?"

"Oh, I guess I do! Like, $(2,3) + (1,2) = (3,5)$?"

"Correct. In school you've probably heard points in the coordinate plane referred to as position vectors. That's what we're talking about here."

Tetra nodded. "Okay, I think I'm starting to get this."

Having already been scolded once, I decided to just sit back and enjoy Miruka and Tetra's back-and-forth. Which was fine by me; I found it quite entertaining.

"So if we're letting the set of vectors V be the set of points in the coordinate plane," Tetra said, "what does it mean to multiply a vector by a scalar?"

"The same thing as multiplying a vector by a real number."

"So, just extending or contracting the vector without changing which way it's pointing?"

"That's right, it's the same thing. We represent real-number multiplication of a vector in the plane as multiplication of the vector by a scalar. But the vector space axioms don't say anything about the orientation or size of vectors, do they? To get that we need to introduce inner products."

"Oh, okay."

"But first, a quick review of the basics regarding vector spaces in the coordinate plane. A vector multiplied by a scalar is a vector, and the sum of two vectors is a vector. Nothing surprising there, it's exactly what we learn about vectors at school. So all these definitions for linear spaces probably sound like a rehashing of things you've heard before, just maybe with some different words."

"Yeah, but this isn't the first time we've started from such humble beginnings," I said. "It's always like this when we start from axioms. They usually aren't super exciting."

Miruka nodded. "Things get interesting when we find other mathematical objects that fulfill the same axioms. For example, when we see that we can treat things that we don't normally consider to be vectors as if they were."

"Like when we saw how we can treat ladder diagrams as groups?" I said.

"Or when we treated the set of rational numbers as a field!" Tetra added.

"Good one!" I said.

Many times, we had assigned abstract names like "group" and "field" to mathematical objects that fulfilled some list of axioms. Every time we did so, we discovered new worlds. It looked like vector spaces were shaping up into a similar experience.

"So what kind of non-vectorish thing can we treat as vectors?" Tetra asked.

"Complex numbers, for example," Miruka replied.

"Complex numbers... are vectors?"

"If you want them to be."

6.2.4 $\mathbb{C}$ as a Vector Space over $\mathbb{R}$

Miruka snagged the nearest notebook and started filling a page with symbols, letters, and diagrams as she continued her lecture.

"So," she began. "We're thinking of examples of vector spaces. Vectors in the coordinate plane are a familiar example. We want to find something less familiar, though, something we can call a vector because it fulfills the axioms for a vector space. We'll start with complex numbers."

"Meaning we're going to call complex numbers vectors too?" Tetra asked.

"That's right. Let's consider the set of all complex numbers $\mathbb{C}$ as a set of vectors, and the set of real numbers $\mathbb{R}$ as a set of scalars. If we consider multiplication of a complex number by a real number as vector multiplication by a scalar, and a sum of complex numbers as a sum of vectors, then we can consider $\mathbb{C}$ as a vector space over $\mathbb{R}$."

Set of all complex numbers $\mathbb{C}$ Set of vectors
Set of all real numbers $\mathbb{R}$ Set of scalars

Considering $\mathbb{C}$ as a vector space over $\mathbb{R}$.

"Very cool!" I said.

"Just look at the complex plane, and it's easy to see how we can consider complex numbers and points in a coordinate plane in the same way, and how we can view both as vector spaces. Let's see that one more time."

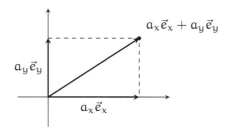 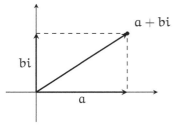

Coordinate plane as a vector space   Complex plane as a vector space

We sat in silence for a moment, considering Miruka's graph. Viewed this way, the coordinate plane and the complex plane were something like different instruments playing the same tune, a piece titled "vector spaces."

After enjoying the melody for a time, Tetra spoke: "Okay, I'm feeling a little more comfortable with vector spaces now, but I'm still not quite sure what the point of all this is. Why would we want to be able to consider the complex plane and the coordinate plane in the same way?"

I realized Tetra was now speaking more sharply than she would have in the past. She was as eager and polite and humble as she'd always been, but she no longer hesitated before speaking her mind.

"Hmph." Miruka crossed her arms. "Okay, another example then."

6.2.5 $\mathbb{Q}(\sqrt{2})$ as a Vector Space over $\mathbb{Q}$

"Let's consider $\mathbb{Q}(\sqrt{2})$ as a vector space over $\mathbb{Q}$," Miruka said. "We'll adjoin $\sqrt{2}$ to $\mathbb{Q}$ to create the extension field $\mathbb{Q}(\sqrt{2})$ and let the set of rational numbers $\mathbb{Q}$ be our set of scalars. That lets us consider $\mathbb{Q}(\sqrt{2})$ to be a vector space over $\mathbb{Q}$. Do you remember $\mathbb{Q}(\sqrt{2})$, Tetra?"

Field $\mathbb{Q}(\sqrt{2})$ (adjunction of $\sqrt{2}$ to the field of rationals)	Set of vectors
Field of rationals $\mathbb{Q}$	Set of scalars

Considering $\mathbb{Q}(\sqrt{2})$ as a vector space over $\mathbb{Q}$.

"I do!" Tetra said. "I've even been reading up on fields!"

"Good for you. So $\mathbb{Q}(\sqrt{2})$ is—"

Tetra leaned forward and held up a hand. "Wait! Let me see if I can describe it, to make sure I understand everything." Miruka nodded her consent and Tetra brightened. "Okay, so $\mathbb{Q}(\sqrt{2})$ is an adjunction of $\sqrt{2}$ to $\mathbb{Q}$, which means it's a field in which we can perform the four basic arithmetic operations. So in this case, we can do arithmetic using rational numbers and $\sqrt{2}$."

"Examples?"

"Huh?"

"Give me some example elements of $\mathbb{Q}(\sqrt{2})$."

"Oh, sure. The key to understanding, right? No problem. Like I said, I've been studying. So here's some example elements we can make by performing arithmetic using the rationals and $\sqrt{2}$."

$$1 \quad 0 \quad 0.5 \quad -\tfrac{1}{3} \quad \sqrt{2} \quad \tfrac{\sqrt{2}}{3} \quad \tfrac{1+3\sqrt{2}}{2-\sqrt{2}}$$

"Well done." Miruka expressed her satisfaction with a nod. "Moving on, an expression created using only arithmetic operations is called a 'rational expression.'"

"Rational expression, got it."

"Also, the value of an integer rational *expression* is a rational *number*."

"Ah, okay. I'll be sure to keep those straight. And we can say that the value of any rational expression we've created using rational numbers is itself a rational number, right?"

"Exactly. Because the field of rational numbers is closed."

Tetra gave a deep nod. "Okay, I'm good."

"A problem, then. Does $\sqrt{\sqrt{2}} \in \mathbb{Q}(\sqrt{2})$ hold?"

"I wouldn't think so, no, since you can't make $\sqrt{\sqrt{2}}$ using just rational numbers and arithmetic. You could also say that it won't have a value you come up with using only the rationals and $\sqrt{2}$."

"Very good."

I noticed that Tetra was also becoming more confident in her answers to Miruka's questions. She wasn't always right, but that didn't hold her back.

"Here's how I think of it," Tetra said. "There's this guy who has all the rational numbers. We go up to him and say 'Here, this is for you,' and hand him $\sqrt{2}$, an irrational number. He used to only be able to create rational numbers, no matter how hard he tried, but our gift has broadened his horizons—now he can use this $\sqrt{2}$ to make all kinds of new numbers! And he lived happily ever after. The end."

Miruka smiled. "I guess that's the gist of it."

"Such a...romantic take on things," I said.

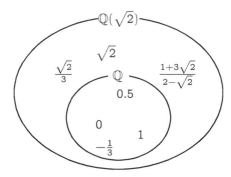

Field of rational numbers ℚ and extension field $\mathbb{Q}(\sqrt{2})$.

"I mean, there's no point in giving him, say, a 0.5, right? That's a rational number, so he already has it."

"That's a good way to think about it," Miruka said, her expression now serious. "The adjunction of a rational number to the field of rational numbers doesn't do anything to change the field. We can write that as a equation like this."

$$\mathbb{Q}(0.5) = \mathbb{Q}$$

"Sure," Tetra said.

"More generally, ℚ(a rational number) = ℚ will always hold. By the way, we can write $\mathbb{Q}(\sqrt{2})$ like this."

$$\mathbb{Q}(\sqrt{2}) = \{p + q\sqrt{2} \mid p \in \mathbb{Q}, q \in \mathbb{Q}\}$$

"Oh, okay."

"Why do you think I didn't write the element as a fraction, like $\frac{p+q\sqrt{2}}{r+s\sqrt{2}}$?"

"Hmm, good point. $\mathbb{Q}(\sqrt{2})$ is a field, which means we can perform all the arithmetic operations in it, so I guess it would make sense to write the generalized form like that. I mean, we have to allow for division, right?"

"But what about rationalization of denominators?"

"Oh... Oh! Of course. Even if we start from a fraction, we can just simplify to get it in the form $p + q\sqrt{2}$. Right?"

"Right. So long as you know you can get from the fractional form to $p+q\sqrt{2}$, confirming that $\mathbb{Q}(\sqrt{2})$ is closed under arithmetic is just a matter of calculating."

[add.] $(p+q\sqrt{2})+(r+s\sqrt{2}) = \underbrace{(p+r)}_{\in \mathbb{Q}}+\underbrace{(q+s)}_{\in \mathbb{Q}}\sqrt{2}$

[sub.] $(p+q\sqrt{2})-(r+s\sqrt{2}) = \underbrace{(p-r)}_{\in \mathbb{Q}}+\underbrace{(q-s)}_{\in \mathbb{Q}}\sqrt{2}$

[mult.] $(p+q\sqrt{2})(r+s\sqrt{2}) = \underbrace{(pr+2qs)}_{\in \mathbb{Q}}+\underbrace{(ps+qr)}_{\in \mathbb{Q}}\sqrt{2}$

[div.]
$$\frac{p+q\sqrt{2}}{r+s\sqrt{2}} = \frac{p+q\sqrt{2}}{r+s\sqrt{2}} \cdot \frac{r-s\sqrt{2}}{r-s\sqrt{2}} \quad \text{Rationalize}$$
$$= \frac{(p+q\sqrt{2})(r-s\sqrt{2})}{(r+s\sqrt{2})(r-s\sqrt{2})}$$
$$= \frac{(pr-2qs)+(qr-ps)\sqrt{2}}{r^2-2s^2}$$
$$= \underbrace{\frac{pr-2qs}{r^2-2s^2}}_{\in \mathbb{Q}}+\underbrace{\frac{qr-ps}{r^2-2s^2}}_{\in \mathbb{Q}}\sqrt{2}$$

"Interesting!" Tetra said. "This alone makes it worth learning how to rationalize denominators. We've used this rational + rational $\sqrt{2}$ form when checking for closure under arithmetic, like when we're confirming axioms. I'm starting to see how this kind of transformation can be useful."

"Hey, Miruka, about $\mathbb{Q}(\sqrt{2})$," I said. "Can I think of this adjunction of $\sqrt{2}$ to $\mathbb{Q}$ as a field that we've expanded out as far as we can using arithmetic operations on $\mathbb{Q}$ and $\sqrt{2}$?"

"I suppose," Miruka said. "But I think it's more common to describe it as the *smallest* field that contains both $\mathbb{Q}$ and $\sqrt{2}$. By smallest, I'm assuming the magnitude relation is based on inclusion. So, a quick problem."

Letting n be a positive integer, when does $\mathbb{Q}(\sqrt{n}) = \mathbb{Q}$?

"Oh, that's easy," Tetra said. "It's when $\sqrt{n}$ is rational! Because if we give a rational to someone who already has all the rationals, we haven't broadened his horizons at all. So $\mathbb{Q}(\sqrt{n}) = \mathbb{Q}$ when n is $1^2, 2^2, 3^2, 4^2, \ldots$, and so on."

"Well done."

6.2.6 Expanse

"So what's so cool about vector spaces, anyway?" Tetra asked.

Miruka shot up from her seat, and Tetra and I looked up toward her. I reflected on the fact that I'd never before had a teacher whose lectures made me want to learn more.

"We can use vector spaces to represent the broadness of things," Miruka said. "Just as the word 'space' implies, a vector space is a kind of expanse. We use vectors to mathematically represent the extent of that expanse."

"I'm...not sure what you mean."

Miruka turned to me. "What do *you* think I mean?"

"Do you mean..." For some reason I heard the clonking of Yuri's wooden clogs. "...dimensions?"

Miruka raised her index finger. "I do," she said. "Which leads us to the question, what is dimension? First off, a set of vectors that can uniquely represent an arbitrary point in a vector space by linear combination is called a *basis* of that linear space. The number of elements in that basis is called its dimension. Good so far?"

Tetra and I nodded.

"Okay," Miruka continued. "You can also consider the number of vectors *necessary and sufficient* to represent any point in the vector space as its dimension. We can uniquely represent an arbitrary point (a_x, a_y) in the coordinate plane as a linear combination $a_x \vec{e}_x + a_y \vec{e}_y$ of vectors $\vec{e}_x$ and $\vec{e}_y$, so this is a two-dimensional vector space. Similarly, we can uniquely represent a given complex number $a + bi$ as a linear combination $a \cdot 1 + b \cdot i$ of vectors 1 and i, so that's a two-dimensional vector space too.

"If that's hard to follow, you can look at $a + bi$ in the same way as we looked at $a_x \vec{e}_x + a_y \vec{e}_y$ before, and see a structure like this."

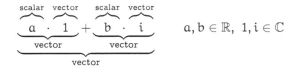

$\mathbb{C}$ as a vector space over $\mathbb{R}$.

"But there's more. We can also uniquely represent an arbitrary element $p + q\sqrt{2}$ of the field $\mathbb{Q}(\sqrt{2})$, using the vectors 1 and $\sqrt{2}$ in the linear combination $p \cdot 1 + q \cdot \sqrt{2}$. Again, two-dimensional. Here's the structure we find when we look at $p + q\sqrt{2}$."

$\mathbb{Q}(\sqrt{2})$ as a vector space over $\mathbb{Q}$.

"Dimension... and basis..." Tetra muttered, scribbling notes.

"While you're writing, add that $\{\vec{e}_x, \vec{e}_y\}$ is one basis we can use to consider the coordinate plane as a vector space over $\mathbb{R}$," Miruka said.

"I see," I said. "And the number of elements in that basis is its dimension."

Miruka nodded. "We usually use $\{1, i\}$ as a basis when considering $\mathbb{C}$ as a vector space over $\mathbb{R}$, but we have other options. We could just as well use $\{-1, i\}$, for example, or $\{100, -20i\}$, or even $\{1 + i, 1 - i\}$. But no matter which basis you use, it will always contain two vectors."

"Because the 'two' in 'two-dimensional' refers to the number of elements in its basis, right," I said.

"Using linear combinations of vectors to represent all of the points in a vector space is called *spanning* the vector space," Miruka said. "Using this terminology, we can say a basis of a vector space spans that vector space."

"Something like the ribs in an umbrella?" I said.

"Or beams in a building?" Tetra said.

"Visualize it however you wish," Miruka said. "The point is, we can use a basis comprising just two vectors to create the vastness of the coordinate plane. It's a finite means to capture the infinite."

"I see," Tetra said.

"Do you? Let's find out."

When considering $\mathbb{Q}(\sqrt{2})$ as a vector space over $\mathbb{Q}$, can you use $\{\sqrt{2}, 2\}$ as a basis?

Tetra's expression became serious. "Hmm, okay. Maybe I spoke too soon... Oh, wait, I think this works!"

"Explain."

"Um, because the numbers we can write as $p \cdot 1 + q \cdot \sqrt{2}$ when the basis is $\{1, \sqrt{2}\}$ can be rewritten as $q \cdot \sqrt{2} + (\frac{p}{2}) \cdot 2$ when the basis is $\{\sqrt{2}, 2\}$. Right?" Miruka nodded, and Tetra beamed. "So any element in $\mathbb{Q}(\sqrt{2})$ can be written as a sum of—"

"As a linear combination of," Miruka corrected.

"Ah, right, as a *linear combination* of $\sqrt{2}$ and 2, meaning we can span the vector space $\mathbb{Q}(\sqrt{2})$ over $\mathbb{Q}$ using $\{\sqrt{2}, 2\}$ as a basis."

"Correct."

"Anyway, since it spans the vector space, $\{\sqrt{2}, 2\}$ can be a basis."

"You forgot some words," Miruka said.

"I did?"

"Being able to span a vector space means you can represent an *arbitrary* point in that vector space. A basis is a set of vectors that can *uniquely* represent *any* point in the linear space as a linear combination. The way you put it, saying 'since it spans the vector space it can be a basis,' won't cut it. Uniqueness is important here."

"So uniqueness is necessary to define a basis?"

"It is. If you leave it out, if you just say 'a set of vectors that can represent any point in a linear space,' then you could use something like $\{1, \sqrt{2}, 2\}$,—a set with three elements—as a basis for $\mathbb{Q}(\sqrt{2})$. Then the example $p \cdot 1 + q \cdot \sqrt{2}$ you just gave could also be represented like this."

$$\frac{p}{3} \cdot 1 + q \cdot \sqrt{2} + \frac{p}{3} \cdot 2 \qquad \text{(linear combination of } 1, \sqrt{2}, 2)$$

"Without uniqueness, there's no limit on the number of vectors, so we wouldn't be able to define dimensions."

"Huh," Tetra said, then fell silent.

Miruka and I did the same, giving her some space to sort things out in her head. I remembered telling Yuri that having four dimensions meant representing a point using four numbers. I now realized that my explanation, while not strictly wrong, was mathematically incomplete. There was a basis lurking behind those four numbers. *And any point can be uniquely represented by a linear combination of elements of that basis, huh?* Multiply each by some scalar, add them up. The four numbers I was thinking of were those scalars.

After a time, Tetra spoke up. "The elements of the basis, they have to be all unrelated, right?"

"Unrelated?" Miruka repeated.

"Well, like you said, if there's a 2 in the basis we can't add a 1, because we'd lose uniqueness of linear combinations. Maybe I'm not saying that right. I mean, if there's a 2 in there, you can already use it to create a 1, so we shouldn't add that. Also, if we've put a 2 and a $\sqrt{2}$ in the basis, we can't add $2+\sqrt{2}$, since we can make that from 2 and $\sqrt{2}$. Again, no uniqueness of linear combinations. So we have to make them, like, you know... Agh! I don't know how to put it in words. Help!"

"Words to describe what?" I asked.

"To describe when you can and can't make things using linear combinations! There's got to be some words, some concept I can use instead of 'can make' and 'can't make.'"

"Don't worry, Tetra," Miruka said. "You caught on to the concept, that's the important thing. The words you're looking for are 'linear dependence' and 'linear independence.'"

"Dependence and independence?"

"Roughly speaking, a vector you can create using a linear combination of a set of vectors is dependent on those vectors. One you can't create in this way is independent of those vectors. Properly defining those terms is our next task."

"Okay, now I think I really do see. The idea is whether a vector 'depends' on another vector, right? Interesting, interesting..." Tetra got a far-off look on her face, as if she'd slipped off into her own world. "Vectors you can create through linear combinations are dependent on each other. Ones you can't create through linear combinations are independent of each other. It's like linearly independent vectors in a vector space find each other irreplaceable. Kind of romantic, really."

"You're getting kind of far from the math here," Miruka said.

6.3 Linear Independence

6.3.1 The Basics

Miruka resumed her lecture. "Linear independence is a very important concept in vector spaces. Starting with any vector, you can just multiply it by a scalar to make infinitely many more vectors. Even so, multiplication by a scalar can't take you everywhere in a multidimensional space. To reach everywhere, you need linear combinations of multiple vectors that are linearly independent. So let's look at a proper description of linear independence.

"Given a vector space over S, we can define linear independence between vectors v and w like this."

Linear independence (two-dimensional)

For a linear space V over S, vectors $v, w \in V$, and scalars $s, t \in S$, v and w are *linearly independent* if

$$sv + tw = 0 \iff s = 0 \overset{\text{(and)}}{\wedge} t = 0.$$

Otherwise, vectors v and w are *linearly dependent*.

"So when considering V as a vector space over S, we show that vectors v and w are linearly independent like this."

$$sv + tw = 0 \iff s = 0 \wedge t = 0 \qquad (s, t \in S)$$

"If we're taking the coordinate plane to be a vector space over $\mathbb{R}$, vectors $\vec{e}_x$ and $\vec{e}_y$ are linearly independent if this is true."

$$a_x \vec{e}_x + a_y \vec{e}_y = 0 \iff a_x = 0 \wedge a_y = 0 \qquad (a_x, a_y \in \mathbb{R})$$

"Or say we're taking $\mathbb{C}$ to be a vector space over $\mathbb{R}$. Then the condition for linear independence appears as a basic proposition using real and complex numbers."

$$a + bi = 0 \iff a = 0 \wedge b = 0 \qquad (a, b \in \mathbb{R})$$

Pillars of the Heavens

"Finally, here's the condition for linear independence when taking $\mathbb{Q}(\sqrt{2})$ as a vector space over $\mathbb{Q}$."

$$p + q\sqrt{2} = 0 \iff p = 0 \land q = 0 \qquad (p, q \in \mathbb{Q})$$

"Whoa!" I shouted. "I used this when solving the angle trisection problem!"

"Like I said, linear independence is important," Miruka cooly stated.

"They're all in the same form," Tetra said.

"They are," I said, trying to keep my cool despite my racing heart. In truth I felt joy, as if I had discovered a familiar melody in a piece played on an instrument I'd never heard before. A melody titled "Linear Independence."

- $sv + tw = 0 \iff s = 0 \land t = 0$ for vector spaces.

- $a_x \vec{e}_x + a_y \vec{e}_y = 0 \iff a_x = 0 \land a_y = 0$ for the coordinate plane.

- $a + bi = 0 \iff a = 0 \land b = 0$ for the field of complex numbers.

- $p + q\sqrt{2} = 0 \iff p = 0 \land q = 0$ for the extension field $\mathbb{Q}(\sqrt{2})$

Using the fact that a complex number $a + bi$ equaling 0 implies that a and b are both 0 is something that showed up on tests all the time. I'd had no idea that had anything to do with vectors, though. But there it was, $a + bi = 0 \iff a = 0 \land b = 0$ as the condition for linear independence when taking $\mathbb{C}$ to be a vector space over $\mathbb{R}$. A proposition stating that 1 and i are linearly independent.

I sat back and thought about this. *If $a + bi = 0$, then $a = -bi$. So if $a \neq 0$, then $1 = -\frac{b}{a} \cdot i$, meaning the vector 1 can be written as a scalar multiple of the vector i. On the other hand, if $b \neq 0$ then $i = -\frac{a}{b} \cdot 1$, so the vector i can be written as a scalar multiple of the vector 1. In other words, the condition $a = 0 \land b = 0$ means we can't write 1 as a linear combination of i, and we can't write i as a linear combination of 1. So linear independence means neither 1 nor i can be written as a linear combination of the other.* Huh.

"But why limit ourselves to two dimensions?" Miruka said. "Let's generalize."

> **Linear independence (generalized)**
>
> For a linear space V over S, vectors $v_k \in V$, and scalars $s_k \in S$, vectors $v_1, v_2, \ldots, v_m$ are *linearly independent* if
>
> $$s_1 v_1 + s_2 v_2 + \cdots + s_m v_m = 0 \iff s_1 = 0 \wedge s_2 = 0 \wedge \cdots \wedge s_m = 0.$$
>
> Otherwise, vectors $v_1, v_2, \ldots, v_m$ are *linearly dependent*.

"So many symbols..." Tetra moaned.

"Then let's get an overhead view." Miruka said. "If you don't have enough vectors, you can't span the entire vector space."

- Real-number multiples of $\vec{e}_x$ can only make a straight line; they cannot create the coordinate plane.

- Real-number multiples of 1 can only create $\mathbb{R}$; they cannot create $\mathbb{C}$.

- Rational-number multiples of 1 can only create $\mathbb{Q}$; they cannot create $\mathbb{Q}(\sqrt{2})$.

"That makes so much sense," I said. "In other words, there's only so much you can do with scalars alone."

"However," Miruka continued, "if you have too many vectors you can't *uniquely* describe a point in a vector space."

"Right, you'd be able to specify a point in more than one way!" Tetra said.

"Which we don't want. We want to be able to *uniquely* specify *any* point in a vector space. To do so, we need a set of necessary and sufficient vectors, in other words, a basis. When an arbitrary point can be uniquely represented, the basis will be a minimal set of vectors spanning the entire vector space, and it will also be a maximal set of linearly independent vectors. When I say 'minimal' and 'maximal' here, I'm talking about the number of elements. There's no one right way of selecting a basis, but regardless of how you do so, the number of elements you end up with will always be the same—it's invariant." Miruka's eyes gleamed. "Invariant things are worth naming. We name this number of vector elements in the basis the *dimension*."

6.3.2 Dimensional Invariance

"I love this idea of naming invariant things," Tetra said.

I nodded in agreement. "As do I. Very cool."

Miruka suddenly looked away from us. "Think of invariance as something like what physicists call 'conservation,' as in conserved quantities or the various conservation laws. They would naturally want to give names to those."

I had never really thought of things in quite this way. I could understand mathematical concepts if someone explained them to me. I could understand books by reading them. I could understand theorems by working with them—if they weren't too far above my head, at least. But Miruka's idea of giving names to invariant things, that was new to me. I wasn't sure what it was, or where it came from.

Tetra clapped her hands. "Miruka! You're talking about things being well-defined, aren't you?"

Miruka cocked her head. "How's that?"

"You taught us about things being well-defined before. What we're talking about here is a concept that can be defined specifically because it doesn't matter how we make a selection. There are various ways of selecting a basis for a vector space, but the number of vectors in the basis you end up choosing will always be the same, which is why we're able to define the concept of dimensions. If that's the case, then the concept of dimensions is one that's well-defined. Or, that's what I thought, at least. Am I wrong?"

Miruka gave a smile so big it seemed it would spill off of her face.

"Tetra, Tetra, my lovely Tetra. Come over here. No, on second thought..."

Miruka spread her arms wide and walked—nearly ran—to where Tetra was sitting. She wrapped her arms around Tetra in an all-encompassing embrace, and planted a kiss right on her cheek.

"I— What?— Uh, Miruka!"

"I love smart girls."

6.3.3 Degree of Extension

With the girls carrying on, I became a little bit worried about Ms. Mizutani making an appearance—disrupting the peace of her library could result in extreme consequences.

Motioning for everyone to be a bit quieter, I asked Tetra, "By the way, didn't you say you were studying about fields?"

Tetra brightened. "I am! I wanted to think a little more about all this $\mathbb{Q}(\sqrt{\text{discriminant}})$ stuff. I tried reading a book about it, but it was just a bunch of theorems and proofs, so I didn't get much out of it."

"Yeah, that happens."

Indeed, it had happened to me, when trying to read a math text in a somewhat narrow field. There were plenty of terms I knew, and I could follow the examples, but most of it was a long series of theorems followed by proofs, which made it hard to follow the overall flow.

"I did understand how if you can't solve a quadratic equation in $\mathbb{Q}$, you can take its discriminant from the quadratic formula and solve it in $\mathbb{Q}(\sqrt{\text{discriminant}})$ instead. I think I understand that, at least. I was also able to see that there's a *lot* more to learn about fields..."

Miruka nodded. "There are many cases where you'll want to consider the field you get from adjoining $\sqrt{\text{discriminant}}$ to $\mathbb{Q}$. So, a quick review. If $\sqrt{\text{discriminant}} \in \mathbb{Q}$, then the adjunction is just $\mathbb{Q}$ itself."

"Nothing changes, right!"

"In the case where $\sqrt{\text{discriminant}} \notin \mathbb{Q}$, however, the adjunction is the field extension $\mathbb{Q}(\sqrt{\text{discriminant}})$."

"Yep! You've given it a present, expanding its horizons!"

"Fair enough. But by *how much* have you expanded its horizons?"

Tetra opened her mouth, then closed it again. "By...how much?"

"That's right. I want to know how big the expansion is."

Miruka gave a mischievous smile I'd often seen before, one she used to provoke a reaction.

"How big the expansion is... I'm not sure—"

Miruka narrowed her eyes. "Maybe I should have said how much broader you've made things."

"How much broader... Are you talking about... dimensions?"

"I am," Miruka said, taking Tetra's hand in hers. "We can capture field extensions through the viewpoint of vector spaces. Specifically, we can use dimensions to describe how much we've extended our field when moving from $\mathbb{Q}$ to its adjunction $\mathbb{Q}(\alpha)$."

Wow. I was so excited I failed to find my voice.

"Let's do this right," Miruka said, taking my hand with her other hand. *So warm.* "Let the field $\mathbb{Q}(\alpha)$ be a vector space over $\mathbb{Q}$. How many linearly independent vectors can we find? In other words, what is the dimension of $\mathbb{Q}(\alpha)$ over $\mathbb{Q}$? To answer that question is to quantitatively capture the extension of the field, and to characterize this element α. It also leads us to a new approach for using vector spaces."

"Yeah?" I said. "What kind of approach?"

"An approach that allows solutions through equations, of course."

My ears perked up. "Equations, did you say?"

"Algebraically solving equations is all about factorization. When factoring, it is necessary to clarify which field we're working in. If it's an extension field that contains all solutions to the equation, the equation can be factored into a product of linear equations. The theory of equations is a theory about fields."

Miruka's voice took on a musical lilt. "So. Back to our original question. When we adjoin an element to a field, by how much have we expanded it? We can use the concept of dimension of the vector space to define that scale of expansion, in other words the degree of extension. It's very interesting to consider how other concepts that we associate with equations—their roots and the number of roots, formulas for finding them, the equation's degree—might be associated with concepts in vector spaces."

Miruka's hand grasped mine more tightly. "This is interesting because a vector space is a bridge between two worlds. It connects the world of equations with the world of fields."

"Again," I squeaked.

Yes, again. So often when we explored mathematics, we discovered pairs of worlds. Fermat's last theorem had brought together algebra and geometry. We had also seen the paired worlds of algebra and analysis. Gödel's incompleteness theorems involved the worlds of formalism and semantics. Working on the angle trisection problem had straddled geometric constructions and numbers. And now, equations and fields. *Maybe mathematicians should be called 'bridge builders.'*

"Another bridge," I said.

Miruka placed her index finger on my lips.

"Don't you love it?" she said.

The concept of vector spaces is significant in its perspective of extracting from among the myriad mathematical objects only the "operational framework" of addition and scalar multiplication.

SHIGA KOJI [12]

CHAPTER 7

The Secrets of Lagrange Resolvents

> The young Prince was all on fire at these words, believing, without weighing the matter, that he could put an end to this rare adventure; and, pushed on by love and honor, resolved that moment to look into it.
>
> CHARLES PERRAULT,
> TRANS. BY A.E. JOHNSON
> *The Sleeping Beauty in the Wood*

7.1 THE CUBIC FORMULA

7.1.1 Tetra

"Hey, you!" Tetra said.

We were in the library where, as had become my custom, I'd headed after my morning summer-session class.

"Funny how often we're running into each other, it being summer vacation and all," I said. I took a peek at her notebook. It was filled with equations, many of them crossed out or marked with corrections. *Pretty advanced stuff.* "Still working hard, I see."

Tetra glumly looked down at what she had been working on. "Math sure can be hard, can't it?"

"You're trying to solve equations?"

"I ran into Mr. Muraki, who said he had some calculation practice for us to work on over summer break."

"Calculation practice? That doesn't sound like him..."

"I know, right? But check this out! *Seven* cards!"

She spread out a rainbow of cards.

- Red card: The Tschirnhaus transformation

- Orange card: Relation between roots and coefficients

- Yellow card: Lagrange resolvents

- Green card: Sums of cubes

- Blue card: Products of cubes

- Indigo card: From coefficients to roots

- Violet card: The cubic formula

"So I guess these are all calculations," Tetra said. "Mr. Muraki said if we work these through in order, we'll end up with the cubic formula!"

"Such a... colorful way to get there," I said.

"I'm still working my way through it all, but I'm starting to get the feeling that the cubic formula's going to be a *lot* more complex than the quadratic formula."

"I'd think so. Honestly, I've never seen it. How far have you gotten?"

"Why, I'd be happy to take you on a little tour! Climb aboard!"

Tetra pulled out a chair, and I sat next to her. Thus began our voyage in search of the cubic formula.

7.1.2 Red Card: The Tschirnhaus Transformation

The first card, the red one, was labeled "The Tschirnhaus transformation."

"First off, what's a Tschirnhaus?" I asked.

"An Ehrenfried, in this case. A German mathematician," Tetra replied.

The Secrets of Lagrange Resolvents

Problem 7-1 (The Tschirnhaus transformation)

Assume a cubic equation in y

$$ay^3 + by^2 + cy + d = 0,$$

where $a \neq 0$. Transforming y as

$$y = x - \frac{b}{3a}$$

gives a cubic equation in x,

$$x^3 + px + q = 0.$$

Find p, q in terms of a, b, c, d.

"Variable transformations, huh? And it says we can solve this by substituting $y = x - \frac{b}{3a}$ into $ay^3 + by^2 + cy + d$. Calculations, sure enough."

Tetra angled her notebook toward me. "Here's what I've got so far."

$ay^3 + by^2 + cy + d$

"We start by substituting $x - \frac{b}{3a}$ into y."

$$= a\left(x - \frac{b}{3a}\right)^3 + b\left(x - \frac{b}{3a}\right)^2 + c\left(x - \frac{b}{3a}\right) + d$$

"Let's expand the cubes and squares."

$$= a\left(x^3 - 3 \cdot \frac{b}{3a}x^2 + 3 \cdot \frac{b^2}{9a^2}x - \frac{b^3}{27a^3}\right)$$
$$+ b\left(x^2 - 2 \cdot \frac{b}{3a}x + \frac{b^2}{9a^2}\right) + c\left(x - \frac{b}{3a}\right) + d$$

"Getting rid of the parentheses... "

$$= ax^3 - 3a \cdot \frac{b}{3a}x^2 + 3a \cdot \frac{b^2}{9a^2}x - a \cdot \frac{b^3}{27a^3}$$
$$+ bx^2 - 2b \cdot \frac{b}{3a}x + b \cdot \frac{b^2}{9a^2} + cx - c \cdot \frac{b}{3a} + d$$

"Now some cleanup."

$$= ax^3 - bx^2 + \frac{b^2}{3a}x - \frac{b^3}{27a^2} + bx^2 - \frac{2b^2}{3a}x + \frac{b^3}{9a^2} + cx - \frac{bc}{3a} + d$$

"Let's bring all those x's together."

$$= ax^3 + (-b + b)x^2 + \left(\frac{b^2}{3a} - \frac{2b^2}{3a} + c\right)x - \frac{b^3}{27a^2} + \frac{b^3}{9a^2} - \frac{bc}{3a} + d$$

"Some more cleanup."

$$= ax^3 - \frac{b^2 - 3ac}{3a}x + \frac{2b^3 - 9abc + 27a^2d}{27a^2}$$

"So now we've rewritten $ay^3 + by^2 + cy + d = 0$ like this."

$$ax^3 - \frac{b^2 - 3ac}{3a}x + \frac{2b^3 - 9abc + 27a^2d}{27a^2} = 0$$

"To get this into $x^3 + px + q = 0$ form we need the x^3 coefficient to be 1, so let's divide both sides by a."

$$x^3 - \frac{b^2 - 3ac}{3a^2}x + \frac{2b^3 - 9abc + 27a^2d}{27a^3} = 0$$

"Now we can find p and q through comparison of the coefficients with $x^3 + px + q = 0$."

$$\begin{cases} p = -\dfrac{b^2 - 3ac}{3a^2} \\ q = \dfrac{2b^3 - 9abc + 27a^2d}{27a^3} \end{cases}$$

"Looks like you did a great job here, Tetra," I said. "Something still bothering you?"

"Well, like you said this transformation is just calculations—substituting, expanding, combining terms... So, like..." Tetra cocked her head and made a pained face. "So what? I was expecting something interesting to happen, but in the end it was just a lot of algebraic shuffling about."

"That's not all it was," I said. "Take a closer look at what you've done here. After the transformation, the squared term has vanished."

$ay^3 + by^2 + cy + d = 0$ equation in y (before)
$\qquad\qquad\qquad\qquad\quad\downarrow$ Tschirnhaus transform
$x^3 \qquad\quad + px + q = 0$ equation in x (after)

"Hey, you're right!"
"So I think the point of this red card was mainly to simplify things. Think of it as... preparation. Getting ready to derive the cubic formula."

Answer 7-1 (The Tschirnhaus transformation)

Substituting $y = x - \frac{b}{3a}$ into a cubic equation $ay^3 + by^2 + cy + d = 0$ to get $x^3 + px + q = 0$, p, q can be written in terms of a, b, c, d as follows:

$$\begin{cases} p &= -\dfrac{b^2 - 3ac}{3a^2} \\ q &= \dfrac{2b^3 - 9abc + 27a^2 d}{27a^3} \end{cases}$$

7.1.3 Orange Card: Relations Between Roots and Coefficients

"Did you work through the second card too?" I asked.
"The orange card, right," Tetra said. "This one."

Problem 7-2 (Relation between roots and coefficients)

Letting $x = \alpha, \beta, \gamma$ be the roots of a cubic equation $x^3 + px + q = 0$, find the relation between the roots and coefficients of the equation.

"This shouldn't be too hard," I said.

"It wasn't! The card tells us that the roots are alpha, beta, and gamma, so we just have to expand $(x-\alpha)(x-\beta)(x-\gamma)$, like this."

$$(x-\alpha)(x-\beta)(x-\gamma)$$
$$= (x^2 - \beta x - \alpha x + \alpha\beta)(x-\gamma)$$
$$= (x^2 - (\alpha+\beta)x + \alpha\beta)(x-\gamma)$$
$$= x^3 - \gamma x^2 - (\alpha+\beta)x^2 + (\alpha+\beta)\gamma x + \alpha\beta x - \alpha\beta\gamma$$
$$= x^3 - (\alpha+\beta+\gamma)x^2 + (\alpha\beta + \beta\gamma + \gamma\alpha)x - \alpha\beta\gamma$$

"We know this is equal to $x^3 + px + q$, so we can just compare coefficients to see what p and q are."

$$\begin{aligned}&x^3 - (\alpha+\beta+\gamma)x^2 + (\alpha\beta+\beta\gamma+\gamma\alpha)x - \alpha\beta\gamma\\ =\,&x^3 \qquad\qquad\qquad\quad + \qquad\qquad\qquad px + q\end{aligned}$$

Answer 7-2 (Relation between roots and coefficients)

Letting $x = \alpha, \beta, \gamma$ be the roots of a cubic equation $x^3 + px + q = 0$,

$$\begin{cases} 0 = \alpha + \beta + \gamma \\ p = \alpha\beta + \beta\gamma + \gamma\alpha \\ q = -\alpha\beta\gamma \end{cases}$$

"So I got that far," Tetra said, making a duck face, "but that's where I'm stuck."

7.1.4 Yellow Card: Lagrange Resolvents

"You're working on the third card now?" I asked.

The Secrets of Lagrange Resolvents

> **Problem 7-3 (Lagrange resolvents)**
>
> Let $x = \alpha, \beta, \gamma$ be the roots of a cubic equation $x^3 + px + q = 0$, and define L and R as
>
> $$\begin{cases} L = \omega\alpha + \omega^2\beta + \gamma \\ R = \omega^2\alpha + \omega\beta + \gamma \end{cases},$$
>
> where ω is a primitive cube root of unity. Represent α, β, γ in terms of L, R.

Tetra nodded. "What I was trying to do just now is set up a system of equations that would allow me to write α, β, γ in terms of L, R, but it isn't going so well. The only equations the problem gives are $L = \omega\alpha + \omega^2\beta + \gamma$ and $R = \omega^2\alpha + \omega\beta + \gamma$, but I'm looking for three unknowns, α, β, and γ. Don't you need three equations to find three unknowns?"

Tetra held her head in her hands. I pulled her notebook closer and started to think.

"Yeah, that's interesting," I said. "So you were trying to get rid of two of α, β, and γ, right?"

"I was, but I need another equation to do that."

"But you have one more, right here."

"Where?"

"In your answer to the orange card! See? You found that $0 = \alpha + \beta + \gamma$."

$$\begin{cases} L = \omega\alpha + \omega^2\beta + \gamma \\ R = \omega^2\alpha + \omega\beta + \gamma \\ 0 = \alpha + \beta + \gamma \quad \text{from Answer 7-2} \end{cases}$$

"Aha!" Tetra folded her arms and gave a deep nod. "Exactly what I needed. Now I can set up the system of equations!"

I held up a hand to restrain her. "Actually, now you can just do it in your head."

"Seriously?"

"Sure. The card says ω is a primitive cube root of unity, so we know this."

$$\omega^3 = 1$$

"Also, ω is a root of the cyclotomic polynomial $\Phi_3(x) = x^2+x+1$, so this must be true."

$$\omega^2 + \omega + 1 = 0$$

"So are you seeing it now?"

"Uh... no?"

"Add up the terms of the three equations."

$$\begin{array}{rrrr}
L = & \omega\alpha + & \omega^2\beta + & \gamma \\
R = & \omega^2\alpha + & \omega\beta + & \gamma \\
+)\quad 0 = & \alpha + & \beta + & \gamma \\
\hline
L+R = & (\omega+\omega^2+1)\alpha + & (\omega^2+\omega+1)\beta + & (1+1+1)\gamma \\
L+R = & 0\alpha + & 0\beta + & 3\gamma \\
L+R = & & & 3\gamma
\end{array}$$

"Oh, look! The alphas and betas went poof!"

"Poof they went. And since we've found that $L+R = 3\gamma$, now we can represent γ in terms of L and R."

$$\gamma = \frac{1}{3}(L+R)$$

"Nice!" Tetra said.

"So just like when we found γ, we should be able to use $\omega^3 = 1$ and $\omega^2+\omega+1 = 0$ to get rid of some letters. Let's see... Ah, right, we can just think about $\omega L + \omega^2 R$."

$$\begin{array}{rrrr}
\omega L = & \omega^2\alpha + & \omega^3\beta + & \omega\gamma \\
\omega^2 R = & \omega^4\alpha + & \omega^3\beta + & \omega^2\gamma \\
+)\quad 0 = & \alpha + & \beta + & \gamma \\
\hline
\omega L + \omega^2 R = & (\omega^2+\omega^4+1)\alpha + & (\omega^3+\omega^3+1)\beta + & (\omega+\omega^2+1)\gamma \\
\omega L + \omega^2 R = & (\omega^2+\omega+1)\alpha + & (1+1+1)\beta + & (\omega+\omega^2+1)\gamma \\
\omega L + \omega^2 R = & 0\alpha + & 3\beta + & 0\gamma \\
\omega L + \omega^2 R = & & 3\beta &
\end{array}$$

"Yay!" Tetra said. "We got $\omega L + \omega^2 R = 3\beta$, which means..."

$$\beta = \frac{1}{3}(\omega L + \omega^2 R)$$

"There ya go. Now we just need α, which we can get from $\omega^2 L + \omega R$."

$$
\begin{array}{rl}
\omega^2 L = & \omega^3 \alpha + \quad \omega^4 \beta + \quad \omega^2 \gamma \\
\omega R = & \omega^3 \alpha + \quad \omega^2 \beta + \quad \omega \gamma \\
+) \quad 0 = & \alpha + \quad \beta + \quad \gamma \\
\hline
\omega^2 L + \omega R = & (\omega^3 + \omega^3 + 1)\alpha + (\omega^4 + \omega^2 + 1)\beta + (\omega^2 + \omega + 1)\gamma \\
\omega^2 L + \omega R = & (1 + 1 + 1)\alpha + (\omega + \omega^2 + 1)\beta + (\omega^2 + \omega + 1)\gamma \\
\omega^2 L + \omega R = & 3\alpha + \quad 0\beta + \quad 0\gamma \\
\omega^2 L + \omega R = & 3\alpha
\end{array}
$$

"Okay," Tetra said, "so this time we got $\omega^2 L + \omega R = 3\alpha$. That gives us α."

$$\alpha = \frac{1}{3}(\omega^2 L + \omega R)$$

"And that's all of them! We've represented α, β, γ in terms of L and R!"

$$
\begin{cases}
\alpha &= \dfrac{1}{3}(\omega^2 L + \omega R) \\
\beta &= \dfrac{1}{3}(\omega L + \omega^2 R) \\
\gamma &= \dfrac{1}{3}(L + R)
\end{cases}
$$

Answer 7-3 (Lagrange resolvents)

Letting $x = \alpha, \beta, \gamma$ be the roots of a cubic equation $x^3 + px + q = 0$, and defining L and R as

$$
\begin{cases}
L &= \omega \alpha + \omega^2 \beta + \gamma \\
R &= \omega^2 \alpha + \omega \beta + \gamma,
\end{cases}
$$

α, β, γ can be written in terms of L and R as

$$
\begin{cases}
\alpha &= \dfrac{1}{3}(\omega^2 L + \omega R) \\
\beta &= \dfrac{1}{3}(\omega L + \omega^2 R) \cdot \\
\gamma &= \dfrac{1}{3}(L + R)
\end{cases}
$$

"Such an interesting use of $\omega^2 + \omega + 1 = 0$!" Tetra said. She was looking at me with a blush of excitement.

"Yeah, it was helpful here. But more to the point, why would we want to represent α, β, γ in terms of L and R?" I paused to think. "Well, I guess representing these L and R in terms of coefficients will give us the cubic formula, so I suppose the next card has something to do with finding L and R?"

Tetra took a look at the next card. "Um... Not quite."

"No?"

"Not L and R. It wants us to find $L^3 + R^3$."

"It... what?!"

7.1.5 The Green Card: Sums of Cubes

Problem 7-4 (Sum of cubes)

Letting $x = \alpha, \beta, \gamma$ be the roots of a cubic equation $x^3 + px + q = 0$, and defining L and R as

$$\begin{cases} L = \omega\alpha + \omega^2\beta + \gamma \\ R = \omega^2\alpha + \omega\beta + \gamma \end{cases},$$

represent $L^3 + R^3$ in terms of p, q.

"So it looks like this next card wants us to find $L^3 + R^3$," Tetra said. "Oh, look! There's more on the back."

A hint (green card, reverse side)

$$(L + R)(L + \omega R)(L + \omega^2 R)$$

"What's this $(L + R)(L + \omega R)(L + \omega^2 R)$?" I said. "I wonder if we should expand it?"

"I'll do it!"

$$(L+R)(L+\omega R)(L+\omega^2 R)$$
$$= (L^2 + \omega LR + LR + \omega R^2)(L + \omega^2 R)$$
$$= L^3 + \omega^2 L^2 R + \omega L^2 R + LR^2 + L^2 R + \omega^2 LR^2 + \omega LR^2 + R^3$$
$$= \text{urgh!}$$

"Okay, this is becoming a mess. How can I clean this up?"

"In cases like this, you generally want to focus on one letter," I said. "Let's say we're going to focus on L. Then we sort things out, looking at L^3, L^2, L, and constant terms."

$$(L+R)(L+\omega R)(L+\omega^2 R)$$
$$= L^3 + \omega^2 L^2 R + \omega L^2 R + LR^2 + L^2 R + \omega^2 LR^2 + \omega LR^2 + R^3$$
$$= \underbrace{L^3}_{L^3 \text{ term}} + \underbrace{(\omega^2 + \omega + 1)RL^2}_{L^2 \text{ term}} + \underbrace{(1 + \omega^2 + \omega)R^2 L}_{L \text{ term}} + \underbrace{R^3}_{\text{constant term}}$$
$$= L^3 + R^3$$

"Whoa!" Tetra said. "$\omega^2 + \omega + 1 = 0$ to the rescue again! $L^3 + R^3$ is all that's left!"

"Interesting. So do you see why the hint is helpful?"

"Hmm... I guess it's telling us that instead of calculating $L^3 + R^3$, we can use the 'hint' expression, $(L+R)(L+\omega R)(L+\omega^2 R)$?"

"That's right. Mr. Muraki's hint is actually an identity."

$$L^3 + R^3 = (L+R)(L+\omega R)(L+\omega^2 R)$$

"An identity, of course. An equation that will hold regardless of what L and R are."

"Are you still good on where we're heading?"

"I think so! The green card said

represent $L^3 + R^3$ in terms of p, q,

"but we've replaced that with

represent $(L+R)(L+\omega R)(L+\omega^2 R)$ in terms of p, q, "so we just need to write

(a) $(L + R)$

(b) $(L + \omega R)$

(c) $(L + \omega^2 R)$

in terms of p, q. We've already found (a) $L + R$, when we worked on the yellow card.[1]"

$$\gamma = \frac{1}{3}(L + R)$$

$$L + R = 3\gamma$$

"So I guess next we want to calculate (b) $L + \omega R$?"
"We got something else from the yellow card too," I said.

$$\beta = \frac{1}{3}(\omega L + \omega^2 R)$$

"And that's helpful?"
"Sure, just multiply both sides by $3\omega^2$."

$$\begin{aligned} 3\omega^2 \cdot \beta &= 3\omega^2 \cdot \frac{1}{3}(\omega L + \omega^2 R) \\ &= \omega^3 L + \omega^4 R \\ &= L + \omega R \qquad \text{because } \omega^3 = 1, \omega^4 = \omega \end{aligned}$$

"In other words, we get this for (b)."

$$3\omega^2 \beta = L + \omega R$$

"Oh, I see!" Tetra said. "And we can do the same kind of thing for (c) $(L + \omega^2 R)$!"

$$\alpha = \frac{1}{3}(\omega^2 L + \omega R)$$

---

[1] Answer 7-3, p. 211

The Secrets of Lagrange Resolvents

"This time we just multiply both sides by 3ω!"

$$3\omega \cdot \alpha = 3\omega \cdot \frac{1}{3}(\omega^2 L + \omega R)$$
$$= \omega^3 L + \omega^2 R$$
$$= L + \omega^2 R \qquad \text{because } \omega^3 = 1$$

I nodded. "Very good. Now we've found $L + \omega^2 R$."

$$3\omega\alpha = L + \omega^2 R$$

Tetra tugged on my arm. "That's all of them! Now we have ⓐ, ⓑ, and ⓒ!"

$$\begin{cases} \text{ⓐ} & L + R & = 3\gamma \\ \text{ⓑ} & L + \omega R & = 3\omega^2 \beta \\ \text{ⓒ} & L + \omega^2 R & = 3\omega\alpha \end{cases}$$

"We aren't done yet, though," I said. "We still have some multiplying to do."

"Yep!"

$$L^3 + R^3 = \underbrace{(L+R)}_{\text{ⓐ}} \underbrace{(L+\omega R)}_{\text{ⓑ}} \underbrace{(L+\omega^2 R)}_{\text{ⓒ}} \qquad \text{from the hint}$$
$$= \underbrace{(3\gamma)}_{\text{ⓐ}} \underbrace{(3\omega^2 \beta)}_{\text{ⓑ}} \underbrace{(3\omega\alpha)}_{\text{ⓒ}} \qquad \text{from the calculations above}$$
$$= 27\omega^3 \alpha\beta\gamma$$
$$= 27\alpha\beta\gamma \qquad \text{because } \omega^3 = 1$$

"We're done!"

"Well, almost. We want to write this in terms of p, q, remember."

"Er, what?"

"Don't forget the relation between roots and coefficients we found, $q = -\alpha\beta\gamma$."

$$L^3 + R^3 = 27\alpha\beta\gamma$$
$$= -27q \qquad \text{because } q = -\alpha\beta\gamma \text{ (see p. 208)}$$

"*Now* we're done with you, green card!" Tetra said.

> **Answer 7-4 (Sum of cubes)**
>
> $$L^3 + R^3 = -27q$$

7.1.6 The Blue Card: Products of Cubes

> **Problem 7-5 (Product of cubes)**
>
> Letting $x = \alpha, \beta, \gamma$ be the roots of a cubic equation $x^3 + px + q = 0$, and defining L and R as
>
> $$\begin{cases} L = \omega\alpha + \omega^2\beta + \gamma \\ R = \omega^2\alpha + \omega\beta + \gamma \end{cases},$$
>
> represent $L^3 R^3$ in terms of p, q.

"So now we want to find L^3 and R^3, and multiply them, right?" Tetra asked.

"Hmmm, yeah. We could just jump straight to expanding out $L^3 = (\omega\alpha + \omega^2\beta + \gamma)^3$, but it might be easier to find LR first."

"Wow, I wish I had your sense for where to go with these transformations..."

$$\begin{aligned} LR &= (\omega\alpha + \omega^2\beta + \gamma)(\omega^2\alpha + \omega\beta + \gamma) \\ &= (\omega^3\alpha^2 + \omega^2\alpha\beta + \omega\gamma\alpha) + (\omega^4\alpha\beta + \omega^3\beta^2 + \omega^2\beta\gamma) \\ &\quad + (\omega^2\gamma\alpha + \omega\beta\gamma + \gamma^2) \\ &= \alpha^2 + \beta^2 + \gamma^2 + (\omega^2 + \omega^4)\alpha\beta + (\omega^2 + \omega)\beta\gamma + (\omega + \omega^2)\gamma\alpha \\ &= \alpha^2 + \beta^2 + \gamma^2 + (\omega + \omega^2)(\alpha\beta + \beta\gamma + \gamma\alpha) \end{aligned}$$

"Oooh! Look!" Tetra said. "We can use $\omega^2 + \omega + 1 = 0$ again, in the form $\omega + \omega^2 = -1$!"

$$\begin{aligned} LR &= \alpha^2 + \beta^2 + \gamma^2 + (\omega + \omega^2)(\alpha\beta + \beta\gamma + \gamma\alpha) \\ &= \alpha^2 + \beta^2 + \gamma^2 - (\alpha\beta + \beta\gamma + \gamma\alpha) \quad \text{because } \omega + \omega^2 = -1 \end{aligned}$$

"Looks like we can also simplify the part in parentheses using the relation between roots and coefficients, $\alpha\beta + \beta\gamma + \gamma\alpha = p$."

$$LR = \alpha^2 + \beta^2 + \gamma^2 - (\alpha\beta + \beta\gamma + \gamma\alpha)$$
$$= \alpha^2 + \beta^2 + \gamma^2 - p \qquad \text{from Answer 7-2 (p. 208)}$$
$$= \text{Uh, oh}\ldots$$

"Uh, oh..." Tetra said. "$\alpha^2 + \beta^2 + \gamma^2$ doesn't show up in the relation between roots and coefficients."

"No, but it does say that $\alpha + \beta + \gamma = 0$. So we can square that to create a second-degree term, and know that the result will be 0."

$$(\alpha + \beta + \gamma)^2 = \alpha^2 + \beta^2 + \gamma^2 + 2(\alpha\beta + \beta\gamma + \gamma\alpha)$$
$$0 = \alpha^2 + \beta^2 + \gamma^2 + 2(\alpha\beta + \beta\gamma + \gamma\alpha)$$
$$\alpha^2 + \beta^2 + \gamma^2 = -2(\alpha\beta + \beta\gamma + \gamma\alpha)$$

"Creating a second-degree term, huh? Interesting..."
"And by using $\alpha\beta + \beta\gamma + \gamma\alpha = p$, we can forge a useful tool."

$$\alpha^2 + \beta^2 + \gamma^2 = -2p \qquad \text{(a tool)}$$

"A tool that fits here perfectly!"

$$LR = \underbrace{\alpha^2 + \beta^2 + \gamma^2} - p$$
$$= \underbrace{-2p} - p \qquad \text{using the tool}$$
$$= -3p$$

"The card is asking for L^3R^3," I said, "so we need to cube LR."

$$L^3R^3 = (LR)^3 = (-3p)^3 = -27p^3$$

Answer 7-5 (Product of cubes)

$$L^3R^3 = -27p^3$$

"That's five out of seven!" I said.
"Yep! Just two more cards to go!"

7.1.7 The Indigo Card: From Coefficients to Roots

> **Problem 7-6 (From coefficients to roots)**
>
> Letting $x = \alpha, \beta, \gamma$ be the roots of a cubic equation $x^3 + px + q = 0$, represent α, β, γ in terms of p, q.

"Interesting," I said. "Writing α, β, γ in terms of p, q means we want to write the roots in terms of the coefficients. In other words, we're going to create a formula giving the roots of $x^3 + px + q = 0$."

"We're getting to the heavy stuff!" Tetra said.

"Well, to be fair, what we've done so far wasn't exactly easy."

"Yeah, but all we've really accomplished is to find $L^3 + R^3$ and L^3R^3."

"Which means we're almost at our destination."

Tetra cocked her head. "We... we are?"

"Don't lose track of our path, now. With the yellow card we represented α, β, γ in terms of L and R, right? That means if we can get a grip on L and R, we can create a formula. In other words, we want to know what L^3 and R^3 are."

"I'm sorry, I'm still not sure why we want to know that."

"Because if we take the cube root of L^3 we'll get L."

"Oh, okay. But, L^3 will have three cube roots, right?"

"It will."

"So what am I missing here?"

"Maybe that these will be the cube roots of L^3?"

$$L, \ \omega L, \ \omega^2 L$$

"They will? How so?"

"Like this."

$$\begin{cases} L^3 & = L^3 \\ (\omega L)^3 & = \omega^3 L^3 = L^3 \\ (\omega^2 L)^3 & = (\omega^2)^3 L^3 = (\omega^3)^2 L^3 = L^3 \end{cases}$$

"See? When we cube each one, we end up with L^3."

"We do indeed. Okay, I'm caught up. So our next step is to use $L^3 + R^3$ and L^3R^3 to find L^3 and R^3, right? How are we going to do that?"

How can we find find L^3 and R^3 from $L^3 + R^3$ and L^3R^3?

"Uh oh, you're slipping," I said. "When you know the sum and product of two numbers, and you want to know what those two numbers are, it's time for a quadratic equation."

"A quadratic? How does that help?"

"Well, to find L^3 and R^3, we can create a quadratic in X like this."

$$X^2 - (L^3 + R^3)X + L^3R^3 = 0$$

"I'm still not quite there."

"Think about the factorization $X^2 - (L^3 + R^3)X + L^3R^3 = (X - L^3)(X - R^3)$. Conveniently, we've already found the sum and product of R^3."

$$L^3 + R^3 = -27q \quad L^3R^3 = -27p^3$$
Green card   Blue card

"So now we just make some substitutions, right?"

"Right. We want to solve this quadratic in X."

$$X^2 + 27qX - 27p^3 = 0$$

"Do we have an easy way of solving this?" Tetra asked.

"Sure, our old friend the quadratic formula."

$$X = \frac{-27q \pm \sqrt{(27q)^2 + 4 \cdot 27p^3}}{2}$$

$$= -\frac{27q}{2} \pm \sqrt{\left(\frac{27q}{2}\right)^2 + 27p^3}$$

"Whoa!"

"So now we have two possibilities for L^3 and R^3."

$$-\frac{27q}{2} + \sqrt{\left(\frac{27q}{2}\right)^2 + 27p^3}, \quad -\frac{27q}{2} - \sqrt{\left(\frac{27q}{2}\right)^2 + 27p^3}$$

Tetra nodded. "Right."

"Let's make some cleanup substitutions."

$$\begin{cases} A = -\dfrac{27q}{2} \\ D = \left(\dfrac{27q}{2}\right)^2 + 27p^3 \end{cases} \quad \text{part inside the } \sqrt{}$$

"Now L^3 and R^3 look like this."

$$A + \sqrt{D},\ A - \sqrt{D}$$

"Hang on," Tetra said. "Which one is L^3?"

"Doesn't matter," I said. "Pick one."

"But... The yellow card—the 'Lagrange resolvents' one—defined L and R, so don't we have to stick with that?"

"We do, but L and R are defined in terms of α, β, γ, which are only there to represent the roots of the cubic equation. Nobody ever said anything about which is which. In fact, binding L, R with α, β, γ is something we can do now. Like this, for example."

$$\begin{cases} L = \sqrt[3]{A + \sqrt{D}} \\ R = \sqrt[3]{A - \sqrt{D}} \end{cases}$$

"So now, using the yellow card,[2] the roots of $x^3 + px + q = 0$ look like this."

$$\begin{cases} \alpha = \dfrac{1}{3}(\omega^2 L + \omega R) = \dfrac{1}{3}\left(\omega^2 \sqrt[3]{A + \sqrt{D}} + \omega \sqrt[3]{A - \sqrt{D}}\right) \\ \beta = \dfrac{1}{3}(\omega L + \omega^2 R) = \dfrac{1}{3}\left(\omega \sqrt[3]{A + \sqrt{D}} + \omega^2 \sqrt[3]{A - \sqrt{D}}\right) \\ \gamma = \dfrac{1}{3}(L + R) = \dfrac{1}{3}\left(\sqrt[3]{A + \sqrt{D}} + \sqrt[3]{A - \sqrt{D}}\right) \end{cases}$$

---

[2] Answer 7-3, p. 211

The Secrets of Lagrange Resolvents 221

Answer 7-6 (From coefficients to roots)

Letting $x = \alpha, \beta, \gamma$ be the roots of a cubic equation $x^3 + px + q = 0$, α, β, γ can be written as

$$\begin{cases} \alpha = \dfrac{1}{3}\left(\omega^2 \sqrt[3]{A + \sqrt{D}} + \omega \sqrt[3]{A - \sqrt{D}}\right) \\ \beta = \dfrac{1}{3}\left(\omega \sqrt[3]{A + \sqrt{D}} + \omega^2 \sqrt[3]{A - \sqrt{D}}\right) \\ \gamma = \dfrac{1}{3}\left(\sqrt[3]{A + \sqrt{D}} + \sqrt[3]{A - \sqrt{D}}\right) \end{cases},$$

where

$$\begin{cases} A = -\dfrac{27q}{2} \\ D = \left(\dfrac{27q}{2}\right)^2 + 27p^3 \end{cases}.$$

Tetra sighed. "I'm starting to see the power of definitional equations like the $A = \cdots$ and $D = \cdots$ here. You taught me about general equations and identity equations before, right? Bringing in even more letters always scared me, like it would just make things more complicated, so I always tried to avoid that if I could. But this is an excellent example of how increasing the number of letters can make things much simpler. Without the As and the Ds, this is a complete mess!"

"It sure is."

"These new letters also help us to see the structure here." Tetra was becoming increasingly excited. "You could even say the letters are proof that we've perceived the structure of the problem!"

"I hadn't thought of it that way, but you're right."

"Sure! We're only able to use the letters because we've noticed things like 'this thing and that thing are the same,' or 'the only difference between the things under these cube roots is their signs.'"

Tetra went on like this for a while.

7.1.8 The Violet Card: The Cubic Formula

"The rest should be easy," I said. "Just grunt work."

Problem 7-7 (The cubic formula)

Letting $x = \alpha, \beta, \gamma$ be the roots of a cubic equation $ax^3 + bx^2 + cx + d = 0$, represent α, β, γ in terms of a, b, c, d.

$$A = -\frac{27q}{2} \qquad \text{by definition of A (p. 221)}$$
$$= -\frac{27}{2} \cdot \frac{2b^3 - 9abc + 27a^2d}{27a^3} \qquad \text{write q as } a, b, c, d \text{ (p.207)}$$
$$= -\frac{2b^3 - 9abc + 27a^2d}{2a^3}$$

$$D = \left(\frac{27q}{2}\right)^2 + 27p^3$$
$$= \left(\frac{27}{2} \cdot \frac{2b^3 - 9abc + 27a^2d}{27a^3}\right)^2 + 27 \cdot \left(-\frac{b^2 - 3ac}{3a^2}\right)^3$$
$$= \left(\frac{2b^3 - 9abc + 27a^2d}{2a^3}\right)^2 - \left(\frac{b^2 - 3ac}{a^2}\right)^3$$
$$= \frac{27 \cdot (27a^2d^2 - 18abcd + 4b^3d + 4ac^3 - b^2c^2)}{4a^4}$$

Answer 7-7 (The cubic formula)

Letting $x = \alpha, \beta, \gamma$ be the roots of a cubic equation $ax^3 + bx^2 + cx + d = 0$, we can write α, β, γ as

$$\begin{cases} \alpha = \dfrac{1}{3}\left(\omega^2 \sqrt[3]{A+\sqrt{D}} + \omega \sqrt[3]{A-\sqrt{D}}\right) - \dfrac{b}{3a} \\ \beta = \dfrac{1}{3}\left(\omega \sqrt[3]{A+\sqrt{D}} + \omega^2 \sqrt[3]{A-\sqrt{D}}\right) - \dfrac{b}{3a} \\ \gamma = \dfrac{1}{3}\left(\sqrt[3]{A+\sqrt{D}} + \sqrt[3]{A-\sqrt{D}}\right) - \dfrac{b}{3a} \end{cases}$$

($-\dfrac{b}{3a}$ is the Tschirnhaus transformation part), where

$$\begin{cases} A = -\dfrac{2b^3 - 9abc + 27a^2 d}{2a^3} \\ D = \dfrac{27 \cdot (27a^2 d^2 - 18abcd + 4b^3 d + 4ac^3 - b^2 c^2)}{4a^4} \end{cases}.$$

7.1.9 A Map of Our Journey

"Thanks so much for your help!" Tetra said. "A whole stack of Mr. Muraki's cards, done! But still..." She squirmed in her seat. "I'm sorry, but there's something about this formula that I'm not super-duper happy about."

"What do you mean?"

"Well, we solved the problem and got the cubic formula we were after, but I still don't really *get it*. I mean, what did we really accomplish here?"

There it is—Tetra's greatest strength. She doesn't feel like she's done just because she got an answer. We've worked our way through seven cards to get where we wanted to go, a series of stepping stones (and one big hint) guiding us along, so of course we got the answer at the end. But now that we've arrived at where the problems were taking us, the biggest question of all awaits: What did we really accomplish here?

"That's an excellent question, Tetra. Any ideas on how we can answer it?"

"I think I'll try to create a map of our journey!" Tetra turned to a new page in her notebook. "Let's see... Our final goal was to obtain the cubic formula—a formula that gives the roots of a cubic equation, based on its coefficients."

$$\text{Coefficients} \xrightarrow{\text{Cubic equation}} \text{Roots}$$

"I'm sure there was method behind Mr. Muraki's rainbow madness, but when you're crunching through the calculations it can be hard to see the big picture, so I've wanted to draw a map like this for a while now."

Red card (The Tschirnhaus transformation):

"Working on this card, we transformed the equation so that we were representing p, q in terms of a, b, c, d. Something like this."

$$a, b, c, d \xrightarrow{\text{Tschirnhaus transformation}} p, q$$

Orange card (Relations between roots and coefficients)

"With this card, we represented p, q in terms of α, β, γ, like this."

$$\alpha, \beta, \gamma \xrightarrow{\text{Relation}} p, q$$

Yellow card (Lagrange resolvents)

"This one I'm still not sure about. We introduced two letters, L and R, but I'm not sure what they do. We also saw an ω, which is, um, a primitive cube root of unity. Anyway, we represented α, β, γ in terms of L and R."

$$L, R \xrightarrow{\text{Lagrange resolvent}} \alpha, \beta, \gamma$$

Green card (Sums of cubes)

"With this one, we represented $L^3 + R^3$ in terms of p, q."

$$p, q \xrightarrow{\text{Sum of cubes}} L^3 + R^3$$

Blue card (Products of cubes)

"Kind of like the green card, but this time we represented $L^3 R^3$ in terms of p, q."

$$p, q \xrightarrow{\text{Product of cubes}} L^3 R^3$$

Indigo card (From coefficients to roots)

"Here we used everything we'd found so far to represent α, β, γ in terms of p, q. Along the way, we also found L^3 and R^3, then L and R."

$$p, q \xrightarrow{\text{Coefficients to roots}} \alpha, \beta, \gamma$$

Violet card (The cubic formula)

"Here we summarized everything up, in a sense, by using a, b, c, d to represent α, β, γ."

$$a, b, c, d \xrightarrow{\text{The cubic formula}} \alpha, \beta, \gamma$$

"If you follow the whole thing all the way, you can see the flow from a, b, c, d to α, β, γ."

$$a, b, c, d \xrightarrow{\text{The Tschirnhaus transformation}} p, q$$
$$\xrightarrow{\text{Sum and product of cubes}} L^3 + R^3, L^3 R^3$$
$$\xrightarrow{\text{Solve quadratic}} L^3, R^3$$
$$\xrightarrow{\text{Find cube root}} L, R$$
$$\xrightarrow{\text{Lagrange resolvent}} \alpha, \beta, \gamma$$

"So anyway, looking at all this, I think we can create a map of our journey. Something like...this?"

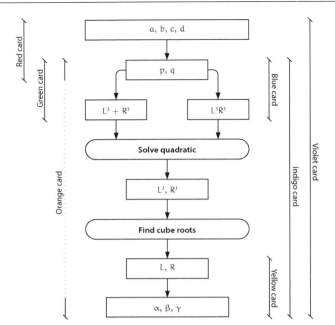

The journey to the cubic equation.

"Very interesting," I said. "Looks like you've got a good grasp on the overall flow of things."

Tetra remained staring at her map. "I think it must be L and R that made this journey possible."

"Yeah? I think Mr. Muraki's hint was super helpful too. The green card, was it?"

"It was helpful, sure, but I think we could have found our way without it, eventually—it basically just saved us a lot of time working through calculations. But without the L and R? No way. I never would have come up with those on my own."

I nodded. "Yeah, I doubt I could either."

"Isn't there something... *unbalanced* about them?"

"Unbalanced? How's that?"

$$\begin{cases} L = \omega\alpha + \omega^2\beta + \gamma \\ R = \omega^2\alpha + \omega\beta + \gamma \end{cases}$$

"Well, the α and β are swapping coefficients, but γ stays exactly the same. Doesn't that feel kinda strange? There's something magical about this L and R. It feels like they're hiding some kind of secret."

I grunted. When I'd first met Tetra, I was always the one doing the teaching. Things were very different now; she was off studying things on her own, and it seemed like I was learning more from her than she was from me. Maybe not about math itself, but about how to approach it.

We sat in silence for a while, staring at the Lagrange resolvent. After a time we both looked up, as if simultaneously recalling our mentor, our mathematical sherpa, who was still writing something over by the window.

7.2 LAGRANGE RESOLVENTS

7.2.1 Miruka

We showed Miruka the seven cards and related the journey they had led us on. This was easily enough to pull her away from what she was writing and put her in lecture mode.

"The sixteenth-century mathematician Niccolò Fontana Tartaglia is often credited with discovering how to solve cubic equations. Today, however, the method is often called the Cardano formula after Gerolamo Cardano, who published it in his own writings."

"I've read about that," I said. "Weren't they in some kind of contest that involved solving math problems?"

"Math battles!" Tetra said, her eyes gleaming.

"Yeah, they used to challenge each other with difficult math problems. The winner would be whoever solved the most. Instead of pistols, they would duel with the mathematical techniques they'd developed."

"Take all that with a grain of salt," Miruka said. "It's likely that Scipione del Ferro found a way to solve cubic equations before Tartaglia did, but del Ferro's disciple Antonio Fiore lost to Tartaglia in a public math contest. Anyway, enough history. Let's get back to the math."

"We were just talking about how the Lagrange resolvent somehow seems to be key to deriving the cubic formula," Tetra said. "Still, I'm not sure what to make of this L and R."

"Joseph-Louis Lagrange, considered by some to be the greatest mathematician of the eighteenth century, studied the methods that Cardano and Euler and others used to solve equations. He was trying to extend known methods for solving third- and fourth-degree equations to figure out how to solve fifth-degree equations, which no one knew how to do. Lagrange's key insight was that solving equations is all about permutations of roots. 'Lagrange resolvents,' which are obviously named after him, are part of that."

"I've gotta say, they seem kind of out-of-the-blue," I said. "Just *poof*, these $\omega\alpha + \omega^2\beta + \gamma$ and $\omega^2\alpha + \omega\beta + \gamma$ show up..."

Miruka snatched up the yellow card. "Take a look. These are Lagrange resolvents."

$$\begin{cases} L = \omega\alpha + \omega^2\beta + \gamma \\ R = \omega^2\alpha + \omega\beta + \gamma \end{cases}$$

"You called these 'unbalanced,' right Tetra?"

"I did. Because the α and the β here get to play 'swap the coefficients,' but poor little γ gets left out of all the fun. Something doesn't feel right about that. It seems... irregular."

"Actually, it isn't coefficients they're swapping, those are roots." Miruka put a finger to her lips and momentarily closed her eyes. "Okay, let's play with some equations and look for regularity. To do so, we should include $\alpha + \beta + \gamma$ in the mix, and forget about L and R for now." She added some new lines in the notebook.

$$\begin{cases} \omega\alpha + \omega^2\beta + \gamma \\ \omega^2\alpha + \omega\beta + \gamma \\ \underwave{\alpha + \beta + \gamma} \end{cases}$$

"We saw $\alpha + \beta + \gamma$ in the relation between roots and coefficients," I said.

"But poor γ is still being left out in the cold," Tetra said.

"Let's set things up so we can see the regularity better," Miruka said. "I'll write coefficients of 1 as ω^3."

$$\begin{cases} \omega^1\alpha + \omega^2\beta + \omega^3\gamma & \text{from } \omega\alpha + \omega^2\beta + \gamma \\ \omega^2\alpha + \omega^1\beta + \omega^3\gamma & \text{from } \omega^2\alpha + \omega\beta + \gamma \\ \omega^3\alpha + \omega^3\beta + \omega^3\gamma & \text{from } \alpha + \beta + \gamma \end{cases}$$

"Ah, right," Tetra said. "Because $\omega^3 = 1$. But if we read off the ω exponents, we get $1, 2, 3$ and $2, 1, 3$ and $3, 3, 3$. I'm still not seeing a pattern here."

$$\begin{cases} \omega^1\alpha + \omega^2\beta + \omega^3\gamma & \omega \text{ exponents are } 1, 2, 3 \\ \omega^2\alpha + \omega^1\beta + \omega^3\gamma & \omega \text{ exponents are } 2, 1, 3 \\ \omega^3\alpha + \omega^3\beta + \omega^3\gamma & \omega \text{ exponents are } 3, 3, 3 \end{cases}$$

"The triple-time rhythm of the omega waltz," Miruka said. "Since $\omega^3 = 1$, we can write ω^1 as ω^4, and we can write ω^3 as ω^6 or ω^9."

$$\begin{aligned} \omega^1 &= \omega^1\omega^3 &&= \omega^{1+3} &&= \omega^4 \\ \omega^3 &= \omega^3\omega^3 &&= \omega^{3+3} &&= \omega^6 \\ \omega^3 &= \omega^3\omega^3\omega^3 &&= \omega^{3+3+3} &&= \omega^9 \end{aligned}$$

Miruka paused to push her glasses up on her nose, then continued writing.

$$\begin{cases} \omega^1\alpha + \omega^2\beta + \omega^3\gamma & \omega \text{ exponents are } 1, 2, 3 \\ \omega^2\alpha + \omega^4\beta + \omega^6\gamma & \omega \text{ exponents are } 2, 4, 6 \\ \omega^3\alpha + \omega^6\beta + \omega^9\gamma & \omega \text{ exponents are } 3, 6, 9 \end{cases}$$

Tetra's mouth dropped open. "Whoa!"

"Definitely some regularity in $1, 2, 3$ and $2, 4, 6$ and $3, 6, 9$. But we can still turn it up a notch."

$$\begin{cases} (\omega^1)^1\alpha + (\omega^1)^2\beta + (\omega^1)^3\gamma \\ (\omega^2)^1\alpha + (\omega^2)^2\beta + (\omega^2)^3\gamma \\ (\omega^3)^1\alpha + (\omega^3)^2\beta + (\omega^3)^3\gamma \end{cases}$$

This time my mouth dropped opened. "Wow..."

"We can also use $\alpha_1, \alpha_2, \alpha_3$ in place of α, β, γ. Let's also name these polynomials $L_3(1), L_3(2), L_3(3)$ while we're at it."

$$\begin{cases} L_3(1) &= (\omega^1)^1 \alpha_1 + (\omega^1)^2 \alpha_2 + (\omega^1)^3 \alpha_3 \\ L_3(2) &= (\omega^2)^1 \alpha_1 + (\omega^2)^2 \alpha_2 + (\omega^2)^3 \alpha_3 \\ L_3(3) &= (\omega^3)^1 \alpha_1 + (\omega^3)^2 \alpha_2 + (\omega^3)^3 \alpha_3 \end{cases}$$

"Now we've got all these nice rows and columns with 1, 2, and 3." Miruka winked. "So I don't think you can claim that there's no regularity in Lagrange resolvents, Tetra."

Lagrange resolvents for cubic equations

$$\begin{cases} L_3(1) &= (\omega^1)^1 \alpha_1 + (\omega^1)^2 \alpha_2 + (\omega^1)^3 \alpha_3 \\ L_3(2) &= (\omega^2)^1 \alpha_1 + (\omega^2)^2 \alpha_2 + (\omega^2)^3 \alpha_3 \\ L_3(3) &= (\omega^3)^1 \alpha_1 + (\omega^3)^2 \alpha_2 + (\omega^3)^3 \alpha_3 \end{cases},$$

where

· ω is a primitive cube root of unity, and

· $\alpha_1, \alpha_2, \alpha_3$ are roots of the cubic equation.

"Okay, I admit there's plenty of regularity there," Tetra said. "But still... so many exponents and subscripts and all that stuff! What a mess!"

Miruka laughed. "Never satisfied, are you. The subscripts make things a bit more complex, sure, but they also reveal patterns."

"They do. I guess the way you write things changes according to what you're trying to show."

"Another advantage of regularity is that it makes it easier to generalize."

"Of course!" I said, taking the pencil from Miruka's hand.

$$L_3(k) = (\omega^k)^1 \alpha_1 + (\omega^k)^2 \alpha_2 + (\omega^k)^3 \alpha_3 \qquad (k = 1, 2, 3)$$

"How about using zetas to represent primitive nth roots of unity?" Miruka suggested.

"Ah, of course!" I said, louder than intended. "Another level of generalization!"

$$L_n(k) = (\zeta_n^k)^1 \alpha_1 + (\zeta_n^k)^2 \alpha_2 + \cdots + (\zeta_n^k)^n \alpha_n \quad (k = 1, 2, 3, \ldots, n)$$

"Exactly," Miruka said. "But look at that long sum. Wouldn't that be better as a sigma? From the laws of exponents $(\zeta_n^k)^j = \zeta_n^{kj}$ will hold, so we can use that to get rid of the parentheses."

$$L_n(k) = \sum_{j=1}^{n} \zeta_n^{kj} \alpha_j \quad (k = 1, 2, 3, \ldots, n)$$

"I think that's about as generalized as we're going to get," Miruka said. "We've uncovered the regularity, and derived the Lagrange resolvent for an equation of degree n."

Tetra let out a whistling sigh.

Lagrange resolvent for an nth-degree equation

$$L_n(k) = \sum_{j=1}^{n} \zeta_n^{kj} \alpha_j,$$

where

- $k = 1, 2, 3, \ldots, n$,
- ζ_n is a primitive nth root of unity, and
- $\alpha_1, \alpha_2, \alpha_3, \ldots, \alpha_n$ are roots of the nth degree equation.

7.2.2 Properties of Lagrange Resolvents

Running her fingers through her long, black hair, Miruka pointed at the "map of our journey"[3] Tetra had created. "Look here. On the way to deriving the cubic formula, you solved two equations. The first was this quadratic."

$$X^2 - (L^3 + R^3)X + L^3 R^3 = 0 \quad \text{quadratic equation in } X$$

---
[3] p. 226

"I remember that," Tetra said.
"From this we can find $X = L^3$ or R^3, then solve two simple cubic equations."

$$Y^3 - L^3 = 0, \; Y^3 - R^3 = 0 \quad \text{cubic equations in } Y$$

"Where did these come from?" Tetra asked.
"These allow us to derive the cube roots of L^3 and R^3," I said. "Specifically, $L, \omega L, \omega^2 L$ and $R, \omega R, \omega^2 R$."

Miruka nodded. "That's right. Before, we focused on ω^k coefficients to reveal the regularity in Lagrange resolvents, but it's more interesting to look at permutations of roots. For example, how about we pull a 'swapper' and exchange α and β? That exchanges L and R."

$$L = \omega\alpha + \omega^2\beta + \gamma$$
$$\updownarrow \quad \text{exchange } \alpha \text{ and } \beta$$
$$R = \omega\beta + \omega^2\alpha + \gamma$$

Tetra raised her hand. "Um...I don't think I quite understand this whole 'permutations of roots' thing."

"Then let's explicitly work out L^3, in other words $L_3(1)^3$."

$$L = \omega\alpha + \omega^2\beta + \gamma$$
$$= \omega\alpha_1 + \omega^2\alpha_2 + \alpha_3$$
$$L^3 = (\omega\alpha_1 + \omega^2\alpha_2 + \alpha_3)^3$$
$$= \alpha_1^3 + \alpha_2^3 + \alpha_3^3 + 6\alpha_1\alpha_2\alpha_3$$
$$+ 3\omega^2(\alpha_1\alpha_2^2 + \alpha_2\alpha_3^2 + \alpha_3\alpha_1^2) + 3\omega(\alpha_1^2\alpha_2 + \alpha_2^2\alpha_3 + \alpha_3^2\alpha_1)$$

Tetra traced a finger along what Miruka had written. "Okay, I can follow this."

"Then let's take a closer look at this expansion of L^3."

$$\alpha_1^3 + \alpha_2^3 + \alpha_3^3 + 6\alpha_1\alpha_2\alpha_3 + 3\omega^2(\alpha_1\alpha_2^2 + \alpha_2\alpha_3^2 + \alpha_3\alpha_1^2) + 3\omega(\alpha_1^2\alpha_2 + \alpha_2^2\alpha_3 + \alpha_3^2\alpha_1)$$

Tetra moved her face closer to the page, explicitly following Miruka's instructions.

Miruka continued. "There are $3! = 6$ ways to permute the three roots, $\alpha_1, \alpha_2, \alpha_3$. Let's work out what each of those permutations would look like." She started writing out a long series of calculations. "I'm going to save some time by using S to represent this part that remains invariant under any permutation."

$$S = \alpha_1^3 + \alpha_2^3 + \alpha_3^3 + 6\alpha_1\alpha_2\alpha_3$$

[1 2 3] leaves L^3 unchanged; a "plopper"

$\quad S + 3\omega^2(\alpha_1\alpha_2^2 + \alpha_2\alpha_3^2 + \alpha_3\alpha_1^2) + 3\omega(\alpha_1^2\alpha_2 + \alpha_2^2\alpha_3 + \alpha_3^2\alpha_1)$
$\quad = L^3$

[1 3 2] exchanges α_2 and α_3 in L^3; a "swapper"

$\quad S + 3\omega^2(\alpha_1\alpha_3^2 + \alpha_3\alpha_2^2 + \alpha_2\alpha_1^2) + 3\omega(\alpha_1^2\alpha_3 + \alpha_3^2\alpha_2 + \alpha_2^2\alpha_1)$
$\quad = S + 3\omega^2(\alpha_2\alpha_1^2 + \alpha_1\alpha_3^2 + \alpha_3\alpha_2^2) + 3\omega(\alpha_2^2\alpha_1 + \alpha_1^2\alpha_3 + \alpha_3^2\alpha_2)$
$\quad = R^3 \quad$ (because this is an exchange of α_2 and α_3 in L^3)

[2 1 3] exchanges α_1 and α_2 in L^3; a "swapper"

$\quad S + 3\omega^2(\alpha_2\alpha_1^2 + \alpha_1\alpha_3^2 + \alpha_3\alpha_2^2) + 3\omega(\alpha_2^2\alpha_1 + \alpha_1^2\alpha_3 + \alpha_3^2\alpha_2)$
$\quad = R^3 \quad$ (because this is an exchange of α_1 and α_2 in L^3)

[2 3 1] rotates α_1 to α_2, α_2 to α_3, α_3 to α_1 in L^3; a "scrambler"

$\quad S + 3\omega^2(\alpha_2\alpha_3^2 + \alpha_3\alpha_1^2 + \alpha_1\alpha_2^2) + 3\omega(\alpha_2^2\alpha_3 + \alpha_3^2\alpha_1 + \alpha_1^2\alpha_2)$
$\quad = S + 3\omega^2(\alpha_1\alpha_2^2 + \alpha_2\alpha_3^2 + \alpha_3\alpha_1^2) + 3\omega(\alpha_1^2\alpha_2 + \alpha_2^2\alpha_3 + \alpha_3^2\alpha_1)$
$\quad = L^3$

[3 1 2] rotates α_1 to α_3, α_2 to α_1, α_3 to α_2 in L^3; a "scrambler"

$\quad S + 3\omega^2(\alpha_3\alpha_1^2 + \alpha_1\alpha_2^2 + \alpha_2\alpha_3^2) + 3\omega(\alpha_3^2\alpha_1 + \alpha_1^2\alpha_2 + \alpha_2^2\alpha_3)$
$\quad = S + 3\omega^2(\alpha_1\alpha_2^2 + \alpha_2\alpha_3^2 + \alpha_3\alpha_1^2) + 3\omega(\alpha_1^2\alpha_2 + \alpha_2^2\alpha_3 + \alpha_3^2\alpha_1)$
$\quad = L^3$

[3 2 1] exchanges α_1 and α_3 in L^3; a "swapper"

$\quad S + 3\omega^2(\alpha_3\alpha_2^2 + \alpha_2\alpha_1^2 + \alpha_1\alpha_3^2) + 3\omega(\alpha_3^2\alpha_2 + \alpha_2^2\alpha_1 + \alpha_1^2\alpha_3)$
$\quad = S + 3\omega^2(\alpha_2\alpha_1^2 + \alpha_1\alpha_3^2 + \alpha_3\alpha_2^2) + 3\omega(\alpha_2^2\alpha_1 + \alpha_1^2\alpha_3 + \alpha_3^2\alpha_2)$
$\quad = R^3 \quad$ (because this is an exchange of α_1 and α_3 in L^3)

"That's so cool!" Tetra said. "Okay, so there are six patterns in which we can change the order of the three roots. But when we actually rearrange the $\alpha_1, \alpha_2, \alpha_3$ roots in L^3, we always end up with either L^3 or R^3, right?"

"That's right," Miruka said. "Furthermore, L^3 and R^3 are yoked together, namely by the quadratic $X^2 - (L^3 + R^3)X + L^3R^3 = 0$. The green and blue cards—sums and products of cubes—are supposed to make you notice that. You should also notice that these sums and products are elements in the coefficient field."

"The... coefficient field?"

"That's right. Let's look at this from the perspective of fields. Take the adjunction of $\sqrt{}$ to the coefficient field, and adjoin $\sqrt[3]{}$ to the resulting extension field. This double adjunction gives us what's called a 'minimal splitting field,' or just a 'splitting field' for short."

"And what's that?"

"The smallest field in which a given cubic polynomial decomposes into linear polynomial factors. For a general cubic, we can start from the coefficient field, adjoin the root operators $\sqrt{}$ and $\sqrt[3]{}$ to it, and that's enough to get a splitting field. This allows us to create a formula that gives the roots to a cubic equation. We adjoin the square root of rational expressions, that would be $\sqrt{D}$, to the coefficient field. Then we take the resulting new field and adjoin to that the cube root of rational expressions—that would be $\sqrt[3]{A + \sqrt{D}}$ and the like. That brings us to the splitting field." She leaned forward and looked at Tetra, then me, straight in the eye. "*Adjoining nth roots until you move from the coefficient field for a polynomial to its splitting field is the algebraic way of solving equations.* When taking the viewpoint of fields, adjunctions that extend fields are important. One we used was $\sqrt{}$. Another was $\sqrt[3]{}$. So this is the essence of Tetra's map."

$$a, b, c, d, \omega \xrightarrow{\sqrt{}} L^3, R^3 \xrightarrow{\sqrt[3]{}} L, R, \alpha, \beta, \gamma$$

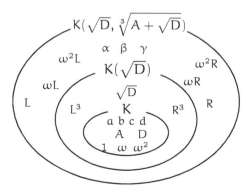

"L_3^3 is invariant for $[1\,2\,3], [2\,3\,1]$, and $[3\,1\,2]$," Miruka said. "Since L_3^3 is invariant in the cyclic group generated by the 'scrambler' $[2\,3\,1]$—"

"Whoa whoa, hold up, Miruka!" Tetra said. "What are these groups doing, creeping into our discussion of fields? Cyclic groups? Where did those come from?"

"Hmph. Tell you what, let's get into the details when Yuri is here."

"Yuri?" I said.

"I hear she's researching the symmetric group S_4."

"Ah, right, she did say something about that." Maybe "researching" was a bit grandiose, but I knew she was doing something with the 24 permutations that comprised S_4.

"Groups, subgroups, normal subgroups, quotient groups... There's no limit to how abstract we can get, but aiming for concreteness will make things easier to understand. Yeah, let's wait for Yuri." Miruka pointed at me. "Bring her."

"Yes, ma'am."

7.2.3 Applications

Following the path laid out by Mr. Muraki's seven cards, we had derived the cubic formula, a formula that gave the roots to a cubic equation. We had fiddled with Lagrange resolvents for cubic equations to find the regularity within them, and we had generalized them to equations of degree n. Which implied that—

"Hey, Miruka," I said. "Since we generalized Lagrange resolvents to degree n, doesn't that mean we should be able to derive a quartic formula in the same way?"

"Not exactly in the same way," Miruka said, "but Lagrange resolvents do make an appearance."

"Hey, no spoilers!" Tetra shouted. "I'm planning on finding the quartic formula on my own!"

"Fair enough. I'll say no more. But save it for homework. Before that, you should take a look at the relation between the quadratic formula and Lagrange resolvents for quadratic equations."

"The quadratic formula? I memorized that one a long time ago."

"Then step one is to forget it. Do you need another kiss to help you forget?" Miruka stood from her chair, and made a move in Tetra's direction.

Tetra waved her hands. "Already forgotten! Quadratic formula? What quadratic formula?!"

Problem 7-8 (Lagrange resolvents for quadratics)

Find the Lagrange resolvents $L_2(1)$ and $L_2(2)$ for a quadratic equation.

7.3 THE QUADRATIC EQUATION

7.3.1 Lagrange Resolvents for Quadratic Equations

After some calculations, Tetra spoke up.

"Okay, the primitive square root of unity—in other words, the number that first becomes 1 when squared—is this."

$$\zeta_2 = -1$$

"Now we use the Lagrange resolvent for a polynomial of degree n."[4]

$$\begin{aligned}
L_2(1) &= \sum_{j=1}^{2} \zeta_2^{1j} \alpha_j \quad \text{Lagrange resolvent with } n=2, k=1 \\
&= \zeta_2^{1 \times 1} \alpha_1 + \zeta_2^{1 \times 2} \alpha_2 \\
&= (-1)^{1 \times 1} \alpha_1 + (-1)^{1 \times 2} \alpha_2 \quad \text{because } \zeta_2 = -1 \\
&= -\alpha_1 + \alpha_2
\end{aligned}$$

$$\begin{aligned}
L_2(2) &= \sum_{j=1}^{2} \zeta_2^{2j} \alpha_j \quad \text{Lagrange resolvent with } n=2, k=2 \\
&= \zeta_2^{2 \times 1} \alpha_1 + \zeta_2^{2 \times 2} \alpha_2 \\
&= (-1)^{2 \times 1} \alpha_1 + (-1)^{2 \times 2} \alpha_2 \quad \text{because } \zeta_2 = -1 \\
&= \alpha_1 + \alpha_2
\end{aligned}$$

---

[4]See p. 231

"Not a terribly exciting result, I guess."

$$\begin{cases} L_2(1) &= -\alpha_1 + \alpha_2 \\ L_2(2) &= \alpha_1 + \alpha_2 \end{cases}$$

Answer 7-8 (Lagrange resolvents for quadratics)

$$\begin{cases} L_2(1) &= -\alpha_1 + \alpha_2 \\ L_2(2) &= \alpha_1 + \alpha_2 \end{cases}$$

"You said something about L and R being key to deriving these formulas for finding the roots to equations, right, Tetra?" Miruka said.

"I did! We started out by finding L^3 and R^3."

"Those were cubes of the Lagrange resolvents for cubic equations. By analogy, lets find the squares of the Lagrange resolvents for quadratic equations."

"Oh, okay. I'll give it a shot!"

$$\begin{cases} L_2(1)^2 &= (-\alpha_1 + \alpha_2)^2 = (\alpha_1 - \alpha_2)^2 \\ L_2(2)^2 &= (\alpha_1 + \alpha_2)^2 = (\alpha_1 + \alpha_2)^2 \end{cases}$$

"Interesting," I said, impressed. We had used the regularity of Lagrange resolvents to generalize them, and applied that to a concrete example. Everything according to theory.

"So $L_2(1)^2$ would be equal to $(\alpha_1 - \alpha_2)^2$, right? Isn't that the key?" Tetra asked.

"That's what I'm asking you," Miruka said. "What do you think?"

Tetra stared at what she'd written, and mumbled, "If I subtract α_2 from α_1... Uh... No, no I just can't do it. Subscripts make me crazy! Okay, enough with the numbers. I'll use α and β instead."

$$\begin{cases} L_2(1)^2 &= (\alpha - \beta)^2 \\ L_2(2)^2 &= (\alpha + \beta)^2 \end{cases}$$

I couldn't resist. "Tetra, the relation between—"

Tetra sharply held up a hand. "Quiet, please!" *Oops.* "I need to end up at the quadratic equation, right? So I'll write the roots as the coefficients of the equation..." I kept silent. "That would be $(\alpha - \beta)^2$ as the coefficients of $ax^2 + bx + c = 0$... So, coefficients, coefficients..."

"Ah..." I started, but stopped myself.

"Oh! Writing this as coefficients means writing it as elementary symmetric polynomials! Sums and products! Right, right. $(\alpha - \beta)^2$ is a symmetric polynomial, because swapping α and β doesn't change its value... And we can write a symmetric polynomial as elementary symmetric polynomials!"

$$L_2(1)^2 = (\alpha - \beta)^2$$
$$= \alpha^2 - 2\alpha\beta + \beta^2$$
$$= (\underbrace{\alpha + \beta}_{\substack{\text{elementary} \\ \text{symmetric} \\ \text{polynomial}}})^2 - 4\underbrace{\alpha\beta}_{\substack{\text{elementary} \\ \text{symmetric} \\ \text{polynomial}}}$$

"Hmph," Miruka grunted.

"Okay, we can write the elementary symmetric polynomials as coefficients, so the rest is easy."

Tetra began writing in her notebook at an impressive speed.

$$\begin{aligned}
L_2(1)^2 &= (\alpha - \beta)^2 \\
&= (\alpha + \beta)^2 - 4\alpha\beta &&\text{as elementary symmetric polynomials} \\
&= \left(-\frac{b}{a}\right)^2 - 4 \cdot \frac{c}{a} &&\text{relation between roots and coefficients} \\
&= \frac{b^2 - 4ac}{a^2}
\end{aligned}$$

"Notice something there?" Miruka asked.

Tetra's eyes grew even larger than usual. "I do I do I do! The $b^2 - 4ac$ showed up! That's the discriminant in the quadratic equation!"

7.3.2 Discriminants

"Exactly," Miruka said in low voice. "When we square Lagrange resolvents for a quadratic, we find the discriminant for the quadratic formula."

$$L_2(1)^2 = \frac{b^2 - 4ac}{a^2} = \frac{\text{discriminant}}{a^2}$$

"That's so cool!" Tetra said.

I was stunned. I had memorized this as part of the quadratic formula so long ago that "b-squared-minus-four-a-c" just rolled off my tongue. What was it doing popping up in the context of Lagrange resolvents? How strange that we could create the discriminant $b^2 - 4ac$ by squaring $L_2(1)$, the Lagrange resolvent for quadratic equations...

$$\frac{-b \pm \sqrt{b^2 - 4ac}}{2a}$$

Miruka continued, calmly. "Letting K be the coefficient field, the roots of a quadratic polynomial are elements in the field

$$K(\sqrt{\text{discriminant}}),$$

"which we can write as

$$K(L_2(1)).$$

"Both $L_2(1)^2 = (\alpha - \beta)^2$ and $L_2(2)^2 = (\alpha + \beta)^2$ are invariant when we swap the roots. In other words, these are symmetric polynomials for the roots. Since they're symmetric polynomials, we can write them as elementary symmetric polynomials, namely, with rational expressions for the coefficients. Lagrange resolvents are definitely useful in finding formulas for polynomial roots. $L_2(1) = \alpha - \beta$ isn't symmetric, so we can't necessarily write it as a rational expression in the coefficients. An adjunction of $L_2(1)$ to the coefficient field solves that problem, though, because the roots will be elements of the resulting field $K(L_2(1))$, which is also $K(\sqrt{\text{discriminant}})$."

Tetra sat thinking about this for a while, nodding. "Okay, I think I'm getting a better grasp on Lagrange resolvents. Won't be long before you're calling them 'Tetragrange resolvents'!"

"Getting a bit ahead of yourself there," Miruka coolly stated.

"Good point—that can come after I've conquered the quartic!"

I half listened to this math girl banter as I let my mind wander. *So, regarding the Lagrange resolvent $L_2(1)$ ...*

- *Adjunction of $L_2(1)$ to the coefficient field gives a field in which the quadratic equation can definitely be solved.*

- *We can create $\sqrt{discriminant}$ from $L_2(1)$.*

- $L_2(1)$ appears in the quadratic formula.

Formulas for finding roots... They look like answers, but really they're questions. They ask if we can find the form of equations. I recalled talking with Yuri about the quadratic formula, how she'd done her best to just memorize it rather than make the effort to really understand it. *It isn't easy, Yuri, but it sure is fun.*

7.4 Roots of Fifth-degree Polynomials

7.4.1 A Quintic Formula?

"Okay, my homework for today is finding a formula for the roots of a fourth-degree equation," Tetra said. "After I've done that, maybe we can work together on using Lagrange resolvents to find a formula for fifth-degree equations?"

"No," Miruka said.

"Er, no?"

"Lagrange resolvents won't help you find a formula for finding the roots of quintic equations, or equations of any higher degree, for that matter."

"Because no such thing exists," I added.

"That's right. There are no formulas that give the roots of equations of order five or more. Put another way, when given an equation of degree five or higher, you *won't necessarily* be able to represent its roots using repeated operations involving basic arithmetic or taking nth roots of its coefficients. That was proven by Paolo Ruffini and Niels Abel."

" 'Won't necessarily,' huh?"

"Indeed. Leading to the obvious question, just what kind of equations *can* we find roots for? It was Évariste Galois who told us that. In fact, he answered this very difficult question just a few years after he started studying mathematics." Miruka paused to scan our faces. "To summarize, Ruffini and Abel proved that fifth-degree equations cannot be solved in a general manner. Galois, meanwhile, showed when we can and when we can't solve fifth-degree equations. Actually, not just fifth-degree equations. Galois gave us the necessary and sufficient conditions for algebraically solving equations of order n."

7.4.2 The Significance of "5"

"Linear, quadratic, cubic, quartic..." Tetra said. "Okay, so those are the kinds of equations that have formulas for finding their roots. But no equations of higher order can have such a thing. That seems so... strange. Something about all this reminds me of when we studied Fermat's last theorem."

The light glinted off of Miruka's glasses. "Does it now?"

"Back then we learned about Wiles's proof, which showed how the equation $x^n + y^n = z^n$ has no natural number solutions for $n \geqslant 3$. So the 'magic number' in that case was 3. Now we're learning that no equation of degree n can have a formula giving its solutions if $n \geqslant 5$, so this time the magic number is 5."

"Magic numbers?" I said, somewhat dubiously.

"Very interesting," Miruka said.

"That's just the thing, though," Tetra said. "I've never really thought of 5 as being a particularly interesting number. I wonder what secrets it might hold?"

> Lagrange also considered several other methods for solving cubic equations, but he noticed that they were all based on the same basic idea; each considered six possible permutations, which produced rational expressions taking just two values and giving three roots, thereby satisfying a quadratic equation.
>
> VICTOR KATZ
> *A History of Mathematics*

My notes

Elementary symmetric polynomials

Elementary symmetric polynomials in α_1
$$\alpha_1$$

Elementary symmetric polynomials in α_1, α_2
$$\alpha_1 + \alpha_2$$
$$\alpha_1 \alpha_2$$

Elementary symmetric polynomials in $\alpha_1, \alpha_2, \alpha_3$
$$\alpha_1 + \alpha_2 + \alpha_3$$
$$\alpha_1 \alpha_2 + \alpha_1 \alpha_3 + \alpha_2 \alpha_3$$
$$\alpha_1 \alpha_2 \alpha_3$$

Elementary symmetric polynomials in $\alpha_1, \alpha_2, \alpha_3, \alpha_4$
$$\alpha_1 + \alpha_2 + \alpha_3 + \alpha_4$$
$$\alpha_1 \alpha_2 + \alpha_1 \alpha_3 + \alpha_1 \alpha_4 + \alpha_2 \alpha_3 + \alpha_2 \alpha_4 + \alpha_3 \alpha_4$$
$$\alpha_1 \alpha_2 \alpha_3 + \alpha_1 \alpha_2 \alpha_4 + \alpha_1 \alpha_3 \alpha_4 + \alpha_2 \alpha_3 \alpha_4$$
$$\alpha_1 \alpha_2 \alpha_3 \alpha_4$$

CHAPTER 8

Building Towers

> If you want to regret your past, go ahead and regret it; the day will come soon enough when you will regret your present.
>
> HIDEO KOBAYASHI
> *My View of Life*

8.1 MUSIC

8.1.1 One Note

"Just one mistake, I think," Miruka said.

"Three," Ay-Ay said, "all on the Bach. It's always Bach that gets me..."

We were in a tea room adjacent to a performance hall, where Miruka and I had come to watch Ay-Ay's recital. We had just joined her for a post-concert lunch of club sandwiches.

"I didn't notice any mistakes at all," I said. "I thought the whole thing was wonderful."

"Nice of you to say so," Ay-Ay said. She attended the same school as Miruka and I, and was in the same grade. She was an amazing pianist, and crazy into music like we were crazy into mathematics. Apparently she'd been studying under the same tutor—who she called "maestro"—since she was a little girl. The recital had featured two pieces each by several of that tutor's students; Ay-Ay had played a Bach piece and one of her own compositions. My seat had allowed me a good view of her hands. To me it had looked like she wasn't using her fingers to hit the keys, it was more like the keys had drawn the correct finger to them. She had looked gorgeous up on stage, her long, wavy hair accentuating a moss-green dress. She was still wearing that dress, and it was hard to keep my eyes off of it.

"So this teacher of yours, does he, like, show you how you should play a piece?" I asked.

Ay-Ay shook her head. "He makes me play a piece, then asks me what I was thinking when I did so. Then he says, 'So you wanted something like this?' and plays the piece exactly as I was imagining it. He also says, 'But this is what you played,' and gives a perfect impression of my poor performance."

"Wow, impressive. What else?"

"Nothing else. Just a period of embarrassment while I'm unable to play in the way I said I wanted to. A time when I'm trying to close the gap between what I want to do and what I'm currently capable of. Maestro playing from time to time to remind me what I'm aiming at. It's maddening, how he can do that."

"Sounds like a kind teacher," Miruka said.

"His words are kind, but beneath that, he's a cruel, cruel man."

Ay-Ay's still riding the natural high from her performance. Eating fast, too.

"How do you describe what you're working on to each other?" I asked. "Do you, like, talk about motifs or something?"

Ay-Ay shook her head. "It's at a lower level than that. It's about how to tie together this sound with that one. About what phrase should come out. About which of the notes that are being hit should come out with the same force. Lots of things like that." Ay-Ay finished off her last sandwich wedge before continuing. "Maestro is always telling me, 'there are no unimportant notes in a score.' You can't create an entire piece all at once, it gets built up note by note. But that's not all. A musical composition isn't just a pile of notes. If you don't understand the entire piece, you can't understand the role that each note plays. Every note supports the piece, and the piece supports every note."

Despite Ay-Ay's claim of her maestro's cruelty, I could tell she'd placed her absolute trust in him.

"That second piece you played," I said, "it started out kind of baroque, but then it shifted to this irregular rhythm... That silence that came just before the shift, and that one note you played to mark it... I can't get that note out of my head."

Ay-Ay laughed. "Then I guess it did its job. All it takes is one note to change the course of things."

8.1.2 One Encounter

After our lunch, Ay-Ay returned to her maestro, leaving Miruka and me in the tea room.

"So is Ay-Ay planning on going pro?" I asked.

"Probably, but she's leaning toward becoming a composer, rather than a performer. A hard row to hoe in either case. She's talking things out with her teacher. I expect she'll be off to Europe after graduation."

"Could be."

"You know what happened to Galois after he graduated *lycée* at fifteen?" Miruka asked. I shrugged. "Failed his college entrance exams."

"No way."

Miruka nodded. "Might have been his interest in mathematics that did it. He'd gotten deep into math just the year before, working his way through Legendre's *Éléments de Géométrie* in just two days, or so the story goes. And so Galois found the world of mathematics, and we came to benefit from his genius."

"Huh."

"Anyway, Galois tried to get into the École Polytechnique, but he failed the oral exam. After that, Louis Richard became his math teacher. A fortunate encounter for Galois, because it was Richard who directed him to Lagrange's work when he learned that Galois was interested in solutions to equations. He also urged Galois to submit his work to mathematics journals. So maybe it was Galois failing to get into the school he was aiming at that led him to the best teacher he could ever have."

"You may be right."

"All it takes is one encounter to change the course of things."

8.2 A LECTURE

8.2.1 In the Library

Several days later I went to the library as usual. As soon as I walked through the door, Tetra latched onto my arm.

"There you are!" she said.

"Whoa, you scared me. What's up?"

"There's something I've been wanting to talk to you about, but... I guess you're busy today?"

"I could make some time. I guess you're ambushing me at the door because you want to go somewhere else?"

"Yeah, Ms. Mizutani's kinda been giving me the evil eye. Okay if we move to the student lounge?"

8.2.2 Degrees of Field Extensions

We relocated to our school's student lounge, where just a few students were hanging out for club activities. Tetra didn't waste any time getting straight to the point: "Remember when Miruka mentioned the degree of a field extension the other day?"

"Mmm... something about how much bigger you've made a field after extending it, I think?"

"Sounds like you need a quick refresher."

In brief, Tetra reminded me that...

- A *field* is a set in which we can perform basic arithmetic (addition, subtraction, multiplication, and division). We can add (or *adjoin*) an element to that set to create a new field, called a *field extension*.

- A *vector space* (or *linear space*) is a set of vectors, for which addition and scalar multiplication are defined. If every vector in a vector space can be uniquely written as a linear combination of a set of vectors, that collection is called a *basis* for the vector space. The number of elements in a basis is called the *dimension* of its vector space.

- A field extension can be viewed as a vector space. When doing so, the dimension of the vector space is called the *degree* of the extension.

"Okay, I remember all this now," I said. "Nice summary, by the way."

Tetra gave a sharp nod. *Something different about her today...*

"So," she continued, "I think the idea was that when we extend a field, the degree of the extension tells us, in a sense, how much larger the field has become. I wanted to create some examples to make sure I understand all this, examples being the key to understanding and all. So I've been studying some books about field theory I borrowed from the library. As best I can, at least. Mind if I ask you some questions?"

Of course, there was no way I could say no. Not when Tetra was so excited about what she was learning. This is how Tetra started on a lecture that took us to places we'd never dreamed of.

8.2.3 Extension Fields and Subfields

"I started with fields that we know well," Tetra began, "like the field of rational numbers $\mathbb{Q}$, the field of real numbers $\mathbb{R}$, and the field of complex numbers $\mathbb{C}$.

"When we perform arithmetic using rational numbers, the result is of course a rational number. Similarly, arithmetic on real numbers gives real numbers, and arithmetic on complex numbers gives complex numbers. This is because $\mathbb{Q}$, $\mathbb{R}$, and $\mathbb{C}$ are each fields. They are also in an inclusion relation, like this."

$$\mathbb{Q} \subset \mathbb{R} \subset \mathbb{C}$$

"This means that $\mathbb{Q}$ is a subset of $\mathbb{R}$, while $\mathbb{R}$ is a subset of $\mathbb{C}$. Furthermore, $\mathbb{Q}$ is a subset of $\mathbb{C}$. Arithmetic operations among these sets also extend in a natural way, in that given a sum of rational numbers $a + b$, the result doesn't change if we decide we want to think of a and b as being real numbers instead. So operations using rational numbers are the same as operations using real numbers, meaning that $\mathbb{Q}$ isn't just a sub*set* of $\mathbb{R}$, it's also a sub*field*. Similarly, $\mathbb{Q}$ and $\mathbb{R}$ are subfields of $\mathbb{C}$. Looked at in the opposite direction, $\mathbb{C}$ is an extension field of $\mathbb{Q}$ and $\mathbb{R}$, and $\mathbb{R}$ is an extension field of $\mathbb{Q}$. Oh, and we can consider any field to be a subfield of itself, and also an extension field of itself.

"I looked in several books to see how to use symbols to show that $\mathbb{C}$ is an extension field of $\mathbb{R}$, in other words that $\mathbb{R}$ is a subfield of $\mathbb{C}$. Some used set notation to show an inclusion relation..."

$$\mathbb{C} \supset \mathbb{R}$$

"...while some put a slash between the field names."

$$\mathbb{C}/\mathbb{R}$$

"That's read '$\mathbb{C}$ over $\mathbb{R}$,' by the way. Here are some examples."

$\mathbb{C}/\mathbb{R}$ ······ $\mathbb{C}$ is an extension field of $\mathbb{R}$ ($\mathbb{R}$ is a subfield of $\mathbb{C}$)
$\mathbb{R}/\mathbb{Q}$ ······ $\mathbb{R}$ is an extension field of $\mathbb{Q}$ ($\mathbb{Q}$ is a subfield of $\mathbb{R}$)
$\mathbb{C}/\mathbb{Q}$ ······ $\mathbb{C}$ is an extension field of $\mathbb{Q}$ ($\mathbb{Q}$ is a subfield of $\mathbb{C}$)

"To show relations among three or more fields, I saw some books write this..."

$$\mathbb{C}/\mathbb{R}/\mathbb{Q}$$

"...but this was more common."

$$\mathbb{C} \supset \mathbb{R} \supset \mathbb{Q}$$

"I also found some new terminology to describe this. Depending on the book, a chained relation between extension fields and subfields like $\mathbb{C} \supset \mathbb{R} \supset \mathbb{Q}$ is called a *tower of fields*, or a *sequence of field extensions*, or an *ascending chain of fields*."

8.2.4 $\mathbb{Q}(\sqrt{2})/\mathbb{Q}$

"I guess that pretty much covers the basics," Tetra said.

"Nicely presented," I said. "Very easy to understand. You're a good teacher." I realized how often she had paid me the same compliment in the past, and reflected on how once again what had once been our customary roles had now swapped.

Tetra blushed. "C'mon, cut it out. Anyway, I worked out this problem for practice."

Building Towers

Problem 8-1 (Degree of a field extension)

Find the degree of the field extension $\mathbb{Q}(\sqrt{2})/\mathbb{Q}$.

"Interesting," I said.
"I started out by making sure I understood all the notations."

$\mathbb{Q}$ ······ the field of rational numbers
$\mathbb{Q}(\sqrt{2})$ ······ adjunction of $\sqrt{2}$ to $\mathbb{Q}$
$\mathbb{Q}(\sqrt{2})/\mathbb{Q}$ ······ $\mathbb{Q}(\sqrt{2})$ as an extension of $\mathbb{Q}$

"The degree of the extension $\mathbb{Q}(\sqrt{2})/\mathbb{Q}$ is, let's see, the dimension of $\mathbb{Q}(\sqrt{2})$ when viewing it as a vector space over $\mathbb{Q}$. Put another way, it's the number of elements in the basis when viewing $\mathbb{Q}(\sqrt{2})$ as a vector space over $\mathbb{Q}$."

"Right."

"Oh, one other thing I saw in the books, we can write the degree of the field extension for $\mathbb{Q}(\sqrt{2})/\mathbb{Q}$ like this."

$$[\mathbb{Q}(\sqrt{2}) : \mathbb{Q}]$$

Tetra grimaced. "Not the most elegant notation, if you ask me."

"Yeah, but having notation like that allows us to use this concept in mathematical expressions."

"I guess. Anyway, we can write $\mathbb{Q}(\sqrt{2})$ like this."

$$\mathbb{Q}(\sqrt{2}) = \{p + q\sqrt{2} \mid p \in \mathbb{Q}, q \in \mathbb{Q}\}$$

"Sure, I see that."

"This means we can write any number belonging to $\mathbb{Q}(\sqrt{2})$ as $p + q\sqrt{2}$, using the basis $\{1, \sqrt{2}\}$."

"In other words, $p + q\sqrt{2} = p \cdot \underline{1} + q \cdot \underline{\sqrt{2}}$," I said.

"Right! Also, there are two elements in the basis $\{1, \sqrt{2}\}$, so the degree of the field extension $[\mathbb{Q}(\sqrt{2}) : \mathbb{Q}]$ is 2."

$$[\mathbb{Q}(\sqrt{2}) : \mathbb{Q}] = 2$$

"And there's the answer!"

Answer 8-1 (Degree of a field extension)

$$[\mathbb{Q}(\sqrt{2}) : \mathbb{Q}] = 2$$

"Well done," I said.
"An extension field of degree is 2 is called a *quadratic extension*. So $\mathbb{Q}(\sqrt{2})/\mathbb{Q}$ is an example of a quadratic extension."

8.2.5 A Problem

"A problem for you!" Tetra said, doing her best Miruka impersonation. She laughed and stuck out her tongue.

What is the value of $[\mathbb{Q}(\sqrt{3}) : \mathbb{Q}]$?

"That's easy," I said. "We're just replacing the $\sqrt{2}$ in your example with $\sqrt{3}$. That makes the basis $\{1, \sqrt{3}\}$, and we can write the extension field as $\mathbb{Q}(\sqrt{3}) = \{p + q\sqrt{3} \mid p \in \mathbb{Q}, q \in \mathbb{Q}\}$, so once again the degree is 2."

$$[\mathbb{Q}(\sqrt{3}) : \mathbb{Q}] = 2$$

"Correct! $\mathbb{Q}(\sqrt{3})/\mathbb{Q}$ is another quadratic extension."

My turn to do an impression. I raised my hand and said, "I have a question! Is this a true statement?"

$[\mathbb{Q}(\sqrt{n}) : \mathbb{Q}] = 2$ for every nonnegative integer n.

"It is!" Tetra said without hesitation.
"Uh, oh. Didn't expect you to be fooled so easily."
"I...what? Oh, wait! It depends on whether $\sqrt{n}$ is a rational number!"
"That's right. We have to consider different cases."

$$[\mathbb{Q}(\sqrt{n}) : \mathbb{Q}] = \begin{cases} 1 & \text{for } \sqrt{n} \in \mathbb{Q} \\ 2 & \text{for } \sqrt{n} \notin \mathbb{Q} \end{cases}$$

Tetra's next impersonation was of a television quiz show host. "Okay, on to our next problem!"

What is the value of $[\mathbb{Q}(\sqrt{5}) : \mathbb{Q}(\sqrt{5})]$?

"Hmm... Let's see... $[\mathbb{Q}(\sqrt{5}) : \mathbb{Q}(\sqrt{5})]$ is the degree of $\mathbb{Q}(\sqrt{5})/\mathbb{Q}(\sqrt{5})$, so the extension field is $\mathbb{Q}(\sqrt{5})$ itself. So we could use a basis like $\{1\}$. Just one element there, so the degree is 1."

$$[\mathbb{Q}(\sqrt{5}) : \mathbb{Q}(\sqrt{5})] = 1$$

"Exactly. So we can write $\mathbb{Q}(\sqrt{5})$ like this."

$$\mathbb{Q}(\sqrt{5}) = \{p \cdot 1 \mid p \in \mathbb{Q}(\sqrt{5})\}$$

8.2.6 $\mathbb{Q}(\sqrt{2}, \sqrt{3})/\mathbb{Q}$

"Here's the next problem," Tetra said, turning her notebook toward me.

Problem 8-2 (Degree of a field extension)

Find the degree of the field extension $\mathbb{Q}(\sqrt{2}, \sqrt{3})/\mathbb{Q}$.

"Ah, interesting," I said.

"$\mathbb{Q}(\sqrt{2}, \sqrt{3})$ is the adjunction of $\sqrt{2}$ and $\sqrt{3}$ to $\mathbb{Q}$. So what do you think I did to find $[\mathbb{Q}(\sqrt{2}, \sqrt{3}) : \mathbb{Q}]$?"

"Well I guess you found its basis?"

"I did, but that's where I immediately went astray. At first, I thought the basis would be $\{1, \sqrt{2}, \sqrt{3}\}$."

$$\mathbb{Q}(\sqrt{2}, \sqrt{3}) = \{p + q\sqrt{2} + r\sqrt{3} \mid p \in \mathbb{Q}, q \in \mathbb{Q}, r \in \mathbb{Q}\} \quad (?)$$

"Er, that's what I was going to say. That's not right?"

Tetra's expression grew serious. "It isn't. $\{1, \sqrt{2}, \sqrt{3}\}$ is not a basis for $\mathbb{Q}(\sqrt{2}, \sqrt{3})/\mathbb{Q}$."

I sat back to think. Tetra was claiming that there was some number in the field $\mathbb{Q}(\sqrt{2}, \sqrt{3})$ that could not be written in the form $p + q\sqrt{2} + r\sqrt{3}$ $(p, q, r \in \mathbb{Q})$. *What kind of number might that be...? Oh!*

"Got it!" I said. "If we use $\{1, \sqrt{2}, \sqrt{3}\}$ as the basis, we can't represent the product of $\sqrt{2}$ and $\sqrt{3}$, namely $\sqrt{2}\sqrt{3} = \sqrt{6}$, as a linear combination. In other words, there are no rational numbers p, q, r satisfying this equation."

$$\sqrt{2}\sqrt{3} = p + q\sqrt{2} + r\sqrt{3}$$

"I knew you'd find that a lot quicker than I did. Took me forever!" "So I guess the basis for $\mathbb{Q}(\sqrt{2}, \sqrt{3})/\mathbb{Q}$ would be this?"

$$\{1, \sqrt{2}, \sqrt{3}, \sqrt{6}\}$$

"Exactly. So the degree is 4."

$$[\mathbb{Q}(\sqrt{2}, \sqrt{3}) : \mathbb{Q}] = 4 \qquad \text{degree of } \mathbb{Q}(\sqrt{2}, \sqrt{3})/\mathbb{Q}$$

"Well what do you know," I said.

"I took some time to figure out where I went wrong here, and realized I hadn't been paying attention to how you make numbers using linear combinations, versus how you make numbers by adjoining numbers to fields."

"I'm not exactly sure what you mean."

"I mean, like..." Tetra blinked several times while carefully choosing her words. "Multiplication in a linear combination only occurs between a scalar and a vector. The scalars in $\mathbb{Q}(\sqrt{2}, \sqrt{3})/\mathbb{Q}$ are all rational numbers, so the vectors that are elements of the basis will only ever be multiplied by rationals."

"Well, sure. Scalar multiplication of vectors is at the heart of vector spaces."

"But look at the field extension $\mathbb{Q}(\sqrt{2}, \sqrt{3})$. That's a field in which we can freely perform multiplication between all of its elements! I had only been thinking about multiplication with rational numbers, so I completely overlooked the possibility of multiplying $\sqrt{2}$ and $\sqrt{3}$. I'm pretty sure that even now I don't really understand the relation between linear combinations in vector spaces and arithmetic in fields. So I'm just going to stick a pin in that and add it to my 'not quite there' list."

I laughed. "That's an interesting list to make."

"An important one! It's easy to lose track of all the things I don't quite get yet. So, like I said, stick a pin in it."

"Fair enough," I said. "Very impressive, in fact. You have a good grasp on what you don't understand, you think about why you don't understand it, and you even keep track of what you're still working on understanding."

"Oh, cut it out."

"Seriously. Here's another list, things that make you a great learner."

- You don't pretend to understand when you really don't.
- You find the borderline between what you understand and what you don't.
- You never assume you've completely understood something.
- You try to figure out why you don't understand something.
- You keep track of what you're sure you don't yet understand.

"Don't forget the 'We know nothing' game!"
"How could I?"
We both laughed.

Answer 8-2 (Degree of a field extension)

$[\mathbb{Q}(\sqrt{2}, \sqrt{3}) : \mathbb{Q}] = 4$

"By the way," Tetra said, "there's a little bit more to this problem."

"Is there, now?"

8.2.7 Products of Degrees of Extension

Tetra had always been an energetic girl, but she had really cranked things up a notch today. Her extracurricular studies seemed to be stoking her engine.

"On to the next part, then," she said. "One of the books I read used an equality like this in its explanation of how to find the degree of $[\mathbb{Q}(\sqrt{2}, \sqrt{3}) : \mathbb{Q}]$."

$$\mathbb{Q}(\sqrt{2}, \sqrt{3}) = \mathbb{Q}(\sqrt{2})(\sqrt{3})$$

"Huh. What's with the $\mathbb{Q}(\sqrt{2})(\sqrt{3})$ on the right?"

"That's the adjunction of $\sqrt{3}$ to the field $\mathbb{Q}(\sqrt{2})$, which is in turn the adjunction of $\sqrt{2}$ to $\mathbb{Q}$."

"Ah, okay."

"Then the degree of $[\mathbb{Q}(\sqrt{2}, \sqrt{3}) : \mathbb{Q}]$ would be this."

$$[\mathbb{Q}(\sqrt{2}, \sqrt{3}) : \mathbb{Q}]$$
$$= [\mathbb{Q}(\sqrt{2})(\sqrt{3}) : \mathbb{Q}] \qquad \text{because } \mathbb{Q}(\sqrt{2}, \sqrt{3}) = \mathbb{Q}(\sqrt{2})(\sqrt{3})$$
$$= \underbrace{[\mathbb{Q}(\sqrt{2}) : \mathbb{Q}]}_{2} \times \underbrace{[\mathbb{Q}(\sqrt{2})(\sqrt{3}) : \mathbb{Q}(\sqrt{2})]}_{2} \qquad \text{multiplicativity formula}$$
$$= 2 \times 2$$
$$= 4$$

"The solution uses something called the multiplicativity formula for degrees, which is also called the 'tower rule' or the 'chain rule.' It's very interesting. In words, it says that the degree of a field extension created by adjunction of multiple numbers is equal to the product of the degrees when adjoining the numbers one at a time. In this case, it says that to find the degree resulting from adjunction of $\sqrt{2}$ and $\sqrt{3}$ to $\mathbb{Q}$, we should multiply the degree of the adjunction of $\sqrt{2}$ to $\mathbb{Q}$ by the degree of the adjunction of $\sqrt{3}$ to $\mathbb{Q}(\sqrt{2})$."

$$[\mathbb{Q}(\sqrt{2})(\sqrt{3}) : \mathbb{Q}] = [\mathbb{Q}(\sqrt{2}) : \mathbb{Q}] \times [\mathbb{Q}(\sqrt{2})(\sqrt{3}) : \mathbb{Q}(\sqrt{2})]$$

"There was a proof of this in the book, but there were so many letters involved, well, I kinda got lost. A little more work and I think I'll get through it, though!"

I nodded encouragement, but otherwise kept silent so as not to disturb the flow of her lecture.

"Another mistake I made was assuming that if adjoining one number to $\mathbb{Q}$, like $\sqrt{2}$ for example, resulted in an extension of degree 2, then adjoining two numbers, say $\sqrt{2}$ and $\sqrt{3}$, would result in a degree of 3. But that's wrong." Tetra nodded several times. "Anyway, using this rule, we can easily find degrees for really complicated field extensions, like $\mathbb{Q}(\sqrt{2}, \sqrt{3}, \sqrt{5}, \sqrt{7})/\mathbb{Q}$."

$$
\begin{aligned}
&[\mathbb{Q}(\sqrt{2}, \sqrt{3}, \sqrt{5}, \sqrt{7}) : \mathbb{Q}] \\
&= [\mathbb{Q}(\sqrt{2})(\sqrt{3})(\sqrt{5})(\sqrt{7}) : \mathbb{Q}] \\
&= [\mathbb{Q}(\sqrt{2}) : \mathbb{Q}] \\
&\quad \times [\mathbb{Q}(\sqrt{2})(\sqrt{3}) : \mathbb{Q}(\sqrt{2})] \\
&\quad\quad \times [\mathbb{Q}(\sqrt{2})(\sqrt{3})(\sqrt{5}) : \mathbb{Q}(\sqrt{2})(\sqrt{3})] \\
&\quad\quad\quad \times [\mathbb{Q}(\sqrt{2})(\sqrt{3})(\sqrt{5})(\sqrt{7}) : \mathbb{Q}(\sqrt{2})(\sqrt{3})(\sqrt{5})] \\
&= 2 \times 2 \times 2 \times 2 \\
&= 2^4 \\
&= 16
\end{aligned}
$$

"I think it works something like this, Tetra." I started sketching out a quick diagram.

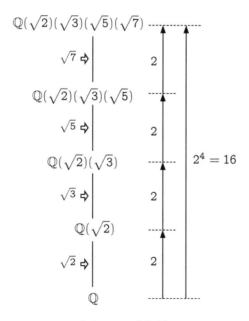

A tower of fields.

"That's right! That's exactly right!" Tetra said.

"It's just the image I got when you called this a 'tower of fields.' It really does look like a tower, built up one adjunction at a time."

$$\mathbb{Q} \subset \mathbb{Q}(\sqrt{2}) \subset \mathbb{Q}(\sqrt{2})(\sqrt{3}) \subset \mathbb{Q}(\sqrt{2})(\sqrt{3})(\sqrt{5}) \subset \mathbb{Q}(\sqrt{2})(\sqrt{3})(\sqrt{5})(\sqrt{7})$$

"It sure does! Math is so much fun..."

"Sorry, didn't mean to change the subject there."

"No worries. Um, right, so anyway $[\mathbb{Q}(\sqrt{2}, \sqrt{3}, \sqrt{5}, \sqrt{7}) : \mathbb{Q}]$ looks like this."

$$[\mathbb{Q}(\sqrt{2}, \sqrt{3}, \sqrt{5}, \sqrt{7}) : \mathbb{Q}] = 2^4 = 16$$

"Right."

"When I saw this I thought, oh, so adjunction of four numbers would result in a degree of 2^4, just like $\mathbb{Q}(\sqrt{2}, \sqrt{3})/\mathbb{Q}$ is an adjunction of two numbers, so the degree is 2^2. But again, I was wrong. Just flipping through a few pages in the book showed me that."

"Oh yeah?"

"Yep. I found an example where adjunction of just one number resulted in an extension of degree 4!"

8.2.8 $\mathbb{Q}(\sqrt{2} + \sqrt{3})/\mathbb{Q}$

Problem 8-3 (Degree of a field extension)

Find the degree of the extension $\mathbb{Q}(\sqrt{2} + \sqrt{3})/\mathbb{Q}$.

"The degree of $\mathbb{Q}(\sqrt{2} + \sqrt{3})/\mathbb{Q}$ is 4?" I asked. "Seriously?"

"Seriously! So maybe the problem should read 'Prove that $[\mathbb{Q}(\sqrt{2} + \sqrt{3}) : \mathbb{Q}] = 4$' instead."

"Okay, this is definitely getting interesting. But still, I think this can be solved in a similar way, working from the definition of the degree of a field extension. Specifically, we consider the field $\mathbb{Q}(\sqrt{2} + \sqrt{3})$ to be a vector space over $\mathbb{Q}$, find a basis, count the number of elements in the basis—"

"Well then, let me rephrase the problem once again: how do we find a basis for $\mathbb{Q}(\sqrt{2} + \sqrt{3})$?"

"Yeah, I guess that's where things get sticky."

"I did spend a day trying to figure this out, but in the end I went back to the books. You'll never believe what I found."

$\mathbb{Q}(\sqrt{2} + \sqrt{3})$
$= \{p + q(\sqrt{2} + \sqrt{3}) + r(\sqrt{2} + \sqrt{3})^2 + s(\sqrt{2} + \sqrt{3})^3 \mid p, q, r, s \in \mathbb{Q}\}$

"Huh. So I guess the basis would be this."

$$\{1,\ \sqrt{2} + \sqrt{3},\ (\sqrt{2} + \sqrt{3})^2,\ (\sqrt{2} + \sqrt{3})^3\}$$

"Four elements in the basis, so the degree is 4," I said.
"That's right!"

Answer 8-3 (Degree of a field extension)

$$[\mathbb{Q}(\sqrt{2} + \sqrt{3}) : \mathbb{Q}] = 4$$

"This basis that was in the book, can I assume it was created as a series of powers, like this?"

$$\{(\sqrt{2} + \sqrt{3})^0,\ (\sqrt{2} + \sqrt{3})^1,\ (\sqrt{2} + \sqrt{3})^2,\ (\sqrt{2} + \sqrt{3})^3\}$$

"You can. Creating the basis in this way allows us to generalize them. Let's see..." Tetra flipped through the pages of her notebook. "Ah. When we consider a field extension $\mathbb{Q}(\theta)/\mathbb{Q}$ to be a vector space over $\mathbb{Q}$, the basis is this."

$$\{1, \theta, \theta^2, \theta^3, \ldots, \theta^{n-1}\}$$

Tetra cocked her head. "I'm not sure why they used thetas, though. This doesn't have anything to do with angles."

"Something's off here," I said. "Are you sure θ can be any number at all? Did you copy the proof in your notes?" I craned my neck to peek at her notebook.

"Not the whole proof, no."

"Uh oh, Tetra. Looks like you skipped over a condition on what you can adjoin to the field."

"I did?"

> **(from Tetra's notes)**
>
> For a complex number θ, a polynomial $p(x)$ with rational coefficients is called the *minimal polynomial* of θ over $\mathbb{Q}$ if it meets the following conditions:
>
> · θ is a root of $p(x)$.
>
> · The coefficient of the highest-degree term in $p(x)$ is 1.
>
> · There is no polynomial of lower degree than $p(x)$ with rational coefficients having θ as a root.
>
> If such a polynomial exists and has degree n, then when considering the field extension $\mathbb{Q}(\theta)/\mathbb{Q}$ as a vector space over $\mathbb{Q}$,
>
> $$\{1, \theta, \theta^2, \theta^3, \ldots, \theta^{n-1}\}$$
>
> is a basis for that vector space.

"I guess you kinda missed this definition of minimal polynomials?" I asked.

"I did, yeah. To tell the truth, I'm still not entirely sure what it says. Not to say that's a good excuse..."

"Don't worry," I said. "I'm not entirely sure what it means either."

8.2.9 Minimal Polynomials

"So how can we parse all this?" Tetra asked. "These three conditions make it seem so hard..."

"Oh, I don't know," I said, taking some time to look carefully at what Tetra had written in her notebook. After a time I said, "Yeah, it isn't so bad if you read closely. Look at the first condition."

· $p(x)$ has θ as a root.

"This just means we're thinking about a polynomial for which $p(\theta) = 0$. It says we're focusing on a polynomial for a specific θ."

"That makes sense. If θ is a root of $p(x)$, then of course $p(\theta) = 0$. Okay, what about the other conditions?"

"I think the second condition, in combination with the first, is to narrow things down to one specific polynomial."

- $p(x)$ has θ as a root.
- The coefficient of the nth-degree term in $p(x)$ is 1.

"How does this limit us to one specific polynomial?" Tetra asked.

"Well, there are infinitely many polynomials with θ as a root, right?"

"There are?"

"Sure. For example, $x^2 - 2$ has $\sqrt{2}$ as root, but so does any multiple of that. Like, I can multiply it by 3 to get $3(x^2 - 2)$, but $\sqrt{2}$ will still be a root. I can even multiply it by another polynomial, like $x(x^2 - 2)$ or $(x^2 + x + 1)(x^2 - 2)$. Again, the result will still have $\sqrt{2}$ as a root."

"Ah, of course."

"Okay, let's add in the last condition for a minimal polynomial."

- $p(x)$ has θ as a root.
- The coefficient of the nth-degree term in $p(x)$ is 1.
- There exists no polynomial of degree less than n with rational coefficients having θ as a root.

"I think this last condition is to uniquely determine the polynomial with θ as a root," I said. "Yeah, it's saying that minimal polynomials are unique."

"It sounds like that would need its own proof."

"Sure. Anything without a proof is just conjecture. I'm sure the book you got this from had something to say about uniqueness."

"I'll be sure to check," Tetra said. "I like how you're able to read this and just know what else must have been in the textbook. It's like... you're making a preemptive strike!"

"Well I'm not exactly declaring war on the problem, but how about if we try to concretely find the minimal polynomial for $\sqrt{2} + \sqrt{3}$?"

"Hold up. Before we do that, I have a question about this part."

If such a polynomial exists, then when considering the field extension $\mathbb{Q}(\theta)/\mathbb{Q}$ as a vector space over $\mathbb{Q}$...

"Why do they go out of their way to specify that this is a vector space 'over $\mathbb{Q}$'?"

"Hmm, good point. No unimportant notes in a score..."

"Huh?"

"Something that Ay-Ay said." I told her about the conversation on music theory that Ay-Ay, Miruka, and I had in the tea room.

"Oh, so you went to Ay-Ay's recital... with Miruka."

I noticed that the tone in Tetra's voice was suddenly lower.

"Uh, yeah. Music sounds so much more powerful in a concert hall than in the school's music room."

"I'm sure it does. Anyway, about this 'over $\mathbb{Q}$' thing..."

"Ah, right. I suspect it's there because when we think about minimal polynomials, we need to be aware of what coefficient field we're working in."

"So if the coefficient field changes, the minimal polynomial will change too?"

"Well of course!" I said, much louder than necessary.

"Eek!" Tetra shouted.

"Uh, sorry about that. But, like, the minimal polynomial for $\sqrt{2}$ over $\mathbb{Q}$ would be $x^2 - 2$, but over $\mathbb{Q}(\sqrt{2})$ it would be $x - \sqrt{2}$."

"Hey, now that you mention it, we had to think about the coefficient field when we factored $x^{12} - 1$. This is the same thing, right? Okay, I'm good. Let's find the minimal polynomial for $\sqrt{2}+\sqrt{3}$ over $\mathbb{Q}$!"

I stopped to think for a minute. *The root operations kind of get in the way if we use rationals for coefficients, so...*

"Ah, okay," I said. "I think we can just do the opposite of solving equations."

"Well please don't just do it all in your head!"

$$x = \sqrt{2} + \sqrt{3}$$ the number for which we want to find the minimal polynomial over $\mathbb{Q}$

$$x - \sqrt{3} = \sqrt{2}$$ prepare to remove the root operation on $\sqrt{2}$

$$(x - \sqrt{3})^2 = (\sqrt{2})^2$$ ⋆ square both sides

$$x^2 - 2\sqrt{3}x + 3 = 2$$ expand both sides

$$x^2 + 3 - 2 = 2\sqrt{3}x$$ prepare to remove the root operation on $\sqrt{3}$

$$x^2 + 1 = 2\sqrt{3}x$$ calculate

$$(x^2 + 1)^2 = (2\sqrt{3}x)^2$$ ⋆⋆ square both sides

$$x^4 + 2x^2 + 1 = 4 \cdot 3x^2$$ expand both sides

$$x^4 + 2x^2 - 12x^2 + 1 = 0$$ calculate

$$x^4 - 10x^2 + 1 = 0$$ clean up

"So the minimal polynomial for $\sqrt{2}+\sqrt{3}$ over $\mathbb{Q}$ is x^4-10x^2+1?" Tetra asked.

"Should be. I'd feel better if we could prove it, though. The coefficient on x^4 is 1, so we're good there, but we need to show that polynomials with rational coefficients having $\sqrt{2} + \sqrt{3}$ as a root cannot be lower than fourth order."

"Right. By the way, what's up with the lines you marked with stars?"

"Oh, those are lines where the transformation only works one way. Like, the first one goes like this."

$$x - \sqrt{3} = \sqrt{2}$$
$$\Downarrow \not\Uparrow$$
$$(x - \sqrt{3})^2 = (\sqrt{2})^2$$

"This works from top to bottom, but not the other way around," I said.

"Because we're squaring here, right?"

"Right. When we square both sides of $x - \sqrt{3} = \sqrt{2}$, we get two solutions mixed in there, $x - \sqrt{3} = \pm\sqrt{2}$. Then we would have gone from a polynomial that has *only* $\sqrt{2}+\sqrt{3}$ as a root to one that has *two* roots, $\sqrt{2}+\sqrt{3}$ and $-\sqrt{2}+\sqrt{3}$."

"I see. And the same thing is happening in the line with two stars, right? We're squaring both sides of $x^2+1 = 2\sqrt{3}x$, which means both $x^2+1 = \pm 2\sqrt{3}x$."

"That's right. The equation $x^2+1 = 2\sqrt{3}x$ has two roots, $\sqrt{2}+\sqrt{3}$ and $-\sqrt{2}+\sqrt{3}$. When we square both sides, we bring in two more roots for $x^2+1 = -2\sqrt{3}x$, namely $\sqrt{2}-\sqrt{3}$ and $-\sqrt{2}-\sqrt{3}$. So in the end, the polynomial $x^4 - 10x^2 + 1$ has four roots."

$$+\sqrt{2}+\sqrt{3},\ -\sqrt{2}+\sqrt{3},\ +\sqrt{2}-\sqrt{3},\ -\sqrt{2}-\sqrt{3}$$

"I get it!" Tetra said. "And the factorization would look like this!"

$$x^4 - 10x^2 + 1$$
$$= \left(x - (+\sqrt{2}+\sqrt{3})\right)\left(x - (-\sqrt{2}+\sqrt{3})\right)$$
$$\left(x - (+\sqrt{2}-\sqrt{3})\right)\left(x - (-\sqrt{2}-\sqrt{3})\right)$$

"Yeah, but it's little cleaner to get rid of those inner parentheses."

$$x^4 - 10x^2 + 1$$
$$= (x - \sqrt{2} - \sqrt{3})(x + \sqrt{2} - \sqrt{3})(x - \sqrt{2} + \sqrt{3})(x + \sqrt{2} + \sqrt{3})$$

"I guess you can write it that way if you want," Tetra said, "but I prefer the parentheses. They make the roots stand out better. That's the message I'm sending through *my* equation!"

8.2.10 A Discovery?

"I can see that," I said. "Okay, so where do we want to go next?"

"Well, now I understand how the minimal polynomial for $\sqrt{2}+\sqrt{3}$ over $\mathbb{Q}$ is $x^4 - 10x^2 + 1$, but frankly... this still feels like another 'so what?'"

"Before we found the basis we saw that the degree of $\mathbb{Q}(\sqrt{2}+\sqrt{3})/\mathbb{Q}$ is 4."

"I guess... But I still don't completely understand." Tetra stopped and bit her lip. "I mean, I know what we're doing. We want the degree of the minimal polynomial for θ to find the degree of $\mathbb{Q}(\theta)/\mathbb{Q}$... Oh, that's interesting, how the word 'degree' shows up in both cases. The *degree* of the minimal polynomial equals the *degree* of the field extension... Huh."

I chuckled. "Getting sidetracked by the vocabulary again?"

"No, wait! This is a huge discovery!"

"Yeah? What have you discovered?"

"Four cows!"

"What on earth...?"

"The yoke! These four numbers..."

$$+\sqrt{2}+\sqrt{3},\ -\sqrt{2}+\sqrt{3},\ +\sqrt{2}-\sqrt{3},\ -\sqrt{2}-\sqrt{3}$$

"...are all yoked together by the equation $x^4 - 10x^2 + 1 = 0$! So these four numbers must be conjugates, right? I may have made my first mathematical discovery!" Tetra's cheeks flushed crimson.

"I don't know that—"

"Something like this: 'All conjugates of $\sqrt{2}+\sqrt{3}$ must be elements in $\mathbb{Q}(\sqrt{2}+\sqrt{3})$, the adjunction of $\sqrt{2}+\sqrt{3}$ to $\mathbb{Q}$.' That would mean we could write these four numbers like this."

$$+\sqrt{2}+\sqrt{3} \in \mathbb{Q}(\sqrt{2}+\sqrt{3})$$
$$-\sqrt{2}+\sqrt{3} \in \mathbb{Q}(\sqrt{2}+\sqrt{3})$$
$$+\sqrt{2}-\sqrt{3} \in \mathbb{Q}(\sqrt{2}+\sqrt{3})$$
$$-\sqrt{2}-\sqrt{3} \in \mathbb{Q}(\sqrt{2}+\sqrt{3})$$

"Yeah... maybe?" But something about this struck me as not quite right.

"It's got to be true." Tetra started biting a fingernail as she thought. "Yeah, this works. Because we can use rational numbers p, q, r, s to write elements of $\mathbb{Q}(\sqrt{2}+\sqrt{3})$ in the form $p +$

$q(\sqrt{2}+\sqrt{3})+r(\sqrt{2}+\sqrt{3})^2+s(\sqrt{2}+\sqrt{3})^3$. So check it out."

$$p+q(\sqrt{2}+\sqrt{3})+r(\sqrt{2}+\sqrt{3})^2+s(\sqrt{2}+\sqrt{3})^3$$
$$=p+q(\sqrt{2}+\sqrt{3})+r\left((\sqrt{2})^2+2\sqrt{2}\sqrt{3}+(\sqrt{3})^2\right)$$
$$+s\left((\sqrt{2})^3+3(\sqrt{2})^2\sqrt{3}+3\sqrt{2}(\sqrt{3})^2+(\sqrt{3})^3\right)$$
$$=p+(q\sqrt{2}+q\sqrt{3})+(2r+2r\sqrt{6}+3r)$$
$$+(2s\sqrt{2}+6s\sqrt{3}+9s\sqrt{2}+3s\sqrt{3})$$
$$=\underbrace{(p+5r)}_{\in\mathbb{Q}}+\underbrace{(q+11s)}_{\in\mathbb{Q}}\sqrt{2}+\underbrace{(q+9s)}_{\in\mathbb{Q}}\sqrt{3}+\underbrace{2r}_{\in\mathbb{Q}}\sqrt{6}$$

"See? Letting $\mathbb{Q}(\sqrt{2}+\sqrt{3})$ be a vector space over $\mathbb{Q}$, we can take the basis $\{1, \sqrt{2}, \sqrt{3}, \sqrt{6}\}$. So..."

$$\mathbb{Q}(\sqrt{2}+\sqrt{3}) = \mathbb{Q}(\sqrt{2}, \sqrt{3}, \sqrt{6}) = \mathbb{Q}(\sqrt{2}, \sqrt{3})$$

"$\mathbb{Q}(\sqrt{2}, \sqrt{3})$ is the adjunction of $\sqrt{2}$ and $\sqrt{3}$ to $\mathbb{Q}$, so $+\sqrt{2}+\sqrt{3}$ and $-\sqrt{2}+\sqrt{3}$ and $+\sqrt{2}-\sqrt{3}$ and $-\sqrt{2}-\sqrt{3}$ are all elements in $\mathbb{Q}(\sqrt{2}, \sqrt{3})$. That means all four of those cows are yoked by $\mathbb{Q}(\sqrt{2}+\sqrt{3})$!" While listening to Tetra, I had been running a different calculation in my head. She was speaking increasingly loudly, however, making that difficult. "Oh, I get it! I see why we want to think about minimal polynomials in field extensions! We aren't using *numbers* to extend fields, we're using *polynomials*! I'm sure that in general, when we consider the minimal polynomial $p(x)$ for θ, all numbers that are yoked together through that $p(x)$ must always belong to $\mathbb{Q}(\theta)$!" Tetra firmly grabbed my arm. "Conjugate numbers always belong to the same field extension! They aren't left all alone, they're always together, sharing the same yoke!"

"Um, I hate to tell you this, but..." I said, separating Tetra from my arm.

"Tell me what?"

"Well, if we set up your prediction as a problem, it would look like this, right?"

Building Towers 265

> **Problem 8-4 (Minimal polynomials and conjugate numbers)**
>
> Is the following proposition always true?
> Proposition: Letting p(x) be the minimal polynomial of θ over $\mathbb{Q}$, all roots of p(x) belong to the field extension $\mathbb{Q}(\theta)$.

"That's right! And I say it *is* true! Such a beautiful proposition..."

"It is, and you're amazing for having come up with it. You clearly know fields far better than I do. But... I've found a counterexample to your proposition."

"A... counterexample?"

"Well, I was thinking about the omega waltz..."

"The what?"

"Let me put that another way. You've been focusing on the $\sqrt{}$ operator in terms like $\sqrt{2}$ and $\sqrt{3}$ and $\sqrt{2} + \sqrt{3}$, right? So I was thinking about what would happen with cube roots, $\sqrt[3]{}$."

"Oh, okay."

"Well, like when we were talking about Lagrange resolvents, given a number L, the three numbers $L, L\omega, L\omega^2$ will be solutions to $x^3 - L^3 = 0$. Here, ω is one of the primitive cube roots of unity, so we'll let $\omega = \frac{-1+\sqrt{3}i}{2}$."

"I'm not sure where this is going."

"Well, let's take the cube root of 2 as an example. The minimal polynomial of $\sqrt[3]{2}$ over $\mathbb{Q}$ is $x^3 - 2$, so we'll look at $x^3 - 2 = 0$. The roots of this equation are this..."

$$\sqrt[3]{2}, \ \sqrt[3]{2}\omega, \ \sqrt[3]{2}\omega^2$$

"...so these are the three conjugates for $\sqrt[3]{2}$. Your three yoked cows, as it were." Tetra was looking increasingly uneasy, but remained silent. "Okay, so think about $\mathbb{Q}(\sqrt[3]{2})$, the adjunction of $\sqrt[3]{2}$ to the field $\mathbb{Q}$. Of course, $\sqrt[3]{2}$ is an element of $\mathbb{Q}(\sqrt[3]{2})$. However, the other two cows in the yoke, $\sqrt[3]{2}\omega$ and $\sqrt[3]{2}\omega^2$, are *not*."

$$\sqrt[3]{2} \in \mathbb{Q}(\sqrt[3]{2})$$
$$\sqrt[3]{2}\omega \notin \mathbb{Q}(\sqrt[3]{2})$$
$$\sqrt[3]{2}\omega^2 \notin \mathbb{Q}(\sqrt[3]{2})$$

"These are counterexamples to your proposition, Tetra."

"Well not so fast, we can't be sure yet! Let me do some calculations to make sure..."

"Actually, there's no need. We can see this straight away. $\sqrt[3]{2}$ is a real number, so all elements in $\mathbb{Q}(\sqrt[3]{2})$ are real numbers. But $\sqrt[3]{2}\omega$ and $\sqrt[3]{2}\omega^2$ aren't real numbers. They'll always include the imaginary unit i."

$$\begin{cases} \sqrt[3]{2}\omega &= \sqrt[3]{2} \cdot \dfrac{-1+\sqrt{3}i}{2} = -\dfrac{\sqrt[3]{2}}{2} + \dfrac{\sqrt[3]{2}\sqrt{3}}{2}i \notin \mathbb{R} \\ \sqrt[3]{2}\omega^2 &= \sqrt[3]{2} \cdot \dfrac{-1-\sqrt{3}i}{2} = -\dfrac{\sqrt[3]{2}}{2} - \dfrac{\sqrt[3]{2}\sqrt{3}}{2}i \notin \mathbb{R} \end{cases}$$

Tetra now had a very different expression from just a few minutes ago. "Oh."

"You see how $\sqrt[3]{2}\omega$ and $\sqrt[3]{2}\omega^2$ can't be in $\mathbb{Q}(\sqrt[3]{2})$, since they aren't real numbers, right?"

Tetra's lips twisted. "I'm such a fool."

"You're doing great, you just—"

"No, I'm foolish and embarrassed, getting all worked up like that. Carrying on about how I'd made some great discovery, while all I did was interrupt your studies."

"Tetra, I—"

"I'm sorry, I think I should go."

Tetra swept her notes into her bag, nodded goodbye, and swiftly left the lounge, leaving me speechless and alone.

> **Answer 8-4 (Minimal polynomials and conjugate numbers)**
>
> The following proposition is *not* always true.
> Proposition: Letting $p(x)$ be the minimal polynomial of θ over $\mathbb{Q}$, all roots of $p(x)$ belong to the field extension $\mathbb{Q}(\theta)$.
> (One counterexample is $\theta = \sqrt[3]{2}, p(x) = x^3 - 2$.)

8.3 A Letter

8.3.1 Going Home

Walking home, I found myself angry. *It isn't my fault*, I thought. *She's the one who dragged me to the student lounge.*

I'd been going to the library every day to study for college entrance exams, but today (again) I'd ended up spending most of the afternoon talking with Tetra. My plans for vacation had been to spend mornings at my summer seminar, afternoons studying at the library, and evenings studying at home. Now summer vacation was already halfway over with, but...

I need to stop going to the library. A senior in high school should spend his summer vacation studying alone. It's not my fault.

I mulled over how Tetra was uncharacteristically taking a poor attitude toward mathematics. Thinking hard about a problem, only to find out you've been wasting your time, is something that happens all the time. You could even say that repeated mistakes are how you learn when studying math on your own.

Tetra had come up with a hypothesis—one worth looking into—but I had provided a counterexample disproving it (and causing her to run out in a snit, no less). Counterexamples are powerful things. A single counterexample is enough to take down the most grandiose of mathematical claims. Tetra had gotten too caught up in words. I could see the appeal in her "sharing a yoke" concept, but math doesn't necessarily work the way you want it to. More to the point...

Why is her being upset making me so upset?

8.3.2 At Home

"I'm home!" I called out.

"Welcome back," my mother answered, coming to the door in an apron. She held out an envelope. "For you."

I took it, expecting school-related junk mail, but it turned out to be a plain white envelope, nothing written on either side.

"What's this?" I asked.

"No idea," Mom said, heading back into the kitchen.

Perplexed, I opened the envelope, releasing a familiar scent. *Citrus*...

8.3.3 The Letter

The letter had no salutation, but from a glance I could see it contained plenty of math.

I sighed. I'd spent the afternoon getting schooled by Tetra, and now that I was home I was going to be schooled by Miruka. *In a letter, no less!*

Resigned, I sat back to read what she'd written.

> Lisa told me I missed you and Yuri on the day of the planning committee meeting for the Galois Festival. Looks like you aren't coming to the school library today, so I'm writing this letter.
>
> $\frac{\pi}{3}$ is a counterexample to the angle trisection problem. When you explain this to Yuri, it might be easier for her if you use $60°$ instead of $\frac{\pi}{3}$. You've probably shown her a proof that $20°$ isn't constructible, but I'm going to give you an overview of how to use degrees of field extensions to show the same thing, so that you two can better enjoy the Festival.

8.3.4 Constructible Numbers

I looked up from the letter. Miruka was right: I had already used proof by mathematical induction to show Yuri that a $60°$ angle couldn't be trisected. *But a proof using degrees of extensions? Seriously?*

I continued reading Miruka's letter.

Starting from two points $(0,0)$ and $(0,1)$ is a restricted form of geometric constructions. It's more common to start from something other than just two points. By adding 0 and 1 you can form $\mathbb{Q}$ through arithmetic, so for a more general construction problem, start from a field resulting from adjunction of some number to $\mathbb{Q}$, and repeatedly perform quadratic extensions.

As you know, figures that are constructible using a straightedge and compass are those involving points having constructible numbers as coordinates. Further, constructible numbers are 0, 1, and any other number that can be created from those through finite repetition of arithmetic and square root operations.

Let's write that using mathematical expressions. The necessary and sufficient conditions for α to be a constructible number are the existence of an integer n and a sequence of real numbers $\sqrt{\alpha_0}, \sqrt{\alpha_1}, \sqrt{\alpha_2}, \ldots, \sqrt{\alpha_{n-1}}$ satisfying the following conditions:

- $K_0 = \mathbb{Q}$
- $K_{k+1} = K_k(\sqrt{\alpha_k}) \quad \sqrt{\alpha_k} \notin K_k, \alpha_k \in K_k \quad (k = 0, 1, 2, \ldots, n-1)$
- $\alpha \in K_n$

This implies the existence of a "tower of fields" like the following.

$$\mathbb{Q} = K_0 \subset K_1 \subset K_2 \subset \cdots \subset K_{n-1} \subset K_n \quad \text{and} \quad \alpha \in K_n$$

Here, we're starting from $\mathbb{Q}$, then creating field extensions by adjunction of $\sqrt{\alpha_k}$ until we arrive at some field K_n that contains α.

We want to use the properties of this tower of fields to give a proof for the angle trisection problem. Thankfully, degrees of field extensions give us everything we need. Each floor of the tower is a field extension— K_{k+1}/K_k, in other words $K_k(\sqrt{\alpha_k})/K_k$.

$$[K_{k+1} : K_k] = [K_k(\sqrt{\alpha_k}) : K_k]$$
$$= 2$$

This means that if α is a constructible number, the degree of $\mathbb{Q}(\alpha)/\mathbb{Q}$ will be 2^n.

$[\mathbb{Q}(\alpha) : \mathbb{Q}]$
$= [K_n : K_0]$
$= [K_1 : K_0] \times [K_2 : K_1] \times \cdots \times [K_n : K_{n-1}]$
$= \underbrace{[K_0(\sqrt{\alpha_0}) : K_0]}_{2} \times \underbrace{[K_1(\sqrt{\alpha_1}) : K_1]}_{2} \times \cdots \times \underbrace{[K_{n-1}(\sqrt{\alpha_{n-1}}) : K_{n-1}]}_{2}$

$\underbrace{\qquad\qquad\qquad\qquad\qquad\qquad\qquad\qquad\qquad}_{n \text{ of these}}$

$= 2^n$

In other words, if α is a constructible number, $\mathbb{Q}(\alpha)/\mathbb{Q}$ is an extension of degree 2^n.

$$[\mathbb{Q}(\alpha) : \mathbb{Q}] = 2^n$$

Saying that one-third of a 60° angle is a constructible number is the same as saying that $\cos 20°$ is a constructible number. However, the minimal polynomial of $2\cos 20°$ over $\mathbb{Q}$ is $x^3 - 3x - 1$, showing that the minimal polynomial of $\cos 20°$ over $\mathbb{Q}$ is a third-degree equation, $x^3 - \frac{3}{4}x - \frac{1}{8}$. Thus, $\mathbb{Q}(\cos 20°)/\mathbb{Q}$ is an extension of degree 3.

$$[\mathbb{Q}(\cos 20°) : \mathbb{Q}] = 3$$

Needless to say, there is no integer $n \geqslant 0$ for which $2^n = 3$, since 3 is not 1, nor is it an even number. So $\cos 20°$ is not a constructible number, meaning you cannot use a straightedge and compass to create a $\cos 20°$ angle, and thus that a 60° angle cannot be trisected using straightedge and compass. So we have 60° as a counterexample for the angle trisection problem, giving us our proof.

Done and done.

8.3.5 Dinner

"Time for dinner!" my mother called out. There was more to Miruka's letter, but I reluctantly set it aside and headed for the dining room.

I ate in a daze, Miruka's letter taking up the better part of my mind. *The degree of a field extension, which we defined using the dimension of a vector space, is equal to the degree of a minimal polynomial. A problem that had started as a geometric construction became a problem in algebra, by way of equations and trigonometric functions and even a dash of number theory. So many topics in mathematics, coming together. Such fun, to see something like "because 3 isn't an even number" pop up in a proof that a $60°$ angle can't be trisected!*

Reading Miruka's letter, it had almost been like she was whispering in my ear. I didn't just hear her. I smelled her citrus scent. I saw her "are you getting this?" expression, and the way she sometimes looked away in embarrassment when I praised her. I could feel her presence the entire time.

After dinner I rushed back to my room.

8.3.6 Solvability of Equations

Further surprises awaited me when I got back to Miruka's letter, which took a sudden turn.

> There is no general formula for the solutions to a fifth-degree equation. The reasons for this are very similar to the reasons why not all angles can be trisected. Namely, there are many structural similarities, strong parallels between these two problems:
> · It not possible to trisect all angles as a geometric construction.
> · It not possible to algebraically solve all fifth-degree equations.
> By the way, it was Dr. Narabikura who decided the overall program for the Galois Festival. Lisa will be the general manager, and she's assigning jobs to the math fans who hang out at the Narabikura Library. I'm refereeing submissions for presentations, under Dr. Narabikura's supervision. Yuri's boyfriend will give a talk about angle trisections. It is precisely because of these strong parallels that we'll be covering construction problems at the Festival.

So let's summarize those parallels.

—*A finite number of restricted operations*
To consider the angle trisection problem, we have to start out by clearly stating exactly what a geometric construction is: "creating a figure using a finite number of operations with a straightedge and compass." Similarly, if we want to think about generalized solutions to fifth-degree equations, we have to be clear about what it means to algebraically solve an equation. Here, we define that as "using a finite number of arithmetic and nth root operations to express the roots of a polynomial."

—*Generalization and special cases*
Not all angles can be trisected using a straightedge and compass, but some can. Similarly, while there exists no general formula that can derive the roots of *every* fifth-degree polynomial, some such equations can be algebraically solved.

—*Existence and constructability*
There exists a trisection of any angle, even if that trisection cannot be derived through geometric construction. Similarly, every fifth-degree polynomial has roots, we just can't necessarily find them algebraically.

—*A tower of fields*
Both of these problems build a tower of fields. In other words, both can be considered as field extension problems. The difference is the direction they take after building that tower. The angle trisection problem is solved by investigating the "size" represented by degrees of extension. However, that isn't enough to address the solvability of fifth-degree polynomials.

8.3.7 Splitting Fields

I was torn between stopping to think deeper about what Miruka had written, or reading on. In the end, I found myself unable to stop.

Let's think a little more about solving equations and field extensions.

We start by adjunction of an nth root to the coefficient field, and set the resulting field as the second floor in our tower. The question then becomes, how far do we have to take those extensions before we can say we've solved the equation? The answer: until we can factor the polynomial into a product of linear expressions. The smallest field that allows such a factorization will be the field resulting from adjunction of all the polynomial's roots to its coefficient field. This is called the "splitting field" for the polynomial.

Algebraically solving a polynomial equation means starting from the coefficient field resulting from adjoining the equation's coefficients to the field of rational numbers, then adjoining various roots to create a tower of fields until we have a field that includes the splitting field. If you can build such a tower for a given polynomial, it can be algebraically solved. If no such tower can be built, it cannot.

Well then, for what kind of polynomial can we build such a tower?

8.3.8 Normal Extensions

Miruka's letter continued.

Let's look at an example of adjunction and factorization. Just like Tetra had fun with $x^{12} - 1$, we'll have some fun with $x^3 - 2$, a frequently used example.

The polynomial $x^3 - 2$ cannot be factored over the field $\mathbb{Q}$, so we call it an irreducible polynomial. It is, however, reducible over the field $\mathbb{Q}(\sqrt[3]{2})$. Specifically, we can factor it into two polynomials, like this:

$$x^3 - 2 = (x - \sqrt[3]{2})(x^2 + \sqrt[3]{2}x + \sqrt[3]{4}) \quad \text{factorization over } \mathbb{Q}(\sqrt[3]{2})$$

The polynomial $x^2 + \sqrt[3]{2}x + \sqrt[3]{4}$ is irreducible over the field $\mathbb{Q}(\sqrt[3]{2})$, so we can't factor it any further. It is reducible over the field $\mathbb{Q}(\sqrt[3]{2}, \omega)$, however, all the way down to a product of linear polynomials.

$x^3 - 2$ irreducible over $\mathbb{Q}$
$= (x - \sqrt[3]{2})(x^2 + \sqrt[3]{2}x + \sqrt[3]{4})$ irreducible over $\mathbb{Q}(\sqrt[3]{2})$
$= (x - \sqrt[3]{2})(x - \sqrt[3]{2}\omega)(x - \sqrt[3]{2}\omega^2)$ irreducible over $\mathbb{Q}(\sqrt[3]{2}, \omega)$

In fact, the field extension $\mathbb{Q}(\sqrt[3]{2}, \omega)/\mathbb{Q}$ has the following property:

> For a given number $\alpha \in \mathbb{Q}(\sqrt[3]{2}, \omega)$, the minimal polynomial of α over $\mathbb{Q}$ can be factored into a product of linear polynomials over $\mathbb{Q}(\sqrt[3]{2}, \omega)$.

In other words, $\mathbb{Q}(\sqrt[3]{2}, \omega)$ contains all numbers that are conjugates of α. An extension like this is called a *normal extension*.

- $\mathbb{Q}(\sqrt[3]{2})/\mathbb{Q}$ is *not* a normal extension.
- $\mathbb{Q}(\sqrt[3]{2}, \omega)/\mathbb{Q}$ *is* a normal extension.

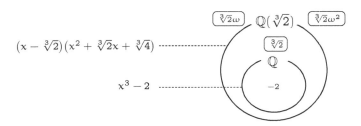

$\mathbb{Q}(\sqrt[3]{2})/\mathbb{Q}$ is not a normal extension.

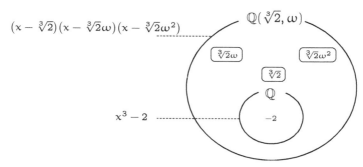

$\mathbb{Q}(\sqrt[3]{2}, \omega)/\mathbb{Q}$ is a normal extension.

$\mathbb{Q}(\sqrt[3]{2}, \omega)$ is equivalent to the field $\mathbb{Q}(\sqrt[3]{2}, \sqrt[3]{2}\omega, \sqrt[3]{2}\omega^2)$, created by adjunction of all roots of $x^3 - 2$ to $\mathbb{Q}$.

$$\mathbb{Q}(\sqrt[3]{2}, \omega) = \mathbb{Q}(\sqrt[3]{2}, \sqrt[3]{2}\omega, \sqrt[3]{2}\omega^2)$$

Generally, a field extension L/K meeting the following requirement is a normal extension:

> For a given number $\alpha \in L$, the minimal polynomial of α over K can be reduced to a product of linear polynomials over L.

Note that we're assuming a field extension with finite degree. We can rephrase the definition for normal extensions like this:

> A field extension for which inclusion of one root of its minimal polynomial implies inclusion of all roots is a *normal extension*.

As Tetra might put it, a normal extension is one that's "neat and tidy," because there's a symmetry to the roots of its minimal polynomial. This neat-and-tidiness makes us want to look into not only the scale of the extension, but also its form. Luckily, we have an excellent tool for mathematical investigations of form: *groups*.

So when thinking about the solvability of polynomial equations, let's build two towers—a *tower of fields* and a *tower of groups*. There is a close correspondence between these towers, one so central to Galois theory that it's named the *Galois correspondence*. With this, we've gotten much closer to the inheritance that Évariste Galois bequeathed to us.

Uh-oh, Ms. Mizutani is on the prowl—time to go. Done and done. For now...

8.3.9 Tackling Truth

Having made it through Miruka's letter, I sat back and took a deep breath.

Wow. Just... wow.

There were big chunks I still didn't understand, and I knew Miruka had skipped over plenty of details. Even so...

Wow.

Field extensions. Towers of fields. Minimal polynomials. Splitting fields. Normal extensions... And now towers of fields and towers of groups, and some correspondence between the two? I was so intrigued I felt dizzy.

I recalled Tetra's "discovery" from earlier in the day.

> Tetra's (incorrect) hypothesis: "If one root of a minimal polynomial belongs to an extension field, then all roots will."

She'd been mistaken at the time, but she really had made a major discovery: "neat and tidy" normal extensions!

> (Alternative) definition for normal extensions: "A field extension for which inclusion of one root of a minimal polynomial implies inclusion of all roots is a *normal extension*."

So what Tetra had actually discovered was a concept, one so important that it already had a name. Had Miruka been there with us, she would have praised Tetra for having discovered that concept before knowing what it's called. In contrast, I had been so preoccupied with looking for counterexamples that I hadn't noticed the significance of what she had found.

I thought about the friends I had made. Miruka, who had written me this letter. Tetra, who had taught me new things leading up to normal extensions... These were irreplaceable encounters in my life, ones that I should never take for granted, no matter what. Even when I considered them to be in the wrong. A partner with whom I could enjoy math was something to treasure.

Together, we were attempting to tackle the great truths of mathematics. I would need all the allies I could find for that endeavor, so I couldn't allow my own pettiness to get in the way of our friendship. No way. There were limits to the time we could study together, and the day when we were forced apart would soon come...

I folded Miruka's letter and started to return it to its envelope. As I did so, I noticed that the envelope also included a small card.

The last committee meeting before the Galois Festival is next Friday, 10:00 AM at the library. Bring Yuri.

—Miruka

Extraordinary students filled the seats in his classroom. This was because he himself was an extraordinary teacher, one capable of perceiving his students' futures and infusing them with the bearing and culture most suited to their essence.

FROM AN OBITUARY FOR LOUIS RICHARD [11]

CHAPTER 9

The Form of Feelings

> What makes a drawing more than just its separate lines?
>
> MARVIN MINSKY
> *The Society of Mind*

9.1 THE SYMMETRIC GROUP S_3

9.1.1 At the Narabikura Library

"...and that's why I call these ploppers, swappers, and scramblers," Yuri said.

We were at the Narabikura Library, seated around an oval table in the conference room called "Beryllium." Yuri was giving Miruka, Tetra, and me a run-down of what she'd been learning. In stark contrast to her brashness when it was just the two of us, she was clearly nervous—possibly because Miruka was there.

It was the day before the Galois Festival. Tomorrow the general public would be coming to see the exhibits, so this was the last chance for preparations and the entire place was abuzz. I didn't really have a grasp on everything that was going on, but a dozen or so high school and college math lovers were at the library, getting ready. They were divided up into several committees, tasked with preparing posters and setting up signs.

Yuri's boyfriend hadn't even graduated junior high yet, but he was off somewhere too, preparing to host a breakout session related to the angle trisection problem. I had gone primarily with the intent of providing moral support, but at some point had gotten entangled in the management side of things. As soon as we had arrived, somewhere around ten o'clock in the morning, Miruka announced that she wanted Yuri to present what she'd been studying over the summer.

"I'm sure Lisa can find a place for Yuri's display while we whip up a poster," Miruka had said, a last-minute change that caused what passed for Lisa as a look of annoyance. She simply groused a bit, however, and headed off to make the necessary arrangements.

By around eleven we were working with Yuri to flesh out her presentation. Our plan was to do that until lunchtime, spend the afternoon making her poster, and get home by evening. *A nice change from studying for entrance exams*, I thought.

Yuri continued her explanation. "I figured giving these things names might make it easier to classify ladder diagrams. Well, I say ladder diagrams, but rather than thinking about the specifics of where the horizontal rungs are, I'm just thinking about how the 1, 2, 3 get lined up after they've passed down the ladder. Then we can just use those three numbers to represent any ladder diagram with three vertical lines. For example, we can use [2 1 3] to represent a swapper that swaps the first two numbers. Mmm, I'm not describing that very well. Here, look at this."

Yuri opened a notebook for us to see.

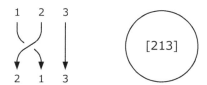

A "swapper" for the two leftmost elements.

"Very nice," Miruka said. "Continue."

"There are $3! = 6$ possible patterns for a ladder diagram with three vertical lines. Like this."

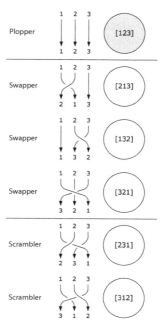

All possible patterns for a three-lined ladder diagram.

"Go on," Miruka said.

"Right. So I tried connecting lots of ladders together, and found out that for example if you put a $[2\,3\,1]$ under a $[2\,1\,3]$, you get a $[1\,3\,2]$, so it's like $[2\,1\,3] \star [2\,3\,1] = [1\,3\,2]$. Like this."

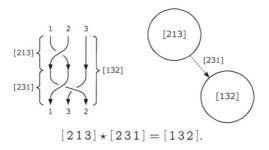

$[2\,1\,3] \star [2\,3\,1] = [1\,3\,2].$

"That looks familiar," I said. "Like with Tetra's equilateral triangles, and Miruka's subgroups."

"The set of all ladder diagrams with three vertical lines forms the symmetric group S_3," Tetra said.

"I guess," Yuri said, "but trying to draw the lines for all these graphs just made a big mess! I can't see the form of the whole symmetric group S_3, so I wanted some way to make it all simpler. To do that I selected one swapper and one scrambler—this [2 1 3] and [2 3 1]—and tried to draw it in a way that made everything as clear as possible. Here's what I came up with. The shaded part is the identity."

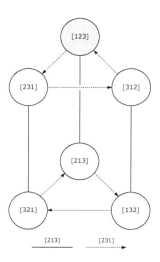

Simplification of the symmetric group S_3.

"Makes things pretty clear, right?" Yuri said. "Looks simple, I know, but this thing took a lot of work! Take the line for the [2 1 3] swapper, for example. It goes back and forth between two ladder diagrams, so it's just a simple line. The scrambler [2 3 1] has a direction, though, so I drew that as an arrow. And check it out! By doing all this I found out there's two groupings, an upper triangle and a lower triangle!"

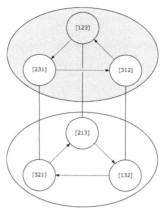

"Upper triangle" and "lower triangle" groupings.

Yuri frowned. "I was kinda disappointed to see that the arrows in the two triangles go in opposite directions, though, clockwise at the bottom and counterclockwise at the top. Would have been cooler if they both went the same way. Well, anyway, that's what I came up with for the symmetric group S_3."

"It's very interesting how those groupings appear like that, though," I said. "I wonder if that's some kind of structure?"

"Of course it is," Miruka said.

"But to be honest, that's about all I got," Yuri said. "This nice pattern of an upper triangle and a lower triangle, with arrows going in different directions."

"You've discovered Cayley graphs," Miruka said. "An excellent way to simplify and capture the overall structure of a group."

Yuri's eyes widened. "This thing has a name?"

"It does. And you've used it to find structure, and furthermore that the upper and lower triangular structures have the same form." Tetra and I nodded. Miruka pointed at me. "Think you can show this relation between the upper and lower triangles using mathematical expressions?"

"This? As equations? I...uh..." I stammered.

Miruka withdrew her finger. "I'll take that as a 'no.' Okay, let's talk a bit about this."

9.1.2 Classifications

Miruka approached the conference room's whiteboard. "The set of all ladder diagrams with three vertical lines is a third-degree symmetric group, so let's call it S_3."

$$S_3 = \{[1\,2\,3], [2\,3\,1], [3\,1\,2], [2\,1\,3], [3\,2\,1], [1\,3\,2]\}$$

"That's the same way you write the elements in a set, right?" Tetra asked.

"It is. The upper triangle Yuri found is a subgroup of S_3. It's a third-order cyclic group, so we'll name that one C_3."

$$C_3 = \{[1\,2\,3], [2\,3\,1], [3\,1\,2]\} \qquad \text{the upper triangle}$$

Yuri nodded. "Okay."

"Do you remember the definition of a cyclic group? It's one that's generated from a single element. For example, $[2\,3\,1]$ generates C_3 here, so we can write it like this."

$$C_3 = \langle [2\,3\,1] \rangle \qquad \text{"upper triangle" generated by } [2\,3\,1]$$

"That's because if you cube a scrambler, you get a plopper!" Yuri nearly shouted.

Miruka held up a finger. "Exactly. When you cube the $[2\,3\,1]$ that Yuri chose, you end up at the identity element. Therefore, $\langle [2\,3\,1] \rangle$ is a cyclic group of order 3."

"Great. I'm totally keeping up with everything you're saying!"

"Glad to hear it. Now that we've named the upper triangle C_3, let's name the lower triangle X_3."

$$C_3 = \{[1\,2\,3], [2\,3\,1], [3\,1\,2]\} \qquad \text{upper triangle}$$
$$X_3 = \{[2\,1\,3], [3\,2\,1], [1\,3\,2]\} \qquad \text{lower triangle}$$

"A problem for you all," Miruka said. "What kind of group is X_3?"

Huh? Miruka's question was met with at least ten seconds of silence.

"Looks like nobody's taking the bait," Miruka said.

"X_3 *isn't* a group... is it?" Tetra said.

"That's right. It doesn't have an identity element, so it can't be a group. Since it isn't a group, it also of course can't be a subgroup of S_3." I heard Yuri take a deep breath. "Even so," Miruka continued, "as you can see, C_3 and X_3 look very similar."

"They do," Tetra said. "They even spin around in kind of the same way."

Miruka began writing on the whiteboard. "The union of C_3 and X_3 is S_3, and the intersection of C_3 and X_3 is the null set."

$$\begin{cases} C_3 \cup X_3 = S_3 & \text{union of } C_3 \text{ and } X_3 \text{ is } S_3 \\ C_3 \cap X_3 = \{\} & \text{intersection of } C_3 \text{ and } X_3 \text{ is the null set} \end{cases}$$

"So when we bring C_3 and X_3 together we get the full group, and there are no common elements between the two. This means we have sorted all the elements in S_3 into one of C_3 or X_3 with no leaks and no dupes. A sorting like this is generally called a *classification*."

"A classification, got it," Yuri said.

"Okay, time for a new symbol."

$$C_3 \star [2\,1\,3]$$

"Oh, the star operator! I was using that!"

"To show the joining of one ladder diagram below another, yes. We're going to expand its meaning, though. Specifically, I want to use this not between ladder diagrams, but between a *set* of ladder diagrams and one specific ladder diagram.

Yuri's face clouded. "I'm not sure why—"

Miruka nodded and spoke calmly. "Let's talk about what this $C_3 \star [2\,1\,3]$ means. Every element in the set C_3 is a ladder diagram. Say we want to connect a $[2\,1\,3]$ beneath each ladder diagram in that set. We can represent that as $C_3 \star [2\,1\,3]$. This isn't something that's theoretically derived, we've just defined things this way. So let's take a detailed look at what kind of set $C_3 \star [2\,1\,3]$ gives us. I'm going to underline the $[2\,1\,3]$'s to make them stand out. The result is something like the distributive law."

$$\begin{aligned} C_3 \star \underline{[2\,1\,3]} &= \{\ [1\,2\,3],\ [2\,3\,1],\ [3\,1\,2]\ \} \star \underline{[2\,1\,3]} \\ &= \{\ [1\,2\,3] \star \underline{[2\,1\,3]},\ [2\,3\,1] \star \underline{[2\,1\,3]},\ [3\,1\,2] \star \underline{[2\,1\,3]}\ \} \\ &= \{\ [2\,1\,3],\ [3\,2\,1],\ [1\,3\,2]\ \} \end{aligned}$$

"So when we append a [2 1 3] beneath each element in C_3, we get [2 1 3], [3 2 1], [1 3 2]. Here's a graph of what's happening."

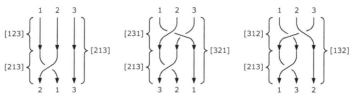

Calculation of $\{[1\,2\,3], [2\,3\,1], [3\,1\,2]\} \star [2\,1\,3]$.

"Notice how that set is the same as X_3, Yuri's 'bottom triangle'?"

$$C_3 \star [2\,1\,3] = \{[2\,1\,3], [3\,2\,1], [1\,3\,2]\}$$
$$X_3 = \{[2\,1\,3], [3\,2\,1], [1\,3\,2]\}$$

"So we can write the upper triangle as C_3 and the bottom triangle as $C_3 \star [2\,1\,3]$. In other words, the lower triangle is the set of all ladder diagrams resulting from appending [2 1 3] to each of the elements in the upper triangle. So here's what we've learned so far, writing S_3 in terms of C_3."

$$\begin{aligned} S_3 &= \text{upper triangle} \cup \text{lower triangle} \\ &= \{[1\,2\,3], [2\,3\,1], [3\,1\,2]\} \cup \{[2\,1\,3], [3\,2\,1], [1\,3\,2]\} \\ &= C_3 \cup X_3 \\ &= C_3 \cup C_3 \star [2\,1\,3] \end{aligned}$$

9.1.3 Cosets

Miruka scanned our faces. "Are we good so far?"

"I'm good," I said.

"I think so," Tetra said.

"Maybe," Yuri said.

Miruka raised her eyebrows. "Maybe? Okay, some actual problems, then. Let's find $C_3 \star a$ for $C_3 = \{[1\,2\,3], [2\,3\,1], [3\,1\,2]\}$ and each $a \in S_3$. In other words…"

$$C_3 \star [1\,2\,3] =$$
$$C_3 \star [2\,3\,1] =$$
$$C_3 \star [3\,1\,2] =$$
$$C_3 \star [2\,1\,3] =$$
$$C_3 \star [3\,2\,1] =$$
$$C_3 \star [1\,3\,2] =$$

We immediately set to work. It wasn't long before we noticed something peculiar.

"Miruka, it looks like we're only getting one of two answers," Tetra said.

$$C_3 \star [1\,2\,3] = \{[1\,2\,3], [2\,3\,1], [3\,1\,2]\} = C_3$$
$$C_3 \star [2\,3\,1] = \{[2\,3\,1], [3\,1\,2], [1\,2\,3]\} = C_3$$
$$C_3 \star [3\,1\,2] = \{[3\,1\,2], [1\,2\,3], [2\,3\,1]\} = C_3$$
$$C_3 \star [2\,1\,3] = \{[2\,1\,3], [3\,2\,1], [1\,3\,2]\} = C_3 \star [2\,1\,3]$$
$$C_3 \star [3\,2\,1] = \{[3\,2\,1], [1\,3\,2], [2\,1\,3]\} = C_3 \star [2\,1\,3]$$
$$C_3 \star [1\,3\,2] = \{[1\,3\,2], [2\,1\,3], [3\,2\,1]\} = C_3 \star [2\,1\,3]$$

Miruka nodded. "When you calculate $C_3 \star a$, there may be some variation on the order of the elements, but the resulting sets will always be one of C_3 or $C_3 \star [2\,1\,3]$."

"This is really cool," Yuri said. "I didn't really get what this star thing was all about when you were talking about it, but after actually using it like this I think I really get it."

"That's why we do this," Miruka said.

"I'm getting a lot more comfortable with this too," Tetra said. "My image of this expression $C_3 \star a$ was something like 'slamming a's onto this subgroup C_3.' But it turns out that the results can only be one of C_3 or $C_3 \star [2\,1\,3]$. This definitely feels like using C_3 to classify the elements of S_3. I guess figures can certainly make things easier to understand, but so can equations, in a different sense."

Miruka nodded in agreement, then continued with her lecture. "These results of classification by C_3, in other words C_3 and $C_3 \star$

[2 1 3], are each *cosets* resulting from division of S_3 by C_3. We'll use $C_3 \backslash S_3$ to represent the set of all cosets resulting from that division. So $C_3 \backslash S_3$ is a set of sets."

$$C_3 \backslash S_3 = \{C_3, C_3 \star [2\,1\,3]\} \qquad \text{set of cosets from dividing } S_3 \text{ by } C_3$$

"Cosets, huh?" Yuri said.

"I'm... still not quite sure I understand what a coset is," Tetra said.

"Think of it as... something like a classification based on the remainder you get after dividing S_3 by C_3."

"Well, that's something else I'm not quite good with. S_3 is a group and C_3 is a subgroup, right? We're dividing groups by subgroups here?"

"That's right. Maybe a picture will make $C_3 \backslash S_3$ a little easier to understand."

Miruka starting drawing on the whiteboard.

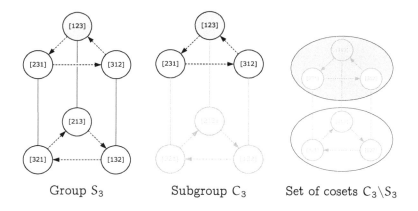

Group S_3 Subgroup C_3 Set of cosets $C_3 \backslash S_3$

"Ah hah!" Tetra said. "Okay, this has made things a lot clearer. We're using the subgroup C_3 as a basis for creating two big chunks, C_3 and $C_3 \star [2\,1\,3]$. I'm still not exactly sure how this is related to 'division,' though."

9.1.4 Clean Forms

We were supposed to be talking about what to do for Yuri's exhibit at the Galois Festival the following day, but had somehow become engrossed in another Miruka lecture. Even so, I was impressed with how much there was to think about regarding the symmetric group S_3, a group with just six elements.

Tetra was looking at her notes with a perplexed expression. "So Miruka, about this division. You said $C_3 \backslash S_3$ was a division of the group S_3 by the subgroup C_3. Can we divide S_3 using *any* subgroup C_3?"

"We can," Miruka said. "If you want a proof—"

Tetra raised a hand to stop her. "We can hold off on a formal proof for now. I just wanted to try dividing S_3 by some other subgroup to see what happens. This second-degree cyclic group C_{2a}, say."

$$C_{2a} = \{[1\,2\,3], [2\,1\,3]\}$$

"Ah, that would be interesting," I said. "C_{2a} would be the middle column in the Cayley graph, so maybe $C_{2a} \backslash S_3$ will give us three pillars?"

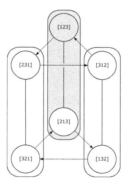

Will $C_{2a} \backslash S_3$ give three pillars?

"That's exactly what I thought," Tetra said. "Let's see..." After some scribbling in her notebook, she turned it to face me. "Well, not what we were expecting. The pillars are tilting over!"

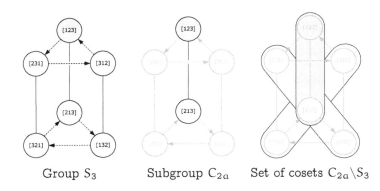

Group S_3 Subgroup C_{2a} Set of cosets $C_{2a}\backslash S_3$

"How did you come up with that?" I asked.

"Well, when we made $C_3\backslash S_3$ before, we calculated $C_3 \star a$ for all $a \in S_3$. So this time I just calculated $C_{2a} \star a$ for all $a \in S_3$."

$$C_{2a} \star [1\,2\,3] = \{[1\,2\,3],[2\,1\,3]\} = C_{2a}$$
$$C_{2a} \star [2\,3\,1] = \{[2\,3\,1],[1\,3\,2]\} = C_{2a} \star [2\,3\,1]$$
$$C_{2a} \star [3\,1\,2] = \{[3\,1\,2],[3\,2\,1]\} = C_{2a} \star [3\,1\,2]$$
$$C_{2a} \star [2\,1\,3] = \{[2\,1\,3],[1\,2\,3]\} = C_{2a}$$
$$C_{2a} \star [3\,2\,1] = \{[3\,2\,1],[3\,1\,2]\} = C_{2a} \star [3\,1\,2]$$
$$C_{2a} \star [1\,3\,2] = \{[1\,3\,2],[2\,3\,1]\} = C_{2a} \star [2\,3\,1]$$

"I see." *She actually worked each one out, just like for the calculation practice we did before. Wow, she's fast.*

"So now we can write the set S_3 in terms of C_{2a}."

$$\begin{aligned} S_3 &= \{[1\,2\,3],[2\,1\,3]\} \cup \{[1\,3\,2],[2\,3\,1]\} \cup \{[3\,1\,2],[3\,2\,1]\} \\ &= \quad\quad C_{2a} \quad\quad \cup \quad C_{2a} \star [2\,3\,1] \quad \cup \quad C_{2a} \star [3\,1\,2] \end{aligned}$$

"So our three pillars are $\{[1\,2\,3],[2\,1\,3]\}$ and $\{[1\,3\,2],[2\,3\,1]\}$ and $\{[3\,1\,2],[3\,2\,1]\}$," I said.

"She's right, our pillars are toppling over!" Yuri said, her eyes moving between Tetra's calculations and her graph.

"Not as pretty as I would have liked," Tetra said.

Miruka nodded. "Indeed. So what we should consider is, what's the difference between sets of cosets $C_3 \backslash S_3$ and $C_{2a} \backslash S_3$? More specifically, what's the difference between subgroups C_3 and C_{2a}?"

"Mmm, give me a minute to sort this all out first."

- We learned how to divide groups by subgroups to make cosets.

- When we divided S_3 by C_3, we got a clean separation into two cosets.

- When we divided S_3 by C_{2a}, we got three cosets, but they weren't separated out so nicely.

- So what's the difference between C_3 and C_{2a}?

9.1.5 Creating Groups

Tetra, Yuri, and I discussed this for a time, but in the end we didn't come to a conclusion regarding the fundamental difference between C_3 and C_{2a}.

"I do know what makes me think the one is prettier than the other, though," Tetra said. "It's that crossing of the columns in the graph with the three cosets."

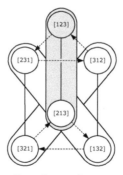

Crossing columns

"Um, Miruka?" Tetra hesitatingly said. "Not to get us sidetracked, but we really write this as $C_3 \backslash S_3$, with a backslash?"

Miruka adjusted her glasses and peered at Tetra. "Something wrong with that?"

"Well, it's just... You said we were dividing groups here, but to show dividing S_3 by C_3, we usually write that with a forward slash, like S_3/C_3. It's just been bugging me, how we're using a backslash instead..."

Miruka snapped her fingers. "You can do it either way," she said. "When dividing a group by a subgroup you can write either S_3/C_3 or $C_3 \backslash S_3$, but they mean different things, depending on whether you're looking for a coset using $a \star C_3$ or $C_3 \star a$."

$$S_3/C_3 = \{\, C_3,\ [2\,1\,3] \star C_3 \,\} \quad \text{set of cosets from } a \star C_3$$
$$C_3 \backslash S_3 = \{\, C_3,\ C_3 \star [2\,1\,3] \,\} \quad \text{set of cosets from } C_3 \star a$$

"Oh, we can use both!" Tetra said. "So do I have this right?"

$a \star C_3$ is the set of all ladder diagrams we get from putting a above each element in C_3.	$C_3 \star a$ is the set of all ladder diagrams we get from putting a below each element in C_3.

"Correct. Because that's how we defined the $\star$ operator."

"Oh! So that would mean... Hang on..."

Tetra began hurriedly writing in her notebook, a task in which she remained immersed for some while. The rest of us remained silent, passing the time in our own ways. Miruka closed her eyes, and was twirling a finger in the air. Yuri was calculating something while referring to her notes. I was doing my best to mentally compare C_3 and C_{2a}.

The biggest difference I noticed was the number of elements, in other words the groups' orders. C_3 had an order of 3, while C_{2a} had an order of 2. I wasn't sure if that had anything to do with how "pretty" the cosets would be, though. Honestly, I wasn't entirely sure what "pretty" had to do with anything. That was too vague a concept for me to consider mathematically. *How can I capture "pretty" in an equation?*

"Got it!" Tetra shouted with that I've-discovered-something look on her face.

"Got what?" I asked.

"The difference between C_3 and C_{2a}, of course! Yuri's Cayley graph shows ladder diagrams being added below ladder diagrams. So I made one where we're adding ladder diagrams *on top* of ladder diagrams. When I did so, I found that the graphs for C_3 come out the same way, whether ladder diagrams are added from the top or the bottom. That's not the case for C_{2a}, though!"

▶ *Cayley graphs for C_3 (groupings are the same)*

ladder diagrams added below | ladder diagrams added above

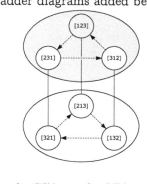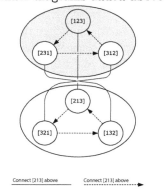

$C_3 \star [213] = \{[213],[321],[132]\}$ | $[213] \star C_3 = \{[213],[321],[132]\}$

▶ *Cayley graphs for C_{2a} (groupings differ)*

ladder diagrams added below | ladder diagrams added above

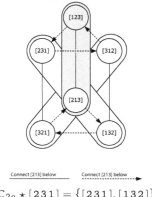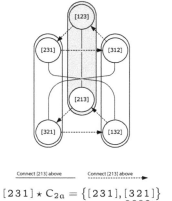

$C_{2a} \star [231] = \{[231],[132]\}$ | $[231] \star C_{2a} = \{[231],[321]\}$

I sat silently, looking at what Tetra had drawn. This time, however, my silence wasn't because I was so enraptured by the math, but because... "I don't get it. I'm sorry, I'm sure this is something amazing, but I'm kinda lost."

"Me too," said Yuri, sounding a little tired.

"Oops, did I say something strange?" Tetra said.

Miruka shook her head. "No, Tetra. You've shown us something essential."

"I...I did? No, I was just showing you something I noticed. I still don't know what it all means."

"I brought up differences in C_3 and C_{2a} in the hopes that someone would notice something, but was starting to think nobody would. Until you did."

"I did? What did I notice?"

"Normal subgroups."

"What's a normal subgroup?" Yuri asked.

"Let's summarize what Tetra just said. Briefly stated, it goes something like this."

$$C_3 \star [2\,1\,3] = [2\,1\,3] \star C_3$$
$$C_{2a} \star [2\,3\,1] \neq [2\,3\,1] \star C_{2a}$$

"Tell me, Tetra. What do these statements assert?"

"That we can...swap things around?"

"Right. It says that if we swap operands in $C_3 \star [2\,1\,3]$ to get $[2\,1\,3] \star C_3$, we haven't changed the result. However, $C_{2a} \star [2\,3\,1]$ and $[2\,3\,1] \star C_{2a}$ are *not* the same. That's exactly what your graphs are saying. In fact, we can make an even more general proof." Miruka pushed her glasses up her nose. "In the case of C_3, this will hold for every element a in S_3."

$$C_3 \star a = a \star C_3$$

"Like you said, it allows for swaps. That's not the case for C_{2a}, or any of the order 2 cyclic subgroups, however. There will exist some element x in S_3, which is not the generator a of the order 2 cyclic subgroup, such that this is true."

$$C_{2a} \star x \neq x \star C_{2a}$$

"This is a very important difference between C_3 and C_{2a}."

"One thing, Miruka," I said. "This equals sign in $C_3 \star a = a \star C_3$, that's an equality between sets, right?"

"It is."

"What's an 'equality between sets'?" Yuri asked.

"Well, $C_3 \star a$ and $a \star C_3$ are both sets, right? So the equals sign here means that the sets on the left and right in $C_3 \star a = a \star C_3$ are the same. It says the set that $C_3 \star a$ creates is equal to the set that $a \star C_3$ creates. But it does *not* mean $x \star a = a \star x$ will be true for each element x in C_3."

"Huh, okay."

Tetra looked dissatisfied. "So, Miruka. I get that C_3 commutes with everything, but—and I hate to say this again, but—so what? We're just saying that C_3 is a normal subgroup, but C_{2a} isn't. Is that all?"

"Yep," Miruka flatly stated.

- C_3 *is* a normal subgroup of S_3.

- C_{2a} is *not* a normal subgroup of S_3.

"Okay," Tetra said. "So what's the big deal about being a normal subgroup of C_3?"

"Hmph." Miruka paused to think for a brief time before saying, carefully selecting each word, "The big deal is that *the set of cosets resulting from division by a normal subgroup is a group*."

"Wait, what?" I said, immediately realizing how silly I sounded.

"Having problems parsing that? How about this: Divide a group by a normal subgroup. The result is a set of cosets. That set of cosets is a group." Uncomfortable silence. "Okay, think about what we do to investigate the structure of groups."

- We count elements to determine the order.

- We search for a generator that creates the group.

- We search for subgroups included in the group.

- We divide groups by subgroups to find cosets.

"That's pretty much our basic investigative toolbox. But another important investigation is looking for normal subgroups. This is important because the sets of cosets we get from dividing the group will show the group's structure in a very natural way. When we consider a set of cosets as a group, it's called a *quotient group*, or a *factor group*. So finding normal subgroups and creating quotient groups are important steps toward finding the structure of a group."

I was—amazed? Astounded? Dazed by the math?

We used logic to gather all these mathematical objects into a *set*. We defined an operator for operations between the elements of that set to give it the structure of a *group*. We searched for subsets that have the structure of a group to find *subgroups*. We used subgroups to divide the group and create sets of *cosets*. Now we've introduced groups into the set of all cosets to create *quotient groups*, which we obtained by dividing groups by normal subgroups...

My head was spinning.

"So $C_3 \backslash S_3$ will be a quotient group, right?" Tetra said.

"Right. C_3 is a normal subgroup, so you can write either $C_3 \backslash S_3$ or S_3 / C_3 for the quotient group, since the two are equivalent."

$$C_3 \backslash S_3 = \{C_3, C_3 \star [2\,1\,3]\}$$
$$= \{C_3, [2\,1\,3] \star C_3\] \quad \text{because } C_3 \star [2\,1\,3] = [2\,1\,3] \star C_3$$
$$= S_3 / C_3$$

"Just what kind of group is this quotient group S_3/C_3?"

"You tell me. What are the elements in S_3/C_3?"

"Oh, uh... It's the set of all cosets, so C_3 and $[2\,1\,3] \star C_3$."

$$S_3/C_3 = \{C_3, [2\,1\,3] \star C_3\}$$

"And how many elements are in S_3/C_3?"

"Just the two, C_3 and $[2\,1\,3] \star C_3$."

"Can you think of another group that has two elements?"

"Uh... Oh! I can! There's only one group in the world with just two elements!"

"That's right. Any group with two elements will be isomorphic to the cyclic group C_2. Which means that applies to S_3/C_3 too."

"Miruka," Yuri weakly called out. "I'm so totally and completely lost."

"How so?" Miruka said.

"First off, what does it mean to $\star$ between cosets?"

"Okay, let's talk about that operator a bit more."

"Hey, Miruka," I said. "All this is truly fascinating, but is this really the time? The Galois Festival is tomorrow. Shouldn't we be focusing on Yuri's poster? Or at least talk about Galois?"

"What do you think we're doing?" Miruka snapped. "Normal subgroups—in other words, subgroups for which the set of all cosets is a group—is at the heart of Galois's work. So much so that on the eve of his death, he took the time to write about them in a letter to his friend Auguste Chevalier. It was Galois who noticed the importance of normal subgroups, in his pursuit of the conditions for a polynomial equation to be algebraically solvable."

"Hold up, Miruka," I said. "You're saying that normal subgroups have something to do with whether we can algebraically find the roots for a polynomial?"

Miruka grinned. "That's exactly what I'm saying. Normal subgroups are one of the most important concepts related to algebraic solvability, in fact. That's why we're learning about them, using the Cayley graphs that Yuri and Tetra drew for us."

"Hey Miruka, okay if we break for some food?" Yuri said.

Miruka looked up at the clock. It was already past two o'clock.

"I guess now's a good time for lunch. Off, then, to Oxygen!"

9.2 Notational Form

9.2.1 At Oxygen

The four of us relocated to Oxygen, the cafe on the library's third floor. It had a lovely outdoor dining patio, but the heat and humidity kept us indoors. A boy who looked like a college student, probably there to prepare for the Galois Festival, approached Miruka and engaged her in some high-level discussion of mathematics, shunting the rest of us off to the sidelines.

"Looks like we're going to have nice weather tomorrow, too," Tetra said to Yuri.

"Meaning *hot*," Yuri replied.

"Were you able to keep up with the discussion of normal subgroups?" I asked her.

Yuri started retying her ponytail. "Mmm, sorta. I understood how the commutative law holds for normal subgroups, so $C_3 \star a = a \star C_3$. I'm not quite there with quotient groups and all that, but I do know that ladder diagrams are lots of fun!"

"They are indeed," I said.

Tetra turned to me. "Speaking of ladder diagrams, don't they basically swap numbers?"

I paused, instinctively feeling that Tetra's question aimed at some fundamental truth.

"What exactly do you mean?"

9.2.2 Swapping Notation

Tetra gave me a rundown of what she was thinking about while we drank tea after lunch.

"I've been thinking that maybe ladder diagrams are number swappers. We've been talking about two patterns, adding ladder diagrams 'above' or 'below' other ladder diagrams. When we draw Cayley graphs, I find it really easy to imagine adding a ladder diagram below another one, because I can picture that in my head. It's harder to imagine adding them from above, though. Preconceived notions or something, I guess.

"So I've been trying to think about this differently. Not as 'attaching a ladder diagram from above,' but in terms of swapping the numbers in the notation. Like, if we add a [2 1 3] beneath a [2 3 1], we aren't really swapping the numbers we see, right? Because if [2 1 3] meant 'swap the 1 and the 2,' then [2 3 1] would become [1 3 2]. Instead, [2 1 3] is telling us to swap the numbers in the first and second *positions* in the notation. The first number in [2 3 1] is a 2, and the second number is a 3, so *those* are the numbers we want to swap. Er, right? Yeah, that's right.

"So anyway, when we place [2 1 3] 'above' [2 3 1], rather than thinking about swapping locations, I find it easier to think about swapping numbers. Specifically, among the numbers in [2 3 1], we

The Form of Feelings

swap the 1 and the 2. Then [2 3 1] becomes [1 3 2], which is exactly the ladder diagram we get when we place [2 1 3] on top of [2 3 1].

"When we studied about symmetric groups, we used a notation where parentheses indicate permutations, which makes perfect sense to me. In the ladder diagram [2 1 3], the 2 comes below the 1, and the 1 comes below the 2. So it's $1 \to 2$, then back to $2 \to 1$. I'll write (12) to show a swap of 1 and 2 like that."

$$[213]\left\{\begin{matrix} 1 & 2 & 3 \\ \diagdown\!\diagup & | \\ 2 & 1 & 3 \end{matrix}\right. \qquad \begin{pmatrix} 1 & 2 & 3 \\ 2 & 1 & 3 \end{pmatrix} \qquad \overbrace{1 \to 2} \qquad (12)$$

[2 1 3] and $\begin{pmatrix} 1 & 2 & 3 \\ 2 & 1 & 3 \end{pmatrix}$ and (12).

"For the ladder diagram [2 3 1], the 2 comes below the 1, the 3 comes below the 2, and the 1 comes below the 3. So it's $1 \to 2 \to 3$, then $3 \to 1$, which I'll write as (123). This is a scramble of $1, 2, 3$."

$$[231]\left\{\begin{matrix} 1 & 2 & 3 \\ & & \\ 2 & 3 & 1 \end{matrix}\right. \qquad \begin{pmatrix} 1 & 2 & 3 \\ 2 & 3 & 1 \end{pmatrix} \qquad \overbrace{1 \to 2 \to 3} \qquad (123)$$

[2 3 1] and $\begin{pmatrix} 1 & 2 & 3 \\ 2 & 3 & 1 \end{pmatrix}$ and (123).

"Here's how my notation compares with Yuri's."

Yuri's notation	[1 2 3]	[2 1 3]	[3 2 1]	[2 3 1]	[3 1 2]	[1 3 2]
Parenthetic notation	()	(12)	(13)	(123)	(132)	(23)

Tetra twisted awkwardly. "I don't think Yuri's notation—writing a ladder diagram like [2 1 3]—is standard mathematics, so when she makes her poster, maybe she should use these parentheses instead? Don't you think?"

Yuri didn't look particularly pleased with this idea.

"Yuri's notation is clearly understandable," Miruka said. Apparently she'd drifted back over after finishing her conversation with the college student. "You're right that [2 1 3] isn't standard notation for a permutation, but so long as she provides a definition that isn't a problem. In fact, I think it provides a more intuitive model for what's happening in a ladder diagram."

"Good point," I said.

"Or, if you prefer, you can say that you've just chosen the lower part of $\left(\begin{smallmatrix} 1 & 2 & 3 \\ 2 & 1 & 3 \end{smallmatrix}\right)$, another standard notation for showing permutations."

"Well I feel much better, then," Yuri said.

"In Galois's first paper, we can see the notation that Galois himself used to represent groups. Letting a, b, c, d be the solutions to a polynomial, he wrote their permutations as $abcd, bacd, cbad, dbca, \ldots$ In Yuri's notation, this would be $[1\,2\,3\,4], [2\,1\,3\,4], [3\,2\,1\,4], [4\,2\,3\,1], \ldots$. So I think this way of writing things is very appropriate for the Galois Festival."

"I concur!" Tetra said. "By the way, that first paper by Galois? What was that about?"

"Necessary and sufficient conditions for algebraically solving polynomial equations," Miruka said. "But let's talk about that tomorrow. I've prepared a poster that will help."

"I can't wait!"

"All notations have their pros and cons. $[2\,1\,3]$ and $\left(\begin{smallmatrix} 1 & 2 & 3 \\ 2 & 1 & 3 \end{smallmatrix}\right)$ work well with ladder diagrams, but it's hard to see what's being swapped with what. Writing (12) is better in that respect, in that it's clear that the 1 and the 2 are what's being swapped. Just use whatever best suits the situation."

"I still want a notation that makes everything clear all the time," Yuri grumbled.

9.2.3 Lagrange's Theorem

"Okay, so I see how there are advantages and disadvantages for various notations," Tetra said. "I guess you could say the same about the difference between figures and equations. A figure can make it easier to see the big picture—"

"Like the groupings in Cayley graphs!" Yuri said.

"—but mathematical expressions can make it easier to see patterns."

"And to give proofs," I added. "Illustrations can be deceiving."

Tetra nodded. "They can. But still, there's a lot to be learned from a good figure. Speaking of proofs, is it true that when you divide a group by a subgroup, the number of elements in each of the resulting cosets will be equal?"

"Meaning what?" Yuri asked.

"Well, the cosets for S_3/C_3 are C_3 and $[2\,1\,3] \star C_3$, and each of those have three elements in them, right?"

"Sure."

"And when we looked at the cosets for S_3/C_{2a} we got C_{2a} and $[2\,3\,1] \star C_{2a}$ and $[3\,1\,2] \star C_{2a}$, and each of those have two elements. That doesn't feel like a coincidence, so it seems like that might be something we can prove."

"We can," Miruka said, already writing on a napkin. "To save time, I'm going to drop the operator. So instead of $[2\,3\,1] \star [2\,1\,3]$ I'll write $[2\,3\,1][2\,1\,3]$, and instead of $[2\,1\,3] \star C_3$ I'll write $[2\,1\,3]C_3$. In general..."

· Write $g \star h$ as gh

· Write $g \star H$ as gH

"That should be clear enough, right?" Miruka said.

"Kind of like multiplication," I said. "Omitting the operator when the context makes the operation clear. Like writing $a \times b$ as ab, or x^1 as x."

"Right. So anyway, here's what Tetra was asking about."

Problem 9-1 (Number of elements in cosets)

Let G/H be the set of all cosets for group G and its subgroup H. Will the number of elements in each coset belonging to G/H equal the number of elements in H?

"To avoid contradictions, we'll assume all these sets have a finite number of elements," Miruka added. "We write the set of all cosets G/H like this."

$$G/H = \{gH \mid g \in G\}$$

"So we can rephrase the question like this: Is the number of elements in the set gH the same as the number of elements in H? Don't forget, the set gH is an abbreviation for $g \star H$, and it's defined like this."

$$gH = \{gh \mid h \in H\}$$

"In other words, we take some element g in G, and gH is the set created by multiplying that g by each element h in H. To show that the number of elements in set gH is the same as the number of elements in H, we need to demonstrate two correspondences."

(1) For any element in set H, there is exactly one corresponding element in set gH.

(2) Conversely, for any element in set gH, there is exactly one corresponding element in set H.

"To start with (1), we can associate element h in set H with element gh in set gH. This associates h with with exactly one corresponding element.

"To show (2), we can go the other way and associate element gh in set gH with element h in set H. Then we'll have associated gh with exactly one corresponding element. This is true because if there is some element $h' \in H$ satisfying $gh = gh'$, then we can multiply both sides of this from the left by g's inverse element g^{-1}, obtaining $g^{-1}gh = g^{-1}gh'$. Since $g^{-1}g$ is the identity element e, we get $eh = eh'$. From the properties of the identity element, this means that $h = h'$, so we see that h and h' are the same element. So we've shown that for a given coset gH belonging to G/H, the number of elements in gH and the number of elements in H are equal. Therefore, the number of elements in all cosets belonging to G/H will be the same as the number of elements in H, which is what we wanted to show. *Quod erat demonstrandum.*"

Answer 9-1 (Number of elements in cosets)

Letting G/H be the set of all cosets for group G and its subgroup H, the number of elements in each coset belonging to G/H equals the number of elements in H.

"This proof shows that there is a bijective—or one-to-one and onto—mapping $f : h \mapsto gh$ between H and gH. Along the way, this proof also shows us that if we divide the number of elements in G by the number of elements in H, we get the number of cosets. This is known as Lagrange's theorem."

Lagrange's theorem

Given a group G and its subgroup H,

$$|G|/|H| = |G/H|,$$

where

- $|G|$ is the order of (number of elements in) group G,
- $|H|$ is the order of subgroup H, and
- $|G/H|$ is the number of elements (cosets) in the set of all cosets G/H.

"We've heard that name before!" Tetra said. "Is that the same Lagrange as in Lagrange resolvents?"

"The very one," Miruka said. "So as an example, the order of group S_3 is $|S_3| = 6$, and the order of its subgroup C_3 is $|C_3| = 3$, so the number of elements in the set of cosets S_3/C_3, in other words the number of cosets, will be $|S_3/C_3| = 2$. Maybe this makes using the slash to indicate division feel a little better?"

$$
\begin{array}{ccc}
|S_3| & / \ |C_3| & = |S_3/C_3| \\
\vdots & \vdots & \vdots \\
6 & / \ \ \ 3 & = \ \ 2
\end{array}
$$

"Interesting!" Tetra said. "And we can do the same thing with S_3/C_{2a}, right?"

$$
\begin{array}{ccc}
|S_3| & / \ |C_{2a}| & = |S_3/C_{2a}| \\
\vdots & \vdots & \vdots \\
6 & / \ \ \ 2 & = \ \ 3
\end{array}
$$

"Very good. Lagrange's theorem also tells us that the order of a subgroup divides the order of its group."

"So, Miruka, would all cosets have the same number of elements?" Yuri asked, scribbling on another napkin. "Would a figure like this be right?"

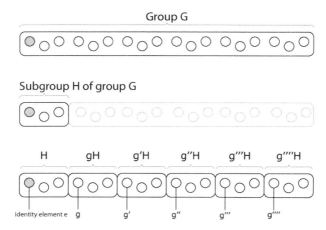

Group G, its subgroup H, and the set of all cosets G/H.

"Now *that* definitely makes things easier to understand," I said.

"The point is that each coset has the same number of dots in it," Yuri said.

Miruka narrowed her eyes, then nodded. "Well done. So Lagrange's theorem is primarily about sets of cosets, but of course it holds for quotient groups as well."

"Makes sense," I said. "Quotient groups are made from sets of cosets, after all."

"Generally, say you have a group G and its normal subgroup H. In other words, G ▷ H. We'll represent the order of quotient group G/H as (G : H), and call this the *index of the subgroup*. For example, the index of the subgroup and the order of the quotient group S_3/C_3 would be this."

$$(S_3 : C_3) = |S_3/C_3| = 2 \quad \text{group index}$$

9.2.4 Notation for Normal Subgroups

"That triangle notion you used," Tetra said. "Does G ▷ H indicate a normal subgroup?"

"It does. I probably should have mentioned that. But anyway, yes, we're going to write G ▷ H to indicate that H is a normal subgroup of G. Here's the standard notation."

Definition of normal subgroups

If
$$gH = Hg$$
holds for a given group G, subgroup H of G, and every element g in G, then H is a *normal subgroup* of G, denoted as

$$G \triangleright H.$$

9.3 THE FORM OF PARTS

9.3.1 $\sqrt[3]{2}$, All Alone

I noticed our table had become covered in napkins, each filled with figures and symbols. There was a general consensus that we should head back to the conference room, so we packed up to leave. Yuri and Miruka were talking about Cayley graphs as they walked to the door. Tetra seemed to be hanging back for some reason.

"Forget something?" I asked.

"There's...something I wanted to talk to you about," she said, pulling me to a corner of the cafe far from any other patrons. She bowed deeply. "I'm so sorry about the other day, leaving you like I did." *Ah, so that's what this is all about.*

"No, I'm the one who should apologize. I didn't mean to make you angry."

"It's not that. Or about the math. It's...something else."

"Namely?"

"The concert."

"What concert?"

"The one I didn't go to. Ay-Ay's concert."

"Oh, her recital?"

"The one you went to with Miruka, yes. I kind of felt like...like $\sqrt[3]{2}$."

"Like...what?"

"Remember when you told me about $\sqrt[3]{2}$? How even if you adjoin $\sqrt[3]{2}$ to $\mathbb{Q}$, there's no $\sqrt[3]{2}\omega$ or $\sqrt[3]{2}\omega^2$ in there with it? Well, that's kind of how I felt. But, that's all on me, not you. You didn't do anything wrong." Tetra briefly met my eyes, then looked back down. "Anyway, I just wanted to say I'm sorry for how I behaved. Enough, let's get back. We don't have much time left to work on Yuri's poster!"

9.3.2 Pursuing Form

Tetra and I walked down the hall toward the conference room, and caught up with Miruka on the way. I saw Lisa up ahead. She'd apparently come from the other direction, pushing a cart and handing out materials to exhibitors in the rooms along the hallway, but Yuri had stopped her to talk about something. We peeked into those rooms as we passed, and saw they were abuzz with people getting ready for the Galois Festival.

"Who'd have thought ladder diagrams could be so much fun?" I said to Tetra, who was walking beside me.

"I know, right?" she said. "I'm also really curious about this investigating groups stuff. Their order and subgroups and all that."

Miruka glanced back at us. "A thing with structure can be broken down into parts."

This struck a chord in me, sending me off on another thread of internal musings. *A thing with structure can be broken down into parts... Not that those parts are unrelated to each other. A picture is not just a collection of lines. A person is not just a collection of cells. And a group is not just a collection of elements; it is precisely the interrelations between those elements that makes it a group. Group theory is what lets us untangle those interrelations. What is the group's order? What subgroups does it have? What normal subgroups does it have? What quotient groups do you get from dividing it by a normal subgroup? All of this is part of unravelling the group's structure.*

We could find normal subgroups within groups, just as we had found regular $1, 2, 3, 4, 6,$ and 12-gons within regular dodecagons. Finding hidden structures—that was what we were after when investigating the form of groups. By outfitting the subject of our inquiry with the structure of a group, we can use the tools of group theory to aid in that investigation.

That's exactly what Galois did...

9.3.3 "Proper" Decompositions

We returned to the Narabikura Library's "Beryllium" conference room.

"You said Galois did a lot of work with normal subgroups, right?" Tetra asked Miruka.

"I did. He didn't call them normal subgroups, but that's what he was using. He called them *décomposition propre*."

"That would be... 'proper decompositions'?"

"Groups can be decomposed into unions of cosets according to their subgroups. For example, if you have $G/H = \{H, aH, a'H, a''H, a'''H\}$, you can also write this."[1]

$$G = H \cup aH \cup a'H \cup a''H \cup a'''H$$

"This is a decomposition of G into the elements of G/H. Or, instead of G/H you could decompose G into the elements of $H\backslash G$, the cosets. If you have $H\backslash G = \{H, Hb, Hb', Hb'', Hb'''\}$, then this holds."

$$G = H \cup Hb \cup Hb' \cup Hb'' \cup Hb'''$$

"Only when H is a normal subgroup, both decompositions are the same. That's what Galois called a 'proper' decomposition. For example, this..."

$$S_3 = C_3 \cup [2\,1\,3]C_3$$

"...is a proper decomposition of group S_3 by its normal subgroup C_3."

---

[1] Note that Galois used $+$ in place of the $\cup$ symbol.

9.3.4 Further Dividing C_3

"I noticed something, thinking about why we're focusing on normal subgroups," Tetra said. "Dividing a group G by a normal subgroup H to create cosets G/H... Doesn't that feel a lot like prime decomposition of an integer to investigate its properties? We've talked about how the prime factors of integers show their structure. Don't normal subgroups similarly show the structure of groups? That's what all this talk of 'dividing' makes me think, anyway, that we can use H and G/H to investigate the properties of group G in the same kind of way."

"Makes sense!" I said.

"You never fail to amaze me, Tetra," Miruka said. "You're exactly right. Until now we've been dividing the group S_3 by its normal subgroup C_3, but we can also divide C_3 itself, by *its* normal subgroups."

Tetra's eyes widened. "What? C_3 has a normal subgroup?"

"It does."

"But C_3 only has three elements, right?"

"It does. A prime number of elements." Miruka winked.

"Is that supposed to tell us something?"

"What are the factors of 3?"

"Well, since it's a prime just 1 and itself."

"So if C_3 has a subgroup, its order will have to be either 3 or 1."

"Oh, right! Lagrange's theorem tells us that."

"Indeed. Actually, C_3 has two subgroups. One is C_3 itself, which has order 3. The other is the identity group, $E_3 = \{[1\,2\,3]\}$, which has order 1. Both of those are normal subgroups of C_3, since for an arbitrary element a in C_3, both $aC_3 = C_3 a$ and $aE_3 = E_3 a$ are true."

Tetra clapped her hands together. "Oh, so the identity group will always be a normal subgroup, right? Just like 1 is a factor of all integers! And the group itself will of course be a normal subgroup, just like any integer is a factor of itself!"

Miruka softly nodded. "Exactly right. So, we've found a chain of normal subgroups, starting from S_3."

- Group S_3 has a normal subgroup C_3, and
- Group C_3 has a normal subgroup E_3.

"In other words..."

$$S_3 \triangleright C_3 \triangleright E_3$$

Miruka walked to the whiteboard. "Let's use a figure to visualize this chain of normal subgroups."

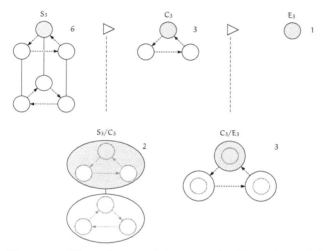

Decomposition of symmetric group S_3 ($S_3 \triangleright C_3 \triangleright E_3$).

"Whoa..." Tetra whispered.

Miruka smiled. "It's like lifting off the ground and taking flight to obtain a bird's-eye view. It lets us ignore the details so that we can see the large-scale structure." Redheaded Lisa entered the room, pushing her cart and expressionless as ever. Miruka glanced her way, but otherwise ignored her. "Investigating what kind of normal subgroups are included in a group is something like taking apart a clock to figure out how it works. We examine its normal subgroups and quotient groups in the same way we would try to figure out how the gears fit together."

It wasn't clear whether Lisa was listening to Miruka. She was busy unloading supplies from her cart: sheets of poster paper, a box of colored markers, a utility knife, mounting boards...

"Once upon a time there was a girl," Miruka said. "When she was in elementary school, she was given a toy clock, a tool for teaching how to tell time. The kind where you spin a knob on the back to move the hands." *What's this all about?* "When she got home, she took apart all the clocks in the house to see how they work. Right, Lisa?" *Aha, okay.*

Lisa continued her work, ignoring Miruka's provocation. Miruka circled behind her and mussed her hair.

Lisa coughed and brushed off Miruka's hand. "Enough!"

"You wanted to see their internal structure, right?"

Lisa regained her composure. "Visual confirmation."

I knew Miruka and Lisa were cousins, but I wasn't quite sure if they were friends.

"Can we use all this stuff?" Tetra asked, sorting through everything Lisa had laid out on the table. "For our poster?"

Lisa nodded.

"Thanks, Lisa," I said. "Looks like everything we need."

Taciturn though she may be, Lisa was never anything short of helpful. She nodded again, and left the room with her cart.

9.3.5 Equivalence with Division

Reviewing her notes, Tetra asked, "But why division?"

"Division for equivalence," Miruka said. "Remainders for classification."

"I'm... not sure what that means."

"We use the same concept when we count days. Say today's Sunday, which we'll call day 0. Then tomorrow is Monday, day 1. So what do you do if you want to know what day of the week day n will be?"

"Oh, you just divide by 7!"

"That's right. You divide n by 7, and look at the remainder. The remainder is your classifier. A 0 means a Sunday, a 1 is a Monday, all the way up to a remainder of 6, which would be Saturday. The result is a set of sets, something like this."

Set of all days = {
$\quad$ { 0, 7, 14, ... }, $\qquad$ ······ Sun.
$\quad$ { 1, 8, 15, ... }, $\qquad$ ······ Mon.
$\quad$ { 2, 9, 16, ... }, $\qquad$ ······ Tue.
$\quad$ { 3, 10, 17, ... }, $\qquad$ ······ Wed.
$\quad$ { 4, 11, 18, ... }, $\qquad$ ······ Thu.
$\quad$ { 5, 12, 19, ... }, $\qquad$ ······ Fri.
$\quad$ { 6, 13, 20, ... } $\qquad$ ······ Sat.
}

"So this division by 7 creates an equivalence for all days separated seven days apart, and the remainder classifies those days into days of the week."

"Okay, I get it now," Tetra said. "Division for equivalence, remainders for classification."

"S_3/C_3 is something similar."

S_3/C_3 = {
$\quad$ {[1 2 3], [2 3 1], [3 1 2]}, $\qquad$ ······ C_3
$\quad$ {[2 1 3], [3 2 1], [1 3 2]} $\qquad$ ······ [2 1 3]C_3
}

"Interesting," Tetra and I synched.

"So this is just another analysis by division," Miruka said, her tempo rising. "G/H is a look at the group G through the granularity of its subgroup H. You can also say we're creating an equivalence for elements belonging to the subgroup H."

"Does this have something to do with quotient groups?" Yuri asked.

Miruka turned to Yuri. "Ah, that brings up an important point. Let's talk about operations between sets. We've defined operations between elements in a group, right, Yuri?"

"Yep!"

"Operations between sets are a natural extension of that. Say you have two elements of a group, a and b. Their product is $ab = c$. What we want is for the product of two cosets, aH and bH, to equal cH. In other words, $(aH)(bH) = cH$. Good so far?"

"That's what we want, meaning that's how we're going to define things?"

"Sure. So long as we fulfill the axioms for group operations, we can define things however we want. Might as well do it in a way that's interesting. In particular, so that if $ab = c$, then $(aH)(bH) = cH$."

Yuri looked uneasy. "Yeah, I think I see that, but still..."

"Hey, Miruka," Tetra cut in. "This is another case of things being 'well-defined,' isn't it!"

"It is," Miruka said.

"What's that?" Yuri asked.

"Well, $ab = c$ is an operation between elements, right? That's a small-scale structure. But $(aH)(bH) = cH$, an operation between cosets, that's a large-scale structure."

Yuri cocked her head. "Huh?"

"When we define an operation between aH and bH, we want the result to belong to cH, regardless of which elements in aH and bH are involved. That's what Miruka means when she says $(aH)(bH) = cH$ when $ab = c$. Right?"

"Well put," Miruka said.

Yuri frowned. "Still not getting it. I thought putting things into equations would make everything clear, but..."

"Think in terms of Cayley graphs, then. The groupings you found before are cosets. We want to take one element from coset aH and another from coset bH, and calculate using those. The result should belong to some coset cH. You're good with that, right?"

"Sure."

"Tetra's 'well-defined' means that no matter which element you take from aH, and no matter which element you take from bH, the results of the operation will always—regardless of how you choose those elements—belong to cH. Then, since the result of the operation will be the same regardless of which elements we pulled from the cosets, we can forget about individual elements. Since we don't care which elements we've selected from the cosets, we can consider this as an operation on cosets."

"Okay, I understand that," Yuri said, "but that thing about not caring which elements we select from the cosets... How can we know that will always work?"

Miruka smiled. "It won't always work. We can't use cosets from just any subgroup. What we can do, though, is make sure we use a subgroup that will produce cosets from which we really don't care which elements we select. Namely—?"

"A normal subgroup!" Tetra shouted.

"Which will provide consistency between the arrows and the groupings in the Cayley graph!" I said.

"Mmm..." Yuri moaned.

"Time for the definition of a normal subgroup," Miruka said. "Let's prove that if H is a normal subgroup of group G, then $(aH)(bH) = (ab)H$ will hold for any elements a, b in G. To prove equality between sets, we need to show that both $\subset$ and $\supset$ hold."

Miruka returned to the whiteboard and sketched out two proofs.

▶ *Proof that* $(aH)(bH) \subset (ab)H$

We can write an arbitrary element in $(aH)(bH)$ as $(ah)(bh')$ (for $h, h' \in H$). From associativity, this element is equal to $a(hb)h'$. Because H is a normal subgroup, $Hb = bH$ holds, and there exists some element h'' in H satisfying $hb = bh''$. Therefore...

$$
\begin{aligned}
(ah)(bh') &= a(hb)h' && \text{change order of operations by associativity} \\
&= a(bh'')h' && \text{there exists } h'' \in H \text{ such that } hb = bh'' \\
&= (ab)(h''h') && \text{change order of operations by associativity} \\
&\in (ab)H && H \text{ is a group, so } h''h' \in H
\end{aligned}
$$

This shows that any element in $(aH)(bH)$ belongs to $(ab)H$, and thus that $(aH)(bH) \subset (ab)H$.

▶ *Proof that* $(aH)(bH) \supset (ab)H$

We can write an arbitrary element in $(ab)H$ as $(ab)h$ (for $h \in H$). From associativity, this element is equal to $a(bh)$. Because $a \in aH$ and $bh \in bH$, the element $(ab)h$ belongs to $(aH)(bH)$.

$$
\begin{aligned}
(ab)h &= a(bh) && \text{change order of operations by associativity} \\
&\in (aH)(bH) && \text{because } a \in aH \text{ and } bh \in bH
\end{aligned}
$$

This shows that any element in $(ab)H$ belongs to $(aH)(bH)$, and thus that $(aH)(bH) \supset (ab)H$.

Because $(aH)(bH) \subset (ab)H$ and $(aH)(bH) \supset (ab)H$,

$$(aH)(bH) = (ab)H.$$

"There you have it," Miruka said. "We can naturally incorporate the structure of a group onto the set of all cosets resulting from dividing a group G by its normal subgroup H. That's the quotient group G/H. This quotient group ignores the small structure within the set of all cosets, focusing instead on the larger structure between cosets. It's a shift from a micro-perspective to a macro-perspective of structure. The forest, not the trees. A panorama from the sky." Miruka paused to survey our little group. "How do you like the view?"

9.4 THE FORM OF SYMMETRIC GROUP S_4

9.4.1 In Beryllium

I glanced at the clock and saw it was already four o'clock.

"So have we decided what to do for Yuri's poster?" I asked. "Maybe a summary of these graphs for S_3 that Yuri and Tetra made?"

"Yuri, have you studied the symmetric group of degree four, S_4?" Miruka asked.

"A little," Yuri answered, digging through her backpack. "I drew a bunch of graphs for it, but I ended up doing a lot of simplification. There are $4! = 24$ patterns, after all!" She pulled out a sheet of paper covered in circles and lines.

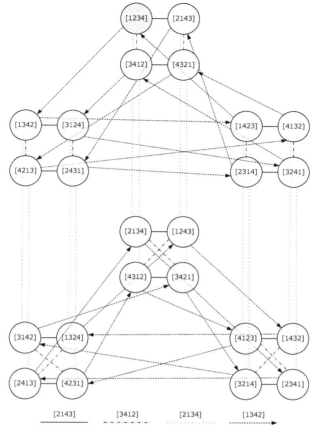

Cayley graph for the symmetric group S_4.

"Its Cayley graph, nice." Miruka said.

"Oh, and I found groupings in this graph of S_4 too! An upper triangle and a lower triangle, like before. But this time, they aren't simple triangles, they're more like triangles made from squares. A grouping of groupings, I guess."

"Oh cool, I see that," I said.

"Yeah, that is cool," Tetra said.

"So Miruka," Yuri said, "We can divide this Cayley graph for S_4 to create cosets, like we did for S_3?"

"Of course. What say we do so? We need to build a tower of normal subgroups, one that shrinks the symmetric group S_4 down to the identity group E_4. That will allow us to more mathematically write out the 'triangles of squares' you found."

We spent a long time sketching out the chain of normal subgroups for S_4.

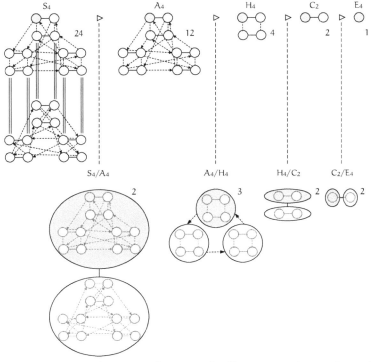

Decomposition of symmetric group S_4 ($S_4 \triangleright A_4 \triangleright H_4 \triangleright C_2 \triangleright E_4$).

When we were finally done, we all worked together to create several posters. The graphs we'd created, definitions of terms, examples of ladder diagrams...

By the time we were done it was pitch dark outside. And what a day it had been. As we got ready to leave, I realized I'd probably never spent so much time at the Narabikura Library. Little did I know what was happening at the bottom of the hill...

9.5 The Form of Feelings

9.5.1 Iodine

"What do you mean, we can't get home?" I said.

"Rail damage," Lisa said. She gave a light cough. "Repairs tomorrow."

"Wait," Tetra said. "You're saying there's something wrong with the train rails, and it won't be fixed until tomorrow?"

Lisa took control of things with a punctiliousness that belied her age. She gathered everyone who had been stranded—some forty or so college and high school students, and five junior high students—in the lecture hall Iodine, where she told everyone about the train problems. All trains on our line were completely stopped, so everyone would have to spend the night at the library. Thankfully it was equipped with rooms for napping, so that wouldn't be too much of a problem. She told minors to notify their parents what was going on, and announced that the Galois Festival would go on tomorrow as planned.

She also proclaimed one rule: "Get some sleep." This was met with a round of booing, but she was adamant. "Lights out at 23:00."

She handed out room assignments she'd printed from her computer, separate quarters for boys and girls. Miruka, Tetra, Yuri, and I helped her distribute pillows and blankets to everyone's room.

9.5.2 Lights Out

Perhaps due to the threat of an enforced bedtime, everything was ready for the conference by nine-thirty. Even Lisa seemed impressed. By ten, everyone had gathered in Oxygen for a late dinner. Our group fell into a discussion of math as we ate. A couple of college students tried to sneak in some beers, but Lisa caught them and sent them off with a stern warning.

At ten forty-five, Lisa used a projector to display an announcement on the wall: "Lights out at 23:00, NO EXCEPTIONS. Work forbidden until 05:00." We set to cleaning up Oxygen, then headed off to our rooms.

Miruka waved me over to the lobby.

"Did you talk to Tetra?" she asked, pushing her glasses up her nose.

"Not about anything in particular," I said, becoming nervous for some reason.

"Never mind then. We're taking a tour tomorrow."

"A tour?"

"Of Galois's first paper."

"Oh, okay."

"Get as much sleep as you can."

"I'll try. Might be hard to shut down my brain, though. Been working it overtime today."

"Hmph... Maybe so."

Miruka tilted her head as if thinking of something. It was a rare thing for me to be alone with her at that hour, making me feel a bit awkward, like there was something I should say. Just then the main lights went out, leaving only the emergency lighting illuminating the room. The darkened nighttime library felt very different from what I was used to. The skylight above us reflected the small emergency lamps amidst the stars above us, creating a fantastical image.

"I guess it's time to—" I started, but Miruka silently and swiftly brought her face to mine. She paused, her eyes filling my field of vision, then pressed her lips against mine for the briefest moment. *So warm...*

"That should help you sleep."

She headed off toward the girls' rooms, leaving behind only a whiff of citrus.

> In other words, when a group G contains another H, the group G can be partitioned into groups, each of which is obtained by operating on the permutations of H with one and the same substitution, so that $G = H + HS + HS' + \cdots$. And also it can be decomposed into groups all of which have the same substitutions, so that $G = H + TH + T'H + \cdots$. These two kinds of decomposition do not ordinarily coincide. When they coincide the decomposition is said to be proper.
>
> ÉVARISTE GALOIS [15]
> *Trans. by Peter M. Neumann*

CHAPTER 10

Galois Theory

> Were I to die in a traffic accident today, the data in my files would be comprehensible by no one. Putting that data in a written form that anyone can understand—in other words, writing papers—is something like creating my last will and testament.
>
> HIROSHI MORI
> *The Silent World of Dr. Kishima*

10.1 THE GALOIS FESTIVAL

10.1.1 A Chronology

"Over here! Over here!" Tetra called out, waving to me.

"You overslept," Yuri complained.

"Good morning, sleepyhead," a clear-eyed Miruka said over her cup of coffee. I could smell fresh croissants. I'd had breakfast with Yuri many times before, but seeing school friends first thing in the morning was a new experience.

We were back in Oxygen, the cafe on the third floor of the Narabikura Library, where we'd been forced to spend the night due to an unforeseen incident. It was the day of the Galois Festival, which the library was hosting, so the place was filled with signs and posters that gave it a festive atmosphere. The festival was planned as a way to teach the general public about Galois theory and was being planned and staffed by the math aficionados that frequented the library.

"Did you see the pamphlet?" Yuri asked, handing me a booklet with "Galois Festival" printed across the top. She opened it to a page featuring a chronology of Galois's life.

Age	Date	
0	Oct 1811	(25th) Évariste Galois is born.
11	Oct 1823	Enters Collège Royal de Louis-le-Grand.
15	Jan 1827	Held back from further studies, begins study of mathematics.
16	Aug 1828	Fails first attempt at entrance exam for École Polytechnique.
	Oct 1828	Meets teacher Louis Richard.
17	May 1829	Submits paper to the French Academy of Sciences. Augustin-Louis Cauchy suggests he apply for the Academy's Grand Prize in Mathematics.
	Jul 1829	Galois's father commits suicide. Fails second attempt at entrance exam for École Polytechnique, disqualifying him from acceptance at that school.
	Aug 1829	Applies to and is accepted at École Normale Supérieure.
18	Feb 1830	Presents a paper for the Academy's Grand Prize in Mathematics (Lost due to death of Fourier, his reviewer).
19	Jan 1831	Expelled for revolutionary activity due to a letter printed in a newspaper. Submits paper to the French Academy of Sciences, rejected in July by reviewers Poisson and Lacroix. (This is the earliest paper by Galois on solvability of equations that still exists.)
	May 1831	Arrested and imprisoned for revolutionary activity.
	Jun 1831	Acquitted of any crime.
	Jul 1831	Re-arrested while crossing the Pont-Neuf bridge, returns to prison.
20	Dec 1831	Found guilty (imprisoned until 29 Apr 1832).
	Mar 1832	Transferred to a sanatorium due to a cholera outbreak. Meets Stéphanie, daughter of a doctor there.
	May 1832	(29th) Writes last letter, to his friend Auguste Chevalier, revising his first paper on solvability of equations.
		(30th) Fights in a duel.
		(31st) Dies in hospital.

Chronology of Galois.

"Such a misfortunate life," Miruka said. "Fails his first shot at the entrance exam for the school he wanted to go to. His father commits suicide just before his second attempt. He fails his second attempt as well. He submits a paper to the French Academy, but it's lost because his reviewer died. He writes a historical paper reviewing the algebraic solvability of equations, but it's rejected. Imprisoned for getting involved in the political activism that followed the French Revolution. And after all that, he dies in a duel."

"It's that dying in a duel thing that really gets me," Tetra said.

"I think it's kind of cool!" Yuri said.

"Yuri!" Miruka snapped. The cafe fell silent, and the other patrons all looked our way. Miruka lowered her voice. "Don't say things like that."

"I'm sorry." Yuri's apology sounded sincere.

After a silence, Miruka spoke again. "Few mathematicians have endured a life like Galois's. There's been a lot of speculation about what his duel was all about. Maybe it had something to do with the revolution. Maybe he was betrayed by a colleague. Possibly a quarrel over a girl. Whatever the reason, one thing remains the same." Miruka closed her eyes. "He didn't live to see his twenty-first birthday."

There was another silence, until it was broken by Lisa's cough. Miruka reopened her eyes.

"Galois is dead, but mathematics lives on. Let's take a tour of his first paper and see how he moved things along."

10.1.2 Galois's First Paper

"You mentioned that paper yesterday, right?" Tetra asked.

Miruka nodded. "Solvability of equations was one of the themes of Galois's research. He intended to summarize his theory of equations in three papers. The first was what he called his *Premier Mémoire*. He submitted that paper for an award, but it was lost. So he rewrote it and resubmitted, but this time it was rejected by Poisson. Still, we refer to it as his first paper. It concerned the 'necessary and sufficient conditions' for algebraically solvable polynomials. In other words, he wanted to take a given equation, see if it meets certain conditions, and thereby determine whether it was solvable through algebraic operations. Of course, actually doing so may be quite involved, but still...

"That first paper was dated January 16th, 1831. Galois continued refining it up until May 29th the following year, the night before his duel. Knowing he might die the following day, he included some refinements in a letter to his friend, Auguste Chevalier."

"Wow, that's... awful," I said.

"Which is why we should repay him by trying to decipher the message he left behind."

"Did he write it in French?" Tetra asked.

"He did, but of course it's been translated into many other languages. But honestly, it's kind of hard to follow. That was Poisson's chief complaint, in fact, and why he didn't understand what Galois was saying, despite being one of the foremost mathematicians of his time. There were many reasons for this. Galois omits too many terms, for one, and didn't even know many of the terms related to groups and fields that we use today. But I think the biggest reason why his paper was hard to follow was because his conditions for solvability weren't given in terms of coefficients, but as substitution groups of solutions."

"That does sound hard," Tetra said.

"It is hard," Miruka acknowledged. "But we have one major advantage over Galois—nearly two centuries of evolution in mathematical terms and concepts. Using those, it isn't all that difficult to make sense of what he wanted to say in his first paper."

"Even for me?" Yuri asked.

"To an extent. That's what I want to do with the display posters I've made. The order and depth of their arguments follow Galois's first paper. I want to respect the historical development of Galois theory, but introduce some modern notation and terms that can make it a little easier to digest. I'm not giving a complete proof, but hope that viewers will at least understand the assertions of the theory."

"To make sure I've gotten this straight," Tetra said, "you said Galois's first paper is about the necessary and sufficient conditions for algebraically solving equations, right?"

"That's right. Its title was *Mémoire sur les conditions de résolubilité des équations par radicaux*, or 'Memoir on the conditions for solvability of equations by radicals.' More specifically it's about one principle, the necessary and sufficient conditions for algebraically solving equations, and one application, the necessary and sufficient conditions for algebraically solving a certain kind of prime degree equation. We're going to read through the 'principle' part."

"We're going to read it?" Tetra asked. "How long is it?"

"Surprisingly short, less than twenty pages in the original. That section defines some terms, then gives four lemmas and five theorems. That's all it takes."

Miruka flipped through the pamphlet, then pointed to a list of bullet items.

- Definition 1: Reducibility and irreducibility

- Definition 2: Substitution groups

- Lemma 1: Properties of irreducible polynomials

- Lemma 2: Creating V from roots

- Lemma 3: Recovering roots from V

- Lemma 4: Conjugates of V

- Theorem 1: Definition of the Galois group of a polynomial

- Theorem 2: Reduction of the Galois group of a polynomial

- Theorem 3: Adjoining all roots of an auxiliary equation

- Theorem 4: Properties of the reduced Galois group

- Theorem 5: Necessary and sufficient conditions for algebraically solving an equation

"Four lemmas and five theorems, huh?" Tetra said, squinting at the list.

"What's a lemma?" Yuri asked.

"Something like a mini-theorem you use to prove the theorem that's your main goal," I said.

"And once we understand these, we'll understand which equations can be solved?" Tetra asked.

"That's the goal," Miruka said, standing. "Let's get started."

We cleared away our plates, and Lisa directed us toward the exhibit rooms.

"I know this won't be too hard, so long as Miruka is here!" Yuri said.

"I know we'll be able to follow the logic, so long as Yuri is here!" Tetra said.

"I'm sure the language won't be a problem, so long as Tetra is here!" I said.

"And so long as you're here—" Miruka started, but didn't finish.

"Everything's fine," Lisa said.

And so we set off on our journey through Galois's first paper, with Miruka as our guide.

"First stop," Lisa said, leading us into the first room.

10.2 Definitions

10.2.1 Definition of Reducibility and Irreducibility

"Galois started off by defining some terms," Miruka said.

Reducibility and irreducibility

Let $f(x)$ be a polynomial in x over coefficient field K. Polynomial $f(x)$ is *reducible* over K if it can be factored over K, and *irreducible* over K otherwise.

"This isn't straight out of Galois's paper. He didn't use the term 'field,' for example. He talked about 'rationality' instead, which he used to represent numbers that can be derived through arithmetic. So he may not have quite conceptualized fields overall, but he was definitely aware of *elements* of fields, which he described using phrases like 'rationally known numbers' or 'rational quantities.' As modern mathematicians, however, we'll talk about fields."

"Wasn't it Dedekind who defined fields?" I asked.

Miruka nodded. "Except that Dedekind called them *Körper*, or 'bodies.'"

"I guess that's where the 'K' comes from," Tetra said.

"In any case," Miruka said, "the important thing here is reducibility and irreducibility. When we talk about whether a polynomial can be factored, we have to be clear about what field we're factoring over."

"Why's that?" Yuri asked.

"Because without specifying that, we can't answer the question. How about this, for example?"

Can the polynomial $x^2 + 1$ be factored?

"It can't!" Yuri said immediately.

"Actually," Tetra said, "we can factor it like this."

$$x^2 + 1 = (x+i)(x-i)$$

"Oh, sure, if you break out the imaginary numbers and all."

"Exactly why we need to say what field we're working with," Miruka said. "If we're limiting ourselves to the field of rationals $\mathbb{Q}$, then no, $x^2 + 1$ can't be factored. It's irreducible, in other words. If we're thinking of coefficients in the field of complex numbers $\mathbb{C}$, however, that's a different story. $x^2 + 1$ can be factored into two polynomials, $x + i$ and $x - i$. It's reducible."

· $x^2 + 1$ is irreducible over the field of rationals $\mathbb{Q}$

· $x^2 + 1$ is reducible over the field of complex numbers $\mathbb{C}$

"Galois himself gives a similar example in his paper. The fact that we can come up with our own example shows that we understand the assertion in that paper." Miruka nodded. "Next, Galois thought about what he called 'known quantities,' more specifically, 'those numbers that can be created from rational expressions involving numbers belonging to the set of all rational numbers adjoined with certain other numbers.' In other words, he was thinking in terms of fields newly created by adjoining numbers to other fields."

"By 'rational expressions,' you mean expressions using arithmetic, right?" Tetra asked.

"I do," Miruka said. "Be sure not to confuse rational *expressions* with rational *numbers*. For example, if we let $\phi(x) = \frac{x+1}{3}$, then $\phi(x)$ is a rational expression in x, and $\phi(1) = \frac{2}{3}$ is a rational number. However, $\phi(\sqrt{2}) = \frac{\sqrt{2}+1}{3}$ is *not* a rational number. We good?" We all nodded. "Okay. So to summarize, we've introduced the terms reducible, irreducible, rational expression, known quantities, and adjoin. These will be the guideposts leading us through the algebraic forest we're about to enter."

"So in the end, Galois was trying to create field extensions?" I asked. "I mean, he used things like the field of rationals $\mathbb{Q}$, and coefficient adjunctions to that field $\mathbb{Q}(a, b)$, and even adjunctions of nth root operations like $\mathbb{Q}(a, b, \sqrt{2}, \sqrt[3]{2})$."

"You could say that. When Galois wrote about using rational expressions to represent numbers, he was basically describing fields. His 'known quantities' are those numbers that can be used as coefficients for each field extension, and we'll be seeing more of those when we take a closer look at field extensions. But before we get to that, we need to talk about substitution groups."

10.2.2 Definition of Substitution Groups

"Galois next introduced the concept of substitution groups, in an attempt to implement Lagrange's hint that he should be using permutations of roots. We'll call some arrangement of a finite number of roots a *permutation*, and a rearrangement of those roots a *substitution*. Galois considered not only specific substitutions that changed the arrangement of a given permutation, but collections of those rearrangements into what he called a '*groupe*,' from which we get the obvious term. Here's essentially how he described it."

> **Galois's definition of substitution groups**
>
> "When we wish to group some substitutions, we make them all begin from one and the same permutation. As the concern is always with questions where the original disposition of the letters has no influence, in the groups that we will consider one must have the same substitutions whichever permutation it is from which one starts. Therefore, if in such a group one has substitutions S and T, one is sure to have the substitution ST."

"This bit about the original arrangement not mattering may be a little confusing, but the substitution groups Galois is thinking about here are what we would describe in modern terminology as subgroups of symmetric groups."

"So Miruka," Yuri said, "these substitution groups are basically the groups we can make using ladder diagrams?"

Miruka nodded. "That's right. The group created by all ladder diagrams with n vertical lines is the symmetric group S_n, and we're referring to a subgroup of S_n as a substitution group."

Symmetric group S_n	The group created from all ladder diagrams with n vertical lines
Substitution group	Subgroup of a symmetric group

"To be more specific," Miruka continued, "Galois considered permutations of the solutions to equations, namely the roots of the polynomial. For example, let's arrange the solutions to a fourth-degree equation, $\alpha_1, \alpha_2, \alpha_3, \alpha_4$."

$$\alpha_1 \; \alpha_2 \; \alpha_3 \; \alpha_4$$

"Of course, that's just one possible permutation. We also could have done it like this."

$$\alpha_1 \; \alpha_3 \; \alpha_4 \; \alpha_2$$

"This rearrangement of permutations from $\alpha_1 \, \alpha_2 \, \alpha_3 \, \alpha_4$ to $\alpha_1 \, \alpha_3 \, \alpha_4 \, \alpha_2$ is a substitution. The subscript 1 stays as 1, while 2 becomes 3, 3 becomes 4, and 4 becomes 2, which looks like this."

"In Yuri's notation, this substitution would be [1 3 4 2]."

$$\alpha_1\, \alpha_2\, \alpha_3\, \alpha_4 \xrightarrow{[1342]} \alpha_1\, \alpha_3\, \alpha_4\, \alpha_2$$

"Applying this substitution to a rational expression created using the roots changes that expression."

$$\frac{\alpha_1 + \alpha_2\alpha_3}{\alpha_1\alpha_4} \xrightarrow{[1342]} \frac{\alpha_1 + \alpha_3\alpha_4}{\alpha_1\alpha_2}$$

"This substitution [1 3 4 2] is one element in the fourth-degree symmetric group S_4. Let's give this substitution a name. We'll call it 'sigma.'"

$$\sigma = [1\,3\,4\,2] = \begin{pmatrix} 1 & 2 & 3 & 4 \\ 1 & 3 & 4 & 2 \end{pmatrix}$$

"So when we apply σ to $\frac{\alpha_1 + \alpha_2\alpha_3}{\alpha_1\alpha_4}$ to rearrange the roots, we can write that like this."

$$\sigma\left(\frac{\alpha_1 + \alpha_2\alpha_3}{\alpha_1\alpha_4}\right) = \frac{\alpha_1 + \alpha_3\alpha_4}{\alpha_1\alpha_2}$$

"This means we've used σ to rearrange the subscripts in the rational expression $\frac{x_1 + x_2 x_3}{x_1 x_4}$, then substituted the roots."

$$\frac{x_1 + x_2 x_3}{x_1 x_4} \xrightarrow{\text{substitution by } \sigma} \frac{x_1 + x_3 x_4}{x_1 x_2} \xrightarrow{\text{substitute roots}} \frac{\alpha_1 + \alpha_3\alpha_4}{\alpha_1\alpha_2}$$

"These are the permutations and substitutions that Galois was interested in."

10.2.3 Two Worlds

Tetra was rapidly nodding. "Okay, okay. So we're substituting the roots used in a rational expression to make another rational expression, right?"

"More to the point," Yuri added, "all the hullabaloo over Galois theory is basically about ladder diagrams?"

"In a sense," Miruka said. "What Galois did is associate ladder diagrams—namely, the theory of substitution groups—to the solvability of equations."

"That's...oddly reassuring," Tetra said.

"Sure, because we're both pretty good with ladder diagrams!" Yuri added.

"Have you noticed the two worlds we're dealing with?" Miruka asked.

"Well, when you put it that way..." Tetra said.

"In the context of Galois's first paper, when we think about the reducibility or irreducibility of polynomials we're in the world of fields. When we consider substitution groups, we're in the world of groups. Galois theory is a bridge between these two worlds." Miruka winked. "Time to move on, to the next part of Galois's first paper."

"This way," Lisa said, guiding us to the next room.

10.3 LEMMAS

10.3.1 Lemma 1: Properties of Irreducible Polynomials

| Lem. 1 | → Lem. 2 → Lem. 3 → Lem. 4 → Th. 1 → Th. 2 → Th. 3 → Th. 4 → Th. 5

"Galois's first paper used a lemma something like this," Miruka said, pointing to a poster displayed in the room we'd entered. "Think of this as a supplementary theorem that we'll use later."

Lemma 1 (Properties of irreducible polynomials)

Let $f(x)$ be a polynomial over field K, and let $p(x)$ be an irreducible polynomial over K. If $f(x)$ and $p(x)$ share a common root, then $p(x)$ evenly divides $f(x)$.

"I don't get it," Yuri said. "What's a polynomial over K?"

"A polynomial whose coefficients are all in field K," I said. "For example, the polynomial $x^2 + 5x + 3$ has coefficients $1, 5, 3$, all of which belong to the field of rational numbers $\mathbb{Q}$. So we can call $x^2 + 5x + 3$ a polynomial over $\mathbb{Q}$."

Galois Theory

"Ah, it's all about the coefficients. Okay, I'm good."

"Of course, we can also consider it as a polynomial over the field $\mathbb{Q}(\sqrt{2})$." Miruka added. "The important thing is that we're conscious of the coefficient field. By the way, this Lemma 1 is similar to this theorem for integers."

> Let P be a prime and N be an integer. If P and N share a common prime factor, then P evenly divides N.

"Wait, you're sure about that?" Yuri asked, cocking her head to think. "Oh, of course! Since P is a prime it can't be split up, so the prime factor is P itself!"

"Kind of a hard explanation to follow," Miruka said. "Can you give us an example, Tetra?"

Tetra's example came immediately. "Let N = 12 and P = 3. Then 3 is a common prime factor between the two, and P evenly divides 12. You're right, that is a lot like this Lemma 1. Irreducible polynomials are like primes, and reducible polynomials are like composite numbers."

World of polynomials	←---→	World of integers
Polynomials	←---→	Integers
Irreducible polynomials	←---→	Primes
Reducible polynomials	←---→	Composite numbers
Common roots	←---→	Common prime factors

"Very good. Yes, irreducible polynomials are a lot like primes. Just like you can't decompose a prime into component integers, you can't decompose an irreducible polynomial into multiple factors. So if an irreducible polynomial P(x) and a polynomial f(x) share a root, then P(x) evenly divides f(x)."

"Sure, I see that," Tetra said.

"Put another way," Miruka continued, "if a polynomial f(x) and an irreducible polynomial p(x) have even one shared root, then f(x) must share *all* the roots of p(x)."

"How strange, for polynomials and integers to be so similar like that," I said.

"Because both are rings with uniqueness of prime factorization."

"I'm good with the integers," Yuri said. "Polynomials, not so much."

"I can give an example," Tetra said. "We'll use $f(x) = x^4 - 1$ as a polynomial over the field of rationals $\mathbb{Q}$, and $p(x) = x^2 + 1$ as the irreducible polynomial." Yuri jotted these polynomials down and nodded for Tetra to continue. "Then $f(x)$ and $p(x)$ have the common root i, since $f(i) = i^4 - 1 = 0$ and $p(i) = i^2 + 1 = 0$. We can decompose $f(x)$ over $\mathbb{Q}$ like this."

$$f(x) = \underbrace{(x^2 + 1)}(x+1)(x-1)$$

"So, see? $f(x)$ has $x^2 + 1$, in other words $p(x)$ itself, as a factor."

"How are you able to just whip out an example like that?" Yuri asked.

"Because we've used this example before, when we learned about cyclotomic polynomials.[1] I remember Miruka saying they work a lot like primes do. Besides, we just used $p(x) = x^2 + 1$ as an example when defining irreducibility." Tetra paused and took a breath. "So many ways to say the same thing..." she said, half to herself.

- The equation $f(x) = 0$ has $x = \alpha$ as a solution.

- The polynomial $f(x)$ has α as a root.

- If we substitute $x = \alpha$ into $f(x)$, we get $f(\alpha) = 0$.

- $f(x)$ has $x - \alpha$ as a factor.

I nodded. "This is what it means to remain conscious of the field that α belongs to when we decompose into factors. Are we working in $\mathbb{Q}$? Or in $\mathbb{Q}(\sqrt{2})$? Or in $\mathbb{R}$ or in $\mathbb{C}$?"

Miruka looked up as if remembering something. "Speaking of which, in Galois's first paper, he assumed we would always be working in a field that contained the field of rational numbers. Specifically, he didn't consider finite fields, fields having only a finite number of elements."

---

[1] See Sec. 4.2.6, p. 116

"Hey Tetra," Yuri said. "Do you study hard math like this in high school? Is that when I'm finally going to understand all this?"

"Mmm... I guess it's not so much getting to a certain point in school as getting to where you can learn on your own. Don't wait for school to teach you, study for yourself."

"Back to Lemma 1," Miruka said. "To summarize, it says that if a polynomial shares *any* root with an irreducible polynomial, then it shares *all* of the roots of that irreducible polynomial. Moving on, we're going to construct a very interesting rational polynomial, V."

"This way," Lisa said, guiding us to the next room.

10.3.2 Lemma 2: Creating V from Roots

Lem. 1 → ⎡Lem. 2⎤ → Lem. 3 → Lem. 4 → Th. 1 → Th. 2 → Th. 3 → Th. 4 → Th. 5

The next room we entered had another of Miruka's posters.

Lemma 2 (Creating V from roots)

Let $f(x)$ be a polynomial over K with no repeated roots, and let $\alpha_1, \alpha_2, \alpha_3, \ldots, \alpha_m$ be the roots of $f(x)$. We can create a function $\phi(x_1, x_2, x_3, \ldots, x_m)$ for which every change of the permutation of the roots in the expression

$$V = \phi(\alpha_1, \alpha_2, \alpha_3, \ldots, \alpha_m)$$

results in a different value. Furthermore, this function can be chosen to be a linear function with integer coefficients $(k_1, k_2, k_3, \ldots, k_m)$:

$$\phi(x_1, x_2, x_3, \ldots, x_m) = k_1 x_1 + k_2 x_2 + k_3 x_3 + \cdots + k_m x_m.$$

"Sooo haaard..." Yuri whimpered.

"Not really," I said. "It just seems that way because it's written to be as general as possible."

"Examples are key to understanding!" Tetra said. "Let me make an example, using $m = 2$, and $x^2 + 1$ as a polynomial $f(x)$ over $\mathbb{Q}$.

Then the permutations of the roots would look like this."

$$\alpha_1 \alpha_2 \quad \text{and} \quad \alpha_2 \alpha_1$$

"There's just two, so we just want to make sure we can create an expression that has a different value when the roots are exchanged. So, let's see... Oh, this would work!"

$$V = \alpha_1 - \alpha_2$$

"So the phi function would be this."

$$\varphi(x_1, x_2) = x_1 - x_2$$

"Did I do that right?" Tetra asked.
Miruka nodded. "Well done."
"All that gobbledygook, just for this?" Yuri said, throwing up her hands.
"That's the gist of it," Miruka said. "Let's dive into the details. For example, when we say this rational expression V takes a different value according to the permutation of the roots, we're saying that if σ (sigma) and τ (tau) are elements of a symmetric group with $\sigma \neq \tau$, then $\sigma(V) \neq \tau(V)$. Let's verify this using Tetra's example. Letting $\sigma_1 = [1\,2], \sigma_2 = [2\,1] \ldots$"

$$\sigma_1(V) = [1\,2](V) = [1\,2](\varphi(\alpha_1, \alpha_2))$$
$$= \varphi(\alpha_1, \alpha_2) = \alpha_1 - \alpha_2 = i - (-i) = +2i$$
$$\sigma_2(V) = [2\,1](V) = [2\,1](\varphi(\alpha_1, \alpha_2))$$
$$= \varphi(\alpha_2, \alpha_1) = \alpha_2 - \alpha_1 = (-i) - i = -2i$$

"And sure enough, $\sigma_1(V) \neq \sigma_2(V)$."
"Okay, I'm getting more comfortable with V and φ now," Tetra said.
"This V is something like the opposite of a symmetric polynomial," Miruka said. "A symmetric polynomial is one that doesn't change when you apply a substitution to its variables. Instead, V is a polynomial that will *always* change when you apply a substitution."

"Huh. So why was Galois interested in it?"

"From the perspective of fields, he wanted to use V as an adjunction to a field in a theorem down the road. From the perspective of groups, V plays a key role in substitutions of roots. What we want to get from Lemma 2 for now, though, is that we can create this polynomial V that will always change in value when the roots of f are substituted. If we're good with that, it's time to move on to Lemma 3."

"This way," Lisa said, already heading out the door.

10.3.3 Lemma 3: Representing Roots in terms of V

Lem. 1 → Lem. 2 → Lem. 3 → Lem. 4 → Th. 1 → Th. 2 → Th. 3 → Th. 4 → Th. 5

The poster in the third room was suspiciously short.

Lemma 3 (Representing roots in terms of V)

We can use the V of Lemma 2 to represent $\alpha_1, \alpha_2, \alpha_3, \ldots, \alpha_m$, the roots of $f(x)$. Specifically, there exist rational functions $\phi_1(x), \phi_2(x), \phi_3(x), \ldots, \phi_m(x)$ over K satisfying

$$\alpha_1 = \phi_1(V), \ \alpha_2 = \phi_2(V), \ \alpha_3 = \phi_3(V), \ \ldots, \ \alpha_m = \phi_m(V).$$

"Let's look at this in terms of Tetra's example for Lemma 2," Miruka said.

"That was $f(x) = x^2 + 1$, with $\alpha_1 = i$ and $\alpha_2 = -i$," Tetra said.

"And $V = \phi(\alpha_1, \alpha_2) = \alpha_1 - \alpha_2 = 2i$," I added.

The three of us turned to Yuri.

"Huh? What?" she sputtered. "Am I supposed to add something?"

"We're using V to create α_1 and α_2, right?" I said.

"Um...okay?" she said. "Oh, wait. Before we used α_1 and α_2 to create V, didn't we! This time we're going the other way around! We're using V to create α_1 and α_2!"

I nodded. "That's right."

Yuri took a moment to write in her notebook, then showed us the result.

$$\alpha_1 = \frac{V}{2}, \ \alpha_2 = -\frac{V}{2}$$

"Not so hard, since $\alpha_1 = i$, $\alpha_2 = -i$, and $V = 2i$," she said.

"Good," Miruka said. "We go the other direction from the V we created using roots, and write the roots as rational functions of V. So $\phi_1(x)$ and $\phi_2(x)$ look like this."

$$\phi_1(x) = \frac{x}{2}, \quad \phi_2(x) = -\frac{x}{2}$$

Tetra sighed. "I understand that we can create V using roots, and we can create roots using V, but I still have no idea where we're going with all this. These lemmas are really from Galois's first paper?"

"We're talking about representing things as rational expressions, but if that becomes your sole focus, it's easy to lose track of the big picture. So let's turn our attention back to fields. If we think in terms of field extensions, we can write out what Lemma 3 says in just one line."

$$K(\alpha_1, \alpha_2, \alpha_3, \ldots, \alpha_m) = K(V)$$

I palmed my forehead. "So *that's* what this means!"

"Uh...it is?" Tetra said.

"It says we can create $K(\alpha_1, \alpha_2, \alpha_3, \ldots, \alpha_m)$ just from the adjunction of V to K."

"This $K(\alpha_1, \alpha_2, \alpha_3, \ldots, \alpha_m)$ is a field?"

"A very important one," Miruka said. "It's the coefficient field K, to which we've adjoined all the roots of some function $f(x)$. So what do you think makes it important?"

Yuri and Tetra shook their heads.

"Because if we're working in this field $K(\alpha_1, \alpha_2, \alpha_3, \ldots, \alpha_m)$, we can factor $f(x)$ as a product of linear expressions," I said. "If we want the highest-degree coefficient to be 1, it looks like this."

$$f(x) = (x - \alpha_1)(x - \alpha_2)(x - \alpha_3) \cdots (x - \alpha_m)$$

Miruka nodded. "Exactly. The field $K(\alpha_1, \alpha_2, \alpha_3, \ldots, \alpha_m)$ is called the 'splitting field' of the polynomial $f(x)$."

My ears perked up at this mention of splitting fields, reminding me of Miruka's letter.[2]

---

[2] See Sec. 8.3.7, p. 272

"So from $K(\alpha_1, \alpha_2, \alpha_3, \ldots, \alpha_m) = K(V)$," Miruka continued, "we see that we can create the splitting field just by adjoining this one element V to the field K."

"I see. So that's why it's worthwhile to consider $K(V)$." Tetra cocked her head. "Well, maybe I don't totally see, but I'm getting there."

"In the next Lemma, number 4, we consider numbers that share a 'yoke' with V."

"This way," Lisa said.

10.3.4   Lemma 4: Conjugates of V

Lem. 1 → Lem. 2 → Lem. 3 → ⌜Lem. 4⌝ → Th. 1 → Th. 2 → Th. 3 → Th. 4 → Th. 5

In the next room, Miruka moved to a whiteboard next to the poster on display.

"We start with the same prerequisites as in Lemmas 1 through 3," she said, writing:

- $f(x)$ is a polynomial over field K with no repeated roots.

- $\alpha_1, \alpha_2, \alpha_3, \ldots, \alpha_m$ are the roots of $f(x)$ (i.e., solutions to the equation $f(x) = 0$).

- V is a rational expression over K created from roots $\alpha_1, \alpha_2, \alpha_3, \ldots, \alpha_m$, and has different values for each permutation of the roots (Lemma 2).

- $\phi_1(x), \phi_2(x), \phi_3(x), \ldots, \phi_m(x)$ are rational functions over K, with $\alpha_k = \phi_k(V)$ (Lemma 3).

Lemma 4 (Conjugates of V)

Let $f_V(x)$ be the minimal polynomial over field K with V as a root, and let $V_1, V_2, V_3, \ldots, V_n$ be the roots of $f_V(x)$. Then, for each $k = 1, 2, 3, \ldots, n$,

$$\phi_1(V_k), \ \phi_2(V_k), \ \phi_3(V_k), \ \ldots, \ \phi_m(V_k)$$

is a permutation of the roots of the polynomial $f(x)$.

"Here we go again," Yuri moaned. "Drowning in a sea of letters..."

"We can do this!" Tetra said. "Let's just work our way through it."

"We've newly created this function $f_V(x)$ as the minimal polynomial with V as a root," Miruka said. "What we're interested in is a field $K(V)$ that's equal to the splitting field. That's why we're interested in the adjoined element V, as well as other elements that 'share a yoke' with V."

"That would be $V_1, V_2, V_3, \ldots, V_n$, the conjugates of V," I said.

"That's right. $f_V(x)$ looks like this."

$$f_V(x) = (x - V_1)(x - V_2)(x - V_3) \cdots (x - V_n)$$

"If we expand this, all the coefficients will be elements in K."

"Right, because $f_V(x)$ is a polynomial over K," I said. "$V_1, V_2, V_3, \ldots, V_n$ are the roots of $f_V(x)$, and one of those is equal to V. For example, we can let $V = V_1$. Then Lemma 3 gives us this."

$$\alpha_1 = \phi_1(V_1), \; \alpha_2 = \phi_2(V_1), \; \alpha_3 = \phi_3(V_1), \; \ldots, \; \alpha_m = \phi_m(V_1)$$

"In other words, we've used the root V_1 of $f_V(x)$ to represent the roots of $f(x)$."

"Okay, sure. I see that," Tetra said.

Miruka continued, "Lemma 4 says that in place of V, we can use any of $V_1, V_2, V_3, \ldots, V_n$ to create the roots $\alpha_1, \alpha_2, \alpha_3, \ldots, \alpha_m$ of $f(x)$. In other words, numbers that are yoked with V can serve as a substitute for V."

"That seems awfully... convenient," Tetra said, doubtfully.

"How about we use Tetra's example to try it for ourselves?" I suggested. "That was $V = 2i$, right? The minimal polynomial over $\mathbb{Q}$ with V as a root would be $x^2 + 4$, and the roots of $x^2 + 4$ are $V_1 = 2i, V_2 = -2i$. So we should make sure that when $x = V_1$ and $x = V_2$ with $\phi_1(x) = \frac{x}{2}, \phi_2(x) = -\frac{x}{2}$, the roots of $f(x)$ pop out."

- If $x = V_1$, we get $\phi_1(V_1) = \frac{2i}{2} = i$, $\phi_2(V_1) = -\frac{2i}{2} = -i$, so we do indeed get the roots for $f(x) = x^2 + 1$. Not surprising, since $V = V_1$.

- If $x = V_2$, we get $\phi_1(V_2) = \frac{-2i}{2} = -i, \phi_2(V_2) = -\frac{-2i}{2} = i$, so once again get the roots for $f(x) = x^2 + 1$. This agrees with Lemma 4.

"Okay, I'm starting to get this," Tetra said. "Once again, clarity through example."

"Let's take a closer look at this example, though," Miruka said. "When V_1 is used, $i, -i$ is the permutation α_1, α_2, and when V_2 is used, $-i, i$ is the permutation α_2, α_1. This is a substitution of the permutation α_1, α_2. Generally, when we use $V_1, V_2, V_3, \ldots, V_n$, we can create a set of n permutations of roots, like this."

- Using V_1, the permutation is $\phi_1(V_1), \phi_2(V_1), \ldots, \phi_m(V_1)$.

- Using V_2, the permutation is $\phi_1(V_2), \phi_2(V_2), \ldots, \phi_m(V_2)$.

- Using V_3, the permutation is $\phi_1(V_3), \phi_2(V_3), \ldots, \phi_m(V_3)$.

- $\vdots$

- Using V_n, the permutation is $\phi_1(V_n), \phi_2(V_n), \ldots, \phi_m(V_n)$.

"This is so...so complex," Yuri said.

"I'm not sure I see where Galois was headed with all this," Tetra said.

"He wanted to use conjugate elements of V to create a set of permutations of roots, to use as a substitution group. He's using conjugate elements of V to create the elements of the substitution group, in other words each individual substitution. This is easy to see using Tetra's example."

- Using V_1, the permutation of roots is α_1, α_2 (corresponding to substitution $[1\,2]$).

- Using V_2, the permutation of roots is α_2, α_1 (corresponding to substitution $[2\,1]$).

"And that does it for the four Lemmas. We're ready to move on to Theorem 1 in Galois's paper."

"This way," Lisa said, heading toward the next room. But...

"Hold up, Lisa," Miruka said. "First, we need to quickly review the theme of that first paper. Do you remember what it was that Galois wanted to show, Yuri?"

"Wasn't it the conditions for solving equations?" Yuri answered.

"Close," Miruka said. "More precisely, the *necessary and sufficient* conditions for algebraically solving equations."

Yuri nodded. "Right."

Miruka pointed at me. "And what have we done in preparation for that?"

I started ticking off concepts on my fingers. "Reducibility and irreducibility. Substitution groups. Irreducible polynomials, and how they're similar to primes. That we can use roots to create V, which takes different values with each substitution. That we can write roots in terms of V. That we can create a splitting field from an adjunction of V. That we can create substitutions of roots by using conjugate elements as a replacement for V ..."

"So Galois needed all these tools to tackle equations?" Tetra asked.

"Indeed," Miruka said. "He didn't tackle them straight on, though."

Tetra cocked her head. "Meaning?"

"Galois's next step was to introduce the concept of the Galois group for a polynomial." Miruka placed a finger alongside her glasses, and spoke slowly, every word clearly enunciated. "Listen up. Understanding this point is extremely important. So, when starting to read Galois's first paper, what might one expect with regards to the necessary and sufficient conditions for algebraically solving an equation?"

No one offered an answer.

"I think," Miruka continued, "they would be hoping for a formula that used a given equation's coefficients to tell them whether it could be solved. Some kind of discriminant for the equation, perhaps. But Galois's first paper gives nothing of the sort. Rather, it uses not coefficients, but *roots* to determine solvability. Specifically, Galois gave the conditions for solvability in terms of the Galois group for the equation. This caused his reviewer Poisson some ... consternation."

"And that's what you were saying is so important? These Galois groups?" Tetra asked.

"Correct. What say we define what those are?"

"This way," Lisa said.

10.4 Theorems

10.4.1 Theorem 1: Definition of the Galois group of a Polynomial

Lem. 1 → Lem. 2 → Lem. 3 → Lem. 4 → $\boxed{\text{Th. 1}}$ → Th. 2 → Th. 3 → Th. 4 → Th. 5

"Galois's Theorem 1 defines the Galois group of a polynomial," Miruka began. "We want to know when we can and when we cannot algebraically solve an equation. Galois's method does not directly investigate that equation. Not directly from its coefficients, at least. It first finds the Galois group for the equation, and considers the equation from that perspective. Galois simply called this the 'group of the equation,' but today we would call it the Galois group for the equation. His Theorem 1 defines that group and, most importantly, guarantees its existence.

"The prerequisites for the theorem are the same as before. We're given a polynomial $f(x)$ that does not have repeated roots, and we consider the equation $f(x) = 0$. We name the roots of $f(x)$ as $\alpha_1, \alpha_2, \alpha_3, \ldots, \alpha_m$. In his Theorem 1, Galois focuses on rational expressions created from these roots. Namely..."

- Is the value of the rational expression known?

- Is the value of the rational expression invariant under substitution of those roots?

"Galois focused on these two points when defining the Galois group of a polynomial. He also gave a concrete construction of a Galois group."

> **Theorem 1 (Definition of the Galois group of a polynomial)**
>
> Given a rational expression r created from the roots of a polynomial $f(x)$ over field K, if the value of r belongs to field K (i.e., $r \in K$), we refer to r as *known*. If a substitution σ of those roots does not change the value of r (i.e., $\sigma(r) = r$), we say that r is *invariant* under σ.
>
> There will always exist a substitution group G for which the following two properties hold for any r:
>
> Property 1 (known if invariant): If the value of r is invariant under all substitutions in G, then the value of r is known.
>
> Property 2 (invariant if known): If the value of r is known, then r is invariant under all substitutions in G.
>
> This substitution group G is called the *Galois group* for the equation $f(x) = 0$ over K.

"Woosh," Tetra said. "Right over my head. I mean, I understand all the words here, but put together like this? Woosh."

"I see," Miruka said. "How about you, Yuri?"

"Well... I can see that the two properties here are kinda opposites, but..."

"But that's about it?" Miruka nodded, then turned to me. "Well?"

"I'm going to need an example before I'm comfortable with this," I said. "I think that should do it, though."

"Then an example you shall have."

We gathered around a circular table near the poster. The table was conveniently supplied with the tools of our trade—pencils and a stack of paper.

10.4.2 The Galois Group for $x^2 - 3x + 2 = 0$

"Okay, then. Let's talk about the Galois group for $x^2 - 3x + 2 = 0$," Miruka said. "We'll let the rational numbers be our field of coefficients. So in terms of Theorem 1 here, we're considering the polynomial $x^2 - 3x + 2 = 0$ over field $K = \mathbb{Q}$. Good so far?"

We all nodded.

"Great. So to cut to the chase, the Galois group for this polynomial is the trivial group, in other words the group containing only the identity element. That is the group that satisfies the properties needed in the Galois group for this equation."

"Why's that?" Yuri asked.

"Well, let's take a look. What are the roots of $x^2 - 3x + 2$?"

"Let's see... We can factor that as $(x-1)(x-2)$, so the roots are 1 and 2!"

"Good. Let's name them. We'll say $\alpha_1 = 1$ and $\alpha_2 = 2$. So Yuri, what does property 1 say?"

"That's the 'known if invariant' one, so, uh..."

Confirming Property 1 (known if invariant)

> If the value of r is invariant under all substitutions in substitution group G, then the value of r is known.

"Remember," Miruka prompted, "you want to think of a rational expression using the roots $\alpha_1 = 1, \alpha_2 = 2$, one whose value won't change under the substitution in the trivial group $E_2 = \{[1\,2]\}$." She pushed a sheet of paper in front of Yuri.

"Okaaay," Yuri said, turning to face it. "So when we move the roots around using this substitution... Hey, wait! The only substitution here is a plopper! Nothing changes!"

"That's right," Miruka said, nodding. "It becomes very clear when you work on an example, doesn't it? The only element in the trivial group E_2 is the identity element $[1\,2]$, so the permutation of roots doesn't change. That means the value of your rational expression won't change, either. For example, let's apply the substitution $\sigma = [1\,2]$ to $\alpha_1 - \alpha_2$."

$$\sigma(\alpha_1 - \alpha_2) = [1\,2](\alpha_1 - \alpha_2) = \alpha_1 - \alpha_2$$

"You're right! Nothing happened!" Yuri said. "$\alpha_1 - \alpha_2 = 1 - 2 = -1$, and $\sigma(\alpha_1 - \alpha_2) = \alpha_1 - \alpha_2 = 1 - 2 = -1$, so the value is -1 in both! It's invariant!"

"So does this 'known if invariant' property hold when we use E_2?"

"Um... Remind me what 'known' means?"

"We're using $\mathbb{Q}$ as our coefficient field, so any rational number is known."

Yuri spoke to herself as she thought. "A rational expression using the roots would be one that uses 1 and 2, which are both rational numbers... so yeah, sure! Those are known! And property 1 is true for E_2!"

"Well done," Miruka said, nodding. "So what about property 2?"

Confirming Property 2 (invariant if known)

> If the value of r is known, then r is invariant under all substitutions in substitution group G.

"Okay, so we want to think about this 'invariant if known' property using E_2, right?" Yuri said. "Which means, if the rational expression we created using 1 and 2 is known, in other words if its value is a rational number... Well *of course* the value of a rational expression using 1 and 2 will be a rational number! And if we substitute the roots using E_2, well the only thing there is a plopper, so nothing will change, so *of course* it will be invariant with the trivial group E_2! So sure, we're good so far as E_2 and property 2 are concerned!"

"And that's all we need," Miruka said. "Properties 1 and 2 both hold, so the trivial group E_2 is the Galois group for the equation $x^2 - 3x + 2 = 0$ over $\mathbb{Q}$."[3]

"Doesn't that seem kind of obvious, though?" I said. "After all, the value of any rational expression with rational roots will definitely be 'known,' and it certainly isn't surprising that the expression will be 'invariant' under the trivial group, since nothing changes, so..."

"So you've caught sight of our goal—rational roots and the fact that roots don't permute under the substitution in the trivial group."

"Uh, how's that our goal?"

"If a root is known, it belongs to the coefficient field. If it belongs to the coefficient field, we can write it using arithmetic operations. If we can do that, we can algebraically solve the equation. So if the

---

[3] Noting that $[2,1](\alpha_1 - \alpha_2) = (\alpha_2 - \alpha_1) = 2 - 1 = 1$ changes the known value -1 to 1 completes the demonstration that the Galois group is no bigger than E_2.

Galois group for an equation is the trivial group, it can be solved algebraically." Everyone remained silent, taking in what Miruka was saying. "So the Galois group for $x^2-3x+2=0$ is the trivial group E_2. This is the simplest example of a Galois group. In the terminology of fields, we can say this is simple because the roots belong to the coefficient field. In the terminology of groups, we can say this is simple because the roots don't permute. All well and good. So let's take a look at the most complex Galois group."

"The most *complex* Galois group? What's that?" Tetra asked.

"A group having all substitutions of the roots. In other words, a symmetric group. Given an equation having m roots, the most complex Galois group for that equation is the symmetric group S_m. Let's look at an example."

10.4.3 The Galois Group for $ax^2+bx+c=0$

"The Galois group for the general quadratic equation $ax^2+bx+c=0$ is the symmetric group S_2," Miruka said.

Tetra raised her hand. "What do you mean by the 'general' quadratic equation?"

"Just the quadratic equation with letters in place of numbers as coefficients. We would normally treat the a, b, c coefficients as numbers, but let's just consider them as letters for now. We can still use a, b, c when performing arithmetic with other numbers, like in $2a$ and b^2-4ac, but they'll just stick around as letters. For example, in the quadratic formula $\frac{-b \pm \sqrt{b^2-4ac}}{2a}$, the a, b, c are just there as letters. In this case, we can consider the coefficient field to be the adjunction of these letters a, b, c to the field of rational numbers $\mathbb{Q}$. In other words, the coefficient field is $\mathbb{Q}(a, b, c)$."

"For some reason that reminds me of raisin bread," Tetra said. "A loaf of $\mathbb{Q}$, with raisins a, b, c baked inside. No matter how you knead the dough, the raisins will be in there, somewhere."

Miruka laughed. "I like it. In any case, the a, b, c here are letters, with no computational relation between them. The quadratic equation with a loaf of raisin bread as its coefficient field is the *general* quadratic equation."

"Got it," Tetra said.

"We can use the quadratic formula to find the solutions to the general quadratic equation, like this."

$$\alpha_1 = \frac{-b + \sqrt{b^2 - 4ac}}{2a}, \ \alpha_2 = \frac{-b - \sqrt{b^2 - 4ac}}{2a}$$

"The Galois group for the general quadratic equation $ax^2 + bx + c = 0$ is the symmetric group S_2. Using Yuri's notation, that would be this."

$$S_2 = \big\{[1\,2], [2\,1]\big\}$$

"Let's make sure that's the case. Once again, we want to confirm 'known if invariant' and 'invariant if known.'"

Confirming Property 1 (known if invariant)

"A rational expression of the roots does not change when there are no swaps. So all we need to think about is a rational expression that's invariant under the S_2 element $[2\,1]$. A rational expression that's invariant after swapping the two roots by $[2\,1]$ would be a symmetric polynomial in those two roots. We can represent any symmetric polynomial in terms of *elementary* symmetric polynomials, so we should investigate whether the sums and products of those elementary symmetric polynomials give known values. To do that, we just need the relation between roots and coefficients."

[+] $\alpha_1 + \alpha_2 = -\frac{b}{a}$ is known, because it belongs to the coefficient field $\mathbb{Q}(a, b, c)$.

[×] Similarly, $\alpha_1 \alpha_2 = \frac{c}{a}$ is known, because it belongs to the coefficient field $\mathbb{Q}(a, b, c)$.

"Summing up, symmetric polynomials can be represented as combinations of elementary symmetric polynomials, and elementary symmetric polynomials are known, so symmetric polynomials are known. Thus, S_2 satisfies property 1."

Confirming Property 2 (invariant if known)

"Saying a rational expression r created from α_1, α_2 has a known value means that its value belongs to the coefficient field $\mathbb{Q}(a, b, c)$. From the relation between roots and coefficients, we know that r is a rational expression that can be expressed as a ratio of symmetric polynomials in α_1 and α_2. In other words, r is a symmetric polynomial in α_1 and α_2, and the value of r is invariant after swapping α_1 and α_2 by [2 1]. Of course, it will be invariant for [1 2] too. Therefore, S_2 also satisfies property 2.

"We've satisfied both properties, so we've shown that S_2 is the Galois group for $ax^2 + bx + c = 0$."

"Still just the first theorem, but I'm already lost," Yuri said.

"Let me summarize it for you then," Miruka said. "In a word, Theorem 1 just asserts that a Galois group exists for any equation. So what kind of group will an equation's Galois group be?"

"Oh, that's straight out of the definition. It will be a group where 'known if invariant' and 'invariant if known' is true."

"Exactly."

"I get that," Tetra said, "but it still isn't quite enough for me to feel like I've become friends with Mssr. Galois."

"Hmph. Okay, let's take a different perspective, then. We can classify the roots of a quadratic equation according to the discriminant $D = b^2 - 4ac$."

"Sure."

"Letting K be the coefficient field, those roots will always belong to the field extension $K(\sqrt{D})$. We know that's true because we can write those roots using arithmetic operations on the coefficients and $\sqrt{D}$."

"Right, using the quadratic formula."

"That's right. Well then, let's compare the coefficient field K and the field extension $K(\sqrt{D})$ to get a big-picture view of where the roots belong."

- If the roots *do not* belong to the coefficient field K, the Galois group *is not* the trivial group. Further, adjunction of $\sqrt{D}$ *will* extend the field.

- If the roots *do* belong to the coefficient field K, the Galois group *is* the trivial group. Further, adjunction of $\sqrt{D}$ *will not* extend the field.

"Huh," Tetra said, nodding slowly.

"Remember, our main interest is whether we can algebraically solve a given equation. The answer to that question depends on what kind of field its roots belong to. That, in turn, depends on what kind of group its Galois group is, which is why we're so interested in Galois groups."

"Okay, this is starting to make sense now. At least, I see how the Galois group for an equation is a really important concept. Still lots of 'don't get it' to clear up, though. About this 'known if invariant' and 'invariant if known' stuff in particular."

I was glad to see Tetra's perseverance was alive and well.

"Hmph." Miruka closed her eyes for several seconds. "Okay, let's talk about how the two properties of the Galois group for an equation captures its essence. Intuitively, at least."

Miruka pulled a fresh sheet of paper from the stack.

"Let G be the Galois group for the general quadratic equation $ax^2 + bx + c = 0$ over a field $K = \mathbb{Q}(a,b,c)$. We'll also say sigma is one element of that group, so $\sigma \in G$. Now we know that these statements hold."

$$\sigma(a) = a, \quad \sigma(b) = b, \quad \sigma(c) = c$$

"Since $K = \mathbb{Q}(a,b,c)$, each of a, b, c are known. Since 'invariant if known,' a substitution σ belonging to the Galois group will just 'pass on' the a, b, c. Now, letting the roots of $ax^2 + bx + c$ be α_1, α_2, we can say this."

$$a\alpha_1^2 + b\alpha_1 + c = 0$$

"Let's apply σ to both sides."

$$\sigma(a\alpha_1^2 + b\alpha_1 + c) = \sigma(0)$$

"This σ just passes on any coefficients, so a little thought gives us this."

$$a\sigma(\alpha_1^2) + b\sigma(\alpha_1) + c = 0$$

"We can also see that $\sigma(\alpha_1^2) = \sigma(\alpha_1 \alpha_1) = \sigma(\alpha_1)\sigma(\alpha_1) = \sigma(\alpha_1)^2$, so we get this."

$$a\sigma(\alpha_1)^2 + b\sigma(\alpha_1) + c = 0$$

"We can do the same thing with any rational expression created from the roots. Like this, for example."

$$\sigma\left(\frac{a\alpha_1 + b\alpha_2\alpha_1}{c\alpha_1^2\alpha_2}\right) = \frac{a\sigma(\alpha_1) + b\sigma(\alpha_2)\sigma(\alpha_1)}{c\sigma(\alpha_1)^2\sigma(\alpha_2)}$$

"Because Galois groups have the properties 'invariant if known' and 'known if invariant,' when we apply substitution σ—which remember is an element of the Galois group—to a rational expression created from the roots, that σ can slide right in, deep into that rational expression. In the end, this shows us which pairs of roots can be swapped. The Galois group contains all that information. The Galois group of a polynomial shows us that polynomial's form. Going one step further, this leads us to Emil Artin's wonderful idea of defining Galois groups as automorphism groups of fields. Artin used vector spaces to reorganize Galois theory, but that's getting ahead of ourselves."

"Oh! Oh!" Tetra shouted, flapping both arms. "I remembered something—when Yuri was talking about S_3, and I saw triangles! This is like that! Flip things and spin them around and you can see equilateral triangles. In the same way, Galois groups bring out the form of a polynomial, showing us which roots can be swapped, right?"

"Correct," Miruka said.

Tetra continued, a smile on her face: "I think I'm getting this, a little at least. Galois groups capture the relationship between polynomial roots. One question, though. How exactly do we find the Galois group of a polynomial?"

"Let's ask Galois."

10.4.4 Creating Galois Groups

"Galois tells us how to create Galois groups in his first paper," Miruka continued. "Let $f(x)$ be a polynomial over K with no repeated roots, and call those roots $\alpha_1, \alpha_2, \alpha_3, \ldots, \alpha_m$. We'll use those roots to construct V.[4] But this time, we're going to consider the minimal polynomial over K having V as a root, and call that $f_V(x)$. We'll use $V_1, V_2, V_3, \ldots, V_n$ to designate the roots of $f_V(x)$. Just be sure not to get confused between the polynomials $f(x)$ and $f_V(x)$."

---

[4] See Lemma 2, p. 333

$f(x)$: A polynomial over K with non-repeating roots $\alpha_1, \alpha_2, \alpha_3, \ldots, \alpha_m$

$f_V(x)$: The minimal polynomial over K with roots $V_1, V_2, V_3, \ldots, V_n$ ($V = V_1$)

"We can use V to write $\alpha_1, \alpha_2, \alpha_3, \ldots, \alpha_m$, like this."

$$\alpha_1 = \phi_1(V), \ \alpha_2 = \phi_2(V), \ \alpha_3 = \phi_3(V), \ \ldots, \ \alpha_m = \phi_m(V)$$

"So let's change this V into the form $V_1, V_2, V_3, \ldots, V_n$ and create a set of n permutations of the roots.[5]"

Using V_1, the permutation becomes $\phi_1(V_1), \phi_2(V_1), \ldots, \phi_m(V_1)$.

Using V_2, the permutation becomes $\phi_1(V_2), \phi_2(V_2), \ldots, \phi_m(V_2)$.

Using V_3, the permutation becomes $\phi_1(V_3), \phi_2(V_3), \ldots, \phi_m(V_3)$.

$$\vdots$$

Using V_n, the permutation becomes $\phi_1(V_n), \phi_2(V_n), \ldots, \phi_m(V_n)$.

"Letting G be the substitution group produced from these n permutations, this G is the Galois group for the equation $f(x) = 0$ over K. In other words, we've created permutations of roots using conjugate elements of V, which we created from the roots of $f(x)$, and the substitution group produced by those permutations is the Galois group. This G has the two properties of the Galois group."

Confirming property 1: Known if invariant

"Letting $F(\alpha_1, \alpha_2, \alpha_3, \ldots, \alpha_m)$ be a rational expression in the roots of f which is invariant under the substitution group G, this will be equal to $F(\phi_1(V), \phi_2(V), \phi_3(V), \ldots, \phi_m(V))$, since $\alpha_k = \phi_k(V)$. Noting that V determines this rational expression, let's define a new rational expression in V, calling it $F'(V)$."

$$F'(V) = F(\phi_1(V), \phi_2(V), \phi_3(V), \ldots, \phi_m(V))$$

---

[5]See Lemma 4, p. 337

"The value of the rational expression $F(\alpha_1, \alpha_2, \alpha_3, \ldots, \alpha_m)$ is invariant in substitution group G, so the value of the rational expression $F'(V)$ will be the same, regardless of which $V_1, V_2, V_3, \ldots, V_n$ we use in place of V."

$$F'(V) = F'(V_1) = F'(V_2) = F'(V_3) = \cdots = F'(V_n)$$

"From here, we can represent $F'(V)$ in terms of $V_1, V_2, V_3, \ldots, V_n$."

$$F'(V) = \frac{1}{n}\left(F'(V_1) + F'(V_2) + F'(V_3) + \cdots + F'(V_n)\right)$$

"This $F'(V)$ is a symmetric polynomial in $V_1, V_2, V_3, \ldots, V_n$. So now we just want to be able say an *elementary* symmetric polynomial in $V_1, V_2, V_3, \ldots, V_n$ is known. Since $V_1, V_2, V_3, \ldots, V_n$ are roots of the minimal polynomial $f_V(x)$, this will hold."

$$f_V(x) = (x - V_1)(x - V_2)(x - V_3) \cdots (x - V_n)$$

"The coefficients when this is expanded will be elementary symmetric polynomials in $V_1, V_2, V_3, \ldots, V_n$, and furthermore since $f_V(x)$ is a polynomial over K, these coefficients are elements in K. Therefore, elementary symmetric polynomials in $V_1, V_2, V_3, \ldots, V_n$ will be in K. In other words, they are known. From this, the value of $F'(V)$ is known, also meaning the value of $F(\alpha_1, \alpha_2, \alpha_3, \ldots, \alpha_m)$ is known.

"And there you go. Property 1, 'known if invariant,' holds."

Confirming Property 2: Invariant if known

"On the other hand, suppose the value of a rational expression $F(\alpha_1, \alpha_2, \alpha_3, \ldots, \alpha_m)$ with coefficient field K is known, in other words an element in K. Let that value be R, with $R \in K$. Then we can create a new $F'(V)$ like this."

$$F'(V) = F(\phi_1(V), \phi_2(V), \phi_3(V), \ldots, \phi_m(V)) - R$$

"Of course, $F'(V) = 0$, so the rational function $F'(x)$ has V as a root. We can get rid of any denominator in the equation $F'(x) = 0$, so the left side becomes a polynomial, and we'll call that polynomial

$F''(x)$. Then V is also a root of $F''(x)$. This polynomial $F''(x)$ will have a shared root V with the minimal polynomial $f_V(x)$. A minimal polynomial is also an irreducible polynomial, so $F''(x)$ will share all roots $V_1, V_2, V_3, \ldots, V_n$ with $f_V(x)$.[6] Therefore this will hold."

$$F''(V) = F''(V_1) = F''(V_2) = F''(V_3) = \cdots = F''(V_n) = 0$$

"The permutations produced by $V_1, V_2, V_3, \ldots, V_n$ become the substitution group G, so this equation shows that $F''(V)$ is invariant under G, and its value is always 0. So the value of $F'(V)$ too is invariant—namely, zero—under G, and the value of $F(\alpha_1, \alpha_2, \alpha_3, \ldots, \alpha_m)$ will be invariant under G, with value R.

"And that's that. Property 2, 'invariant if known,' holds. Since both properties hold, the substitution group G is the Galois group."

"So we use these conjugate elements $V_1, V_2, V_3, \ldots, V_n$ of V to create the Galois group," I said, "with each conjugate element producing the elements of the substitution group. I guess it's the fact that they're yoked together by the irreducible polynomial $f_V(x)$ that lets us say we're producing the structure for the Galois group G?"

"In a sense, I suppose."

"Miruka, this is way too hard for me," Yuri said.

"So this is it?" Tetra asked. "This is how we create a Galois group?"

"This is it," Miruka said. "What say we give it a try?"

10.4.5 The Galois Group for $x^3 - 2x = 0$

"We'll start with something simple," Miruka said. "Let $K = \mathbb{Q}$, and $m = 3$. Now we need a polynomial over $\mathbb{Q}$ with no repeated roots, so we'll use $f(x) = x^3 - 2x$. It's easy to find the roots of $f(x)$, by factorization."

$$x^3 - 2x = x(x - \sqrt{2})(x + \sqrt{2})$$

"That gives us roots $\alpha_1, \alpha_2, \alpha_3$."

$$\alpha_1 = 0, \ \alpha_2 = +\sqrt{2}, \ \alpha_3 = -\sqrt{2}$$

---

[6]See Lemma 1, p. 330

"Next, we use these roots to construct V,[7] which will have different values under substitutions of these three roots. For example, we could come up with a rational expression $\phi(x_1, x_2, x_3)$ like this."

$$\phi(x_1, x_2, x_3) = 1x_1 + 2x_2 + 4x_3$$

"There are infinitely many other possibilities. Note that I'm explicitly writing the x_1 as $1x_1$ to help us stay focused on the coefficients. Let's make sure we get different values for the rational expression from all six permutations of $\alpha_1, \alpha_2, \alpha_3$ in $\phi(x_1, x_2, x_3)$. We'll call these six values $V_1, V_2, V_3, V_4, V_5, V_6$, and calculate them like so."

$$\begin{aligned}
V_1 &= \phi(\alpha_1, \alpha_2, \alpha_3) = 1\alpha_1 + 2\alpha_2 + 4\alpha_3 = 0 + 2\sqrt{2} - 4\sqrt{2} = -2\sqrt{2} \\
V_2 &= \phi(\alpha_1, \alpha_3, \alpha_2) = 1\alpha_1 + 2\alpha_3 + 4\alpha_2 = 0 - 2\sqrt{2} + 4\sqrt{2} = +2\sqrt{2} \\
V_3 &= \phi(\alpha_2, \alpha_1, \alpha_3) = 1\alpha_2 + 2\alpha_1 + 4\alpha_3 = +\sqrt{2} + 0 - 4\sqrt{2} = -3\sqrt{2} \\
V_4 &= \phi(\alpha_2, \alpha_3, \alpha_1) = 1\alpha_2 + 2\alpha_3 + 4\alpha_1 = +\sqrt{2} - 2\sqrt{2} + 0 = -\sqrt{2} \\
V_5 &= \phi(\alpha_3, \alpha_1, \alpha_2) = 1\alpha_3 + 2\alpha_1 + 4\alpha_2 = -\sqrt{2} + 0 + 4\sqrt{2} = +3\sqrt{2} \\
V_6 &= \phi(\alpha_3, \alpha_2, \alpha_1) = 1\alpha_3 + 2\alpha_2 + 4\alpha_1 = -\sqrt{2} + 2\sqrt{2} + 0 = +\sqrt{2}
\end{aligned}$$

"Sure enough, all six values are different. Next, we create $f_V(x)$, the minimal polynomial over K with V as a root. We can select any of the V_1 through V_6 values, so let's just use $V = V_1 = -2\sqrt{2}$. In other words, we're creating $f_V(x)$ as the minimal polynomial over $\mathbb{Q}$ having $-2\sqrt{2}$ as a root. This isn't hard, either. We just need to come up with a lowest-degree polynomial with $\mathbb{Q}$ coefficients and having $-2\sqrt{2}$ as a root, and a coefficient of 1 on the highest-degree term. We can do that like this, using V_1 and V_2."

$$\begin{aligned}
f_V(x) &= (x - V_1)(x - V_2) \\
&= \left(x - (-2\sqrt{2})\right)\left(x - (+2\sqrt{2})\right) \\
&= x^2 - 8
\end{aligned}$$

"So this will hold for the polynomial $x^2 - 8$ over $\mathbb{Q}$."

- It is irreducible over $\mathbb{Q}$.

- It is the lowest-degree polynomial having $V_1 = -2\sqrt{2}$ as a root.

---

[7] See Lemma 2, p. 333

· Its highest-degree term has a coefficient of 1.

"So the minimal polynomial for $V_1 = -2\sqrt{2}$ is this."

$$f_V(x) = x^2 - 8$$

"In Galois's method, n is the degree of $f_V(x)$, so here $n = 2$. Now we can write $\alpha_1, \alpha_2, \alpha_3$ in terms of V.[8] We'll use $\phi_1(x), \phi_2(x), \phi_3(x)$ as our rational functions, like this."

$$\phi_1(x) = 0, \quad \phi_2(x) = -\frac{x}{2}, \quad \phi_3(x) = \frac{x}{2}$$

"Let's confirm we can create roots $\alpha_1, \alpha_2, \alpha_3$ when $x = V_1$."

$$\alpha_1 = \phi_1(V_1) = 0, \quad \alpha_2 = \phi_2(V_1) = +\sqrt{2}, \quad \alpha_3 = \phi_3(V_1) = -\sqrt{2}$$

"Yep, that works. Let's substitute this x with V_1 and V_2 and create a set of $n = 2$ permutations of the roots.[9] These permutations will form the Galois group according to Galois's method."

Using V_1, the permutation of roots will be $\phi_1(V_1), \phi_2(V_1), \phi_3(V_1)$. So the permutation is $0, +\sqrt{2}, -\sqrt{2}$, in other words $\alpha_1, \alpha_2, \alpha_3$.

Using V_2, the permutation of roots will be $\phi_1(V_2), \phi_2(V_2), \phi_3(V_2)$. So the permutation is $0, -\sqrt{2}, +\sqrt{2}$, in other words $\alpha_1, \alpha_3, \alpha_2$.

"From these two permutations $\alpha_1, \alpha_2, \alpha_3$ and $\alpha_1, \alpha_3, \alpha_2$, we can use G to refer to the Galois group for $x^3 - 2x = 0$ over $\mathbb{Q}$, like this."

$$G = \{[1\ 2\ 3], [1\ 3\ 2]\}$$

"In standard notation, we would write it like this."

$$G = \{\begin{pmatrix} 1 & 2 & 3 \\ 1 & 2 & 3 \end{pmatrix}, \begin{pmatrix} 1 & 2 & 3 \\ 1 & 3 & 2 \end{pmatrix}\}$$

Miruka put down her pencil. "In other words, the substitution group created from the identity element and a substitution that swaps α_2 and α_3 is the Galois group for $x^3 - 2x = 0$ over $\mathbb{Q}$."

---

[8] See Lemma 3, p. 335
[9] See Lemma 4, p. 337

"Aha!" I said. "I think I'm starting to see the meaning behind the Galois group of a polynomial. The substitution group G = $\{[1\,2\,3],[1\,3\,2]\}$ shows that among the three roots, only α_2 and α_3 can be swapped."

"Oh, I get it!" Tetra said.

"I don't," Yuri muttered.

"It's like this," I said. "When you use $\alpha_1, \alpha_2, \alpha_3$ to make a rational expression that has a rational value, you can swap $\alpha_2 = +\sqrt{2}$ and $\alpha_3 = -\sqrt{2}$ without changing the value of that expression. Specifically, even when you make that swap, the value of the rational expression won't change. The substitution group G represents this information that α_2 and α_3 form a pair."

Miruka nodded. "That's correct, in a qualitative sense at least. The Galois group for an equation can be viewed as representing a certain sort of symmetry in the roots of that equation. In the case of the polynomial $x^3 - 2x = 0$, we're showing a simple symmetry between α_2 and α_3, but these symmetries will become more complex as the polynomial becomes more complex."

"So Galois groups show symmetries among roots..." Tetra whispered.

"And Galois groups show the form of an expression..." Yuri added.

"As we progress through Galois's first paper, we learn even more properties of Galois groups," Miruka said. "When we create a field extension by adjoining an element to the field of coefficients, the previously irreducible $f_V(x)$ becomes reducible. The minimal polynomial for V changes, so the Galois group will change too. Theorem 2 considers how we can reduce Galois groups."

"This way," Lisa said. She guided us on, to reduction of Galois groups.

10.4.6 Theorem 2: Reduction of the Galois group of a Polynomial

Lem. 1 → Lem. 2 → Lem. 3 → Lem. 4 → Th. 1 → ⟦ Th. 2 ⟧ → Th. 3 → Th. 4 → Th. 5

We strolled through the venue for the Narabikura Library's Galois Festival. Miruka stopped in front of a poster titled "Theorem 2."

"Before we dive in to Theorem 2, let's review the theme of Galois's first paper."

Tetra's hand flew up. "Necessary and sufficient conditions for algebraically solving polynomials!"

Miruka nodded. "Exactly. By the way, Cardano and Lagrange and Euler and all the others went looking for auxiliary equations when trying to algebraically solve equations. Their goal was to solve those auxiliary equations to find roots to adjoin to the field of coefficients."

Yuri turned to me. "Did we use these 'auxiliary equations' for quadratic equations?"

"Sure," I replied. "Remember when we solved [something]^2 = $b^2 - 4ac$?"[10]

"We did, didn't we! More specifically it was [part including x]^2 = [part not including x], right? Our target form!"

"You solved an auxiliary equation for finding the square root of $b^2 - 4ac$," Miruka said, "then adjoined the $\sqrt{b^2 - 4ac}$ you obtained to the coefficient field. A quadratic equation can be solved with the four arithmetic operations in the field $K(\sqrt{\text{discriminant}})$. We can say that from the quadratic formula."

"I see," Yuri said.

Miruka continued. "If an algebraic solution is possible, then the elements we should adjoin will exist. Conversely, if no algebraic solution is possible, no such element will exist. So the key question becomes, under what conditions will there be such an element?"

We all nodded.

"This is where Galois's ideas really shine," Miruka said. "He turned his attention to how Galois groups change when you adjoin the roots of auxiliary equations. Theorem 2, which we're about to read, describes the 'reduction' of the Galois group of a polynomial, which happens when we extend the field of coefficients by adjoining the roots of an auxiliary equation."

"So we extend the field of coefficients by adjoining the roots of an auxiliary equation..." I said.

"...which reduces the Galois group for the polynomial," Yuri said, completing my summary.

---

[10] See p. 34.

"I'm not sure what you mean, this thing about reducing a group," Tetra said. "What's a reduced group?"

"In essence, a reduced group is a subgroup," Miruka replied. "When the field of coefficients changes due to an extension, there will be an accompanying change in the known numbers, and thus in the Galois group as well. This new Galois group will be a subgroup of the one before."

"Okay, a subgroup then. That I can understand," Tetra said.

"Remember," Miruka said, "A Galois group encodes 'invariant if known' and 'known if invariant.' Here, 'known' means belonging to the field of coefficients. In other words, it's a concept from the world of fields. 'Invariant' is instead a concept from the world of groups. So the two characteristics Galois presented in his Theorem 1 show a correspondence between fields and groups, and the Galois group of a polynomial is a bridge between two worlds."

Two worlds... I recalled how Miruka had described the joy she felt when two worlds touched...

"This Theorem 2 we're going to talk about similarly describes a correspondence between field extensions and group reductions. As we'll see, the two have a surprisingly tight relation. When the field is extended the group is reduced, and when the group is reduced the field is extended. Taking advantage of this correspondence, we can use the possibility of a field extension to determine the possibility of reducing a group. And this is very useful, because what we want to find out is whether a field can be extended to contain all the roots for an equation."

"Makes sense," I said.

"Well then, let's start with a review of the prerequisites for Theorem 2."

- $f(x)$ is a polynomial with no repeated roots over field K.

- $\alpha_1, \alpha_2, \alpha_3, \ldots, \alpha_m$ are the roots of $f(x)$ (i.e., solutions to the equation $f(x) = 0$).

- $f_V(x)$ is the minimal polynomial over K having V as a root.

"Now we want to think of some new auxiliary equation $g(x) = 0$ that is irreducible over K. We'll let the roots of $g(x)$ be

$r_1, r_2, r_3, \ldots, r_p$, and in particular let $r = r_1$. The polynomial $f(x)$ over field K can also be considered as a polynomial over field $K(r)$. But extending the coefficient field K to $K(r)$ means the 'known' values have changed, so the Galois group may have changed as well. So here's what we want to know."

> As compared with the Galois group for the equation $f(x) = 0$ <u>over field K</u>, what happens to the Galois group for $f(x) = 0$ <u>over field $K(r)$</u>?

"It's Theorem 2 that answers this question."

Theorem 2 (Reduction of the Galois group for a polynomial)

Let G be the Galois group for an equation $f(x) = 0$ <u>over field K</u>, and let H be the Galois group for $f(x) = 0$ <u>over field $K(r)$</u>. Then, letting r be one solution to the irreducible auxiliary equation $g(x) = 0$ over K, and letting $r_1, r_2, r_3, \ldots, r_p$ be the roots of $g(x)$ (with $r = r_1$), one of the following two cases will hold:

- $G = H$ (i.e., adjoining r does not change the Galois group), or

- $G \supset H$ (i.e., adjoining r reduces the Galois group to a subgroup of itself).

In the case of a reduction, subgroup H will partition group G into p parts as

$$G = \sigma_1 H \cup \sigma_2 H \cup \sigma_3 H \cup \cdots \cup \sigma_p H,$$

where $\sigma_1, \sigma_2, \sigma_3, \ldots, \sigma_p \in G$, and σ_1 equals the identity element e. The number of parts in the partitions will equal the number of roots of $g(x)$.

"In his Theorem 2 Galois wrote that the partition will be into 'p groups,' like it says here. In the nomenclature of modern mathematics, however, we would call these cosets. So $f_V(x)$ is the minimal polynomial over K with root V, and if $f_V(x)$ remains irreducible over $K(r)$, the Galois group hasn't changed. If $f_V(x)$ *is* reducible over $K(r)$, however, then it is a product of p same-degree factors. Galois did not provide a proof for this."

$$f_V(x) = f'_V(x, r_1) \times f'_V(x, r_2) \times f'_V(x, r_3) \times \cdots \times f'_V(x, r_p)$$

"Here, each $f'_V(x, r_k)$ is irreducible over $K(r_k)$. Now, take a look at how $f'_V(x, r_1)$ factors. Say $V_1, V_2, V_3, \ldots, V_q$ are the roots of $f'_V(x, r_1)$. In other words, this is a minimal polynomial over $K(r)$."

$$f'_V(x, r_1) = (x - V_1)(x - V_2)(x - V_3) \cdots (x - V_q)$$

"Note that I'm rewriting the subscripts for the conjugates of V comprising $f'_V(x, r_1)$ as small numbers."

Tetra groaned. "So... many... letters..."

"How about an example, then," Miruka said. "Let's see what happens when $n = 12, p = 3$, and $q = 4$."

$$\begin{aligned}
f_V(x) = &\underbrace{(x - V_1)(x - V_2)(x - V_3)(x - V_4)}_{f'_V(x, r_1) \text{ irreducible over } K(r_1)} \\
&\times \underbrace{(x - V_5)(x - V_6)(x - V_7)(x - V_8)}_{f'_V(x, r_2) \text{ irreducible over } K(r_2)} \\
&\times \underbrace{(x - V_9)(x - V_{10})(x - V_{11})(x - V_{12})}_{f'_V(x, r_3) \text{ irreducible over } K(r_3)}
\end{aligned}$$

"We used $V_1, V_2, V_3, \ldots, V_n$ to construct the Galois group G for a polynomial over field K.[11] After adjoining r, we can use $V_1, V_2, V_3, \ldots, V_q$ to construct the Galois group H for the polynomial over $K(r)$. The order of G is n, and the order of H is q. In the case of the example above, G is constructed from $V_1, V_2, V_3, \ldots, V_{12}$, and after adjoining r, we've reduced G to H, which is a subgroup of G constructed from V_1, V_2, V_3, V_4."

10.4.7 Galois's Error

"Actually, there's a small error in this Theorem 2 of Galois's first paper," Miruka said.

---

[11]See Theorem 1, p. 341

"Galois made a mistake?" I said.

"The day before his duel, he obtained a copy of his first paper and erased a bit that said 'and of prime degree p.' Technically speaking, that correction was an error. The number of parts in the partition may be not p, but some divisor of p. So far as solvability of the equation is concerned, it would be fine if p were prime, but when Galois was revising his first paper, he probably noticed this loosening of the condition that p be prime would make for a better argument. Unfortunately, he didn't have time to make the needed corrections. After all, he was scheduled to fight in a duel the next day. He spent that last night alone, scribbling notes in the margins of his paper. *Il y a quelque chose à completer dans cette démonstration. Je n'ai pas le temps*, he wrote. 'There is something to be completed in this proof. I do not have the time.'"

"Oh, that's so sad," Tetra said, her voice trembling.

"So Galois knew his proof still needed some work," Miruka said. "Given more time he surely could have completed it, but alas... Undoubtedly there was still a lot of mathematics in his head, math he never got a chance to put on paper. Fate allotted him far too short a life. That makes me so... *angry!*"

And with that, Miruka kicked the wall. We all gasped. She soon regained her composure, however, and continued.

"That same night, Galois wrote a letter to his friend Chevalier, where he outlined how his theory of equations could be developed in three papers. In the end, however, he had to leave that task to future mathematicians." Miruka closed her eyes for a time before resuming. "Thankfully, he *was* able to leave us the necessary and sufficient conditions for algebraically solving equations. His first paper gave us the first fruits from the newly born field of Galois theory."

We all fell silent as a brief tribute to Galois.

"Back to the math," Miruka said. "So Theorem 1 defines what a Galois group is, and Theorem 2 tells us how Galois groups get reduced when we adjoin a root r of an auxiliary equation. By adjoining an r having p conjugate elements, we obtained small irreducible factors $f'_V(x, r)$ for $f_V(x)$, so the number of conjugate elements of V creating the Galois group fell to $\frac{1}{p}$ of what it was. That means the order of the Galois group went from n to $q = \frac{n}{p}$. The next Theorem 3 will show us what happens when we adjoin *all* the roots of an auxiliary equation. This will be useful when adjoining roots."

"This way," Lisa said, guiding us toward the next room to learn about adjoining roots.

10.4.8 Theorem 3 (Adjoining all Roots of an Auxiliary Equation)

Lem. 1 → Lem. 2 → Lem. 3 → Lem. 4 → Th. 1 → Th. 2 → ⟦ Th. 3 ⟧ → Th. 4 → Th. 5

"Interesting," I said. "These posters we've been reading as we tour the rooms... It's like we're walking through Galois's first paper."

Miruka nodded. "Theorem 2 showed us how a Galois group may be reduced when we adjoin any *one* root r of an irreducible auxiliary equation to the field of coefficients. So what happens if we adjoin *all* the roots? That's what Theorem 3 tells us."

Theorem 3 (Adjoining all roots of an auxiliary equation)

Let G be the Galois group for an equation $f(x) = 0$ over a field K, and let $r_1, r_2, r_3, \ldots, r_p$ be roots of an irreducible polynomial $g(x)$ over K. When considering $f(x) = 0$ as an equation over field $K(r_1, r_2, r_3, \ldots, r_p)$, the Galois group for the equation is reduced to a normal subgroup H of G.

"See how we're extending field K to field $K(r_1, r_2, r_3, \ldots, r_p)$?"

"Huh?" I said, noticing something. "A field extension with all roots of an irreducible polynomial. Would that be... a normal extension?"

"It would," Miruka said. "A normal field extension is one in which all roots of an irreducible polynomial have been adjoined."

Tetra placed her hands on her head. "Hold up, now. We've been talking so much about fields. We're supposed to be talking about Galois *groups*, right?"

"Tetra, Tetra," Miruka said. "Don't forget, we've built a bridge spanning two worlds, the world of fields and the world of groups.

We can see both. Theorem 3 says there's a correspondence between normal extensions and normal subgroups."

$$K \subset K(r_1, r_2, r_3, \ldots, r_p) \quad \text{(a normal extension of K)}$$
$$\vdots \quad \vdots$$
$$G \triangleright H \quad \text{(a normal subgroup of G)}$$

Theorem 2 says— Adjoining *one* root of an auxiliary equation creates a field extension, and correspondingly reduces the Galois group for an equation to a subgroup.

Theorem 3 says— Adjoining *all* roots of an auxiliary equation creates a *normal* field extension, and correspondingly reduces the Galois group for an equation to a *normal* subgroup.

"What a curious correspondence!" Tetra said. "It's kind of beautiful."

Miruka nodded. "Beautiful indeed. Curious as well. Also, it doesn't just create a correspondence between normal field extensions and normal subgroups. It also says the degree of a normal extension equals the group index of the normal subgroup. In other words, asking how big the field grew after it was extended is the same as asking how small the group became after dividing it by its normal subgroup. In more mathy terms, letting L be the field resulting from a normal extension of K, the degree of the normal extension L/K equals the order of the quotient group G/H. In other words..."

$$[L : K] = (G : H)$$

"It was mathematicians after Galois who cleaned things up with this notation, but this correspondence allowed us to look for the properties of fields by investigating groups."

Yuri broke her long silence. "Miruka, there's a lot here I still don't understand, but this thing about using groups to study fields... I guess that's helpful somehow?"

"Hmm, good question," Miruka said, a finger on her lips. "I guess the best answer would be 'not always.' If your goal is to study the solvability of equations, however, groups are very helpful indeed."

"How's that?"

"When we consider equations in the world of fields, we can search for auxiliary equations. It's difficult to select an appropriate auxiliary equation from among infinitely many candidates. It may even be the case that no appropriate auxiliary equation exists. On the other hand, the world of groups lets us consider Galois groups. Galois groups are substitution groups of finite order, so they'll only have a finite number of normal subgroups. In principle, it's far easier to look for normal subgroups than it is to find a good auxiliary equation. That being said, examining every possibility would still be an enormous amount of work, so it helps to have some tricks."

10.4.9 Repeated Reductions

"Let's summarize our journey so far," Miruka said.

- We want to know whether an equation $f(x) = 0$ is algebraically solvable over a field K.

- Use roots $\alpha_1, \alpha_2, \alpha_3, \ldots, \alpha_m$ of $f(x)$ to construct V.

- Write roots $\alpha_1, \alpha_2, \alpha_3, \ldots, \alpha_m$ in terms of V.

- Let $f_V(x)$ be the minimal polynomial over K having V as a root, and consider conjugate elements $V_1, V_2, V_3, \ldots, V_n$ of V.

- Consider the auxiliary equation $g(x) = 0$.

- When we adjoin to K all roots $r_1, r_2, r_3, \ldots, r_p$ of $g(x)$, the Galois group G for $f(x) = 0$ is reduced to a normal subgroup of G.

- $f_V(x)$ is factored into p irreducible same-degree factors over $K(r_1, r_2, r_3, \ldots, r_p)$.

"Yay! We've factored it!" Tetra squealed.

"We still haven't reached our goal, though," Miruka said.

"No?"

"We've factored $f_V(x)$, yes, but not necessarily into a product of linear expressions. We need to keep reducing the Galois group.

Whether we can keep reducing that group tells us whether the equation can be algebraically solved."

"Halt condition?" Lisa asked, surprising us with her uncharacteristic question.

"Interesting that you immediately want to know where to stop iterating," Miruka replied. "But anyway, our goal is the trivial group. We stop reducing the Galois group when we end up with a group having only one element, the identity element. Do you see why?"

"Known root," Lisa mumbled.

Miruka nodded. "Exactly. If the Galois group for an equation is the trivial group, any rational expression of the roots will be invariant. In other words, any α_k will be invariant, and so from 'known if invariant,' α_k will also be known. If all roots are known, the equation in question can be algebraically solved. Therefore, if the Galois group for an equation is the trivial group, we can algebraically solve that equation.

"Summing up, to determine whether an equation can be solved algebraically, we find the Galois group for that equation, investigate whether the Galois group reduces, and keep reducing it as far as we can, stopping if the Galois group becomes the trivial group."

"Not sure if this will make any sense," Yuri said, "but I have a question. Isn't it $f(x)$ that we're trying to factor into linear expressions, not $f_V(x)$?"

"Same thing. We can construct the roots of $f(x)$ from V.[12] Since $\alpha_k = \phi_k(V)$, if V is known then α_k is known, and we can factor $f(x)$ into linear expressions."

"Okay. Gotta admit, though, I'm getting kinda tired."

"Hang in there. We've almost reached our destination."

"This way," Lisa said, leading us to the next room.

10.4.10 Theorem 4: Properties of Reduced Galois Groups

Lem. 1 → Lem. 2 → Lem. 3 → Lem. 4 → Th. 1 → Th. 2 → Th. 3 → ⬛Th. 4⬛ → Th. 5

The next room had yet another poster.

---

[12]See Theorem 3, pg. 335

Theorem 4 (Properties of reduced Galois groups)

Let r be an element of field $K(\alpha_1, \alpha_2, \alpha_3, \ldots, \alpha_m)$. Let G be the Galois group for $f(x) = 0$ <u>over K</u>. Let H be the Galois group for $f(x) = 0$ over $K(r)$. Then H consists only of substitutions under which the value of r is invariant.

"In Theorem 4, we create a rational expression r using roots $\alpha_1, \alpha_2, \alpha_3, \ldots, \alpha_m$. The Galois group reduces when we adjoin this r to the coefficient field, and the reduced Galois group is a collection of substitutions under which the value of the rational expression r is invariant. This Theorem 4 is used to explicitly construct an adjunction element that reduces the Galois group. Specifically, we'll use Theorem 4 in Theorem 5, our climax—the necessary and sufficient conditions for whether an equation can be algebraically solved."

"I'm so excited!" Tetra said.

"Finally!" Yuri moaned.

"This way," Lisa said, pointing to our final destination.

10.5 Theorem 5: Necessary and Sufficient Conditions for an Equation to be Algebraically Solved

Lem. 1 → Lem. 2 → Lem. 3 → Lem. 4 → Th. 1 → Th. 2 → Th. 3 → Th. 4 → $\boxed{\text{Th. 5}}$

10.5.1 Galois's Question

"We've arrived at the heart of Galois's first paper," Miruka said, scanning our faces. "Here's the question he posed."

> Question: Under what conditions can we algebraically solve an equation?

"In his own words, *Dans quels cas une équation est-elle soluble par de simples radicaux?* Or, Under what circumstances is an equation solvable by simple radicals?"

"If we limit ourselves to first- through fourth-degree equations, the answer is simple."

> Q: Under what conditions are first-degree equations solvable?
>
> A: They're always solvable.
>
> Q: Under what conditions are second-degree equations solvable?
>
> A: They're always solvable.
>
> Q: Under what conditions are third-degree equations solvable?
>
> A: They're always solvable.
>
> Q: Under what conditions are fourth-degree equations solvable?
>
> A: They're always solvable.

"Tell me, Yuri. Why can we make these assertions with such confidence?"

"Uh... Because we have formulas for those?"

"Exactly. There are well-known formulas for finding the solutions to first- through fourth-degree equations. We know we can always find those solutions, so no need to think any further. *But*—" Miruka held up a finger. "As Ruffini and Abel proved, there exists no formula for fifth- or higher-degree equations. If you're given a fifth-degree equation, well, maybe you can solve it, maybe not. For example, Gauss showed that an equation in the form $x^p = 1$ can always be solved algebraically. Generally speaking, however, this was the best they could say for fifth-degree equations."

> Q: Under what conditions are fifth-degree equations solvable?
>
> A: Uh... some?

"Needless to say, this is not a satisfactory answer for most mathematicians. Abel showed that if we can represent each root α_k of an equation as the value of a rational expression $\phi_k(\alpha)$ in one of the roots α, and furthermore that $\phi_k(\phi_j(\alpha)) = \phi_j(\phi_k(\alpha))$, then the equation can be solved. This condition asserts that the Galois group of an equation satisfies the commutative law, and that's why groups satisfying the commutative law are today called abelian groups.

"I remember those!" Tetra said.

"Anyway, mathematicians before Galois had gotten as far as being able to say 'equations in this form are algebraically solvable.' In other words, they had found one *sufficient* condition for algebraic solvability."

Yuri cocked her head. "Meaning?"

"Meaning some particular form could indicate solvability, but that doesn't necessarily mean an equation that's not in that form is not solvable. There may be other equations in other forms that are still solvable."

Miruka pointed to the poster.

> Question: Under what conditions can we algebraically solve an equation?

"Galois was the first human in history to completely answer that question. He found the necessary and sufficient conditions for algebraic solvability, regardless of whether the equation is of fifth degree or sixth degree or one-hundredth degree."

"So how are necessary and sufficient conditions different, specifically?" Yuri asked.

"They say that if an equation *is* in this form it *can* be solved, and if it is *not* in this form it *cannot* be solved. To give this more complete answer, Galois learned the methods of Lagrange, which focused on substitutions of roots, and used that as the foundation for his bridge to the world of fields." I noticed Miruka's cheeks start to flush as she continued. "Galois had to write his papers using the yet-unfinished tools of fields and groups. That's one reason why his first paper is so hard to read. Many later mathematicians put a lot of effort into developing those tools, and thanks to them we now have a much fuller toolbox.

"But let's continue following the path that Galois took, weaving among fields and groups, to pull together the necessary and sufficient conditions for algebraic solvability."

10.5.2 What Does it Mean to Algebraically Solve an Equation?

Miruka continued her lecture.

"There are various ways to describe what we're after."

- The equation can be solved algebraically
- The equation is algebraically solvable
- The equation is solvable by simple radicals
- The roots of the equation can be expressed using only arithmetic operations and radicals

"Using the language of fields, we get this."

"Algebraic solvability" in the language of fields

An equation is algebraically solvable when we can start with the field of coefficients for that equation and extend that field by adjoining radicals until all roots are known.

"In the language of groups, we get this."

"Algebraic solvability" in the language of groups

An equation is algebraically solvable when we can start with the Galois group for the equation and reduce it while fulfilling certain conditions until we obtain the trivial group.

"What we want to know is the exact conditions that make this possible. So in turn, we want to know the conditions in which the Galois group for an equation can be reduced to the trivial group. Whether we're dealing with field extensions or group reductions, we need to be sure to proceed step-by-step. To do that, we'll build two towers—a tower of fields and a tower of groups.

"In the world of fields, it goes like this. We extend the field of coefficients by adjoining a radical. We want to repeat this until we've created a field extension to which all roots of the equation belong.

"In the world of groups, we instead start with the Galois group for the equation, and reduce it. Then we reduce that reduced group, and so on. We repeat this until we end up with the trivial group—a group with only one element, the identity element. When doing so, however, we must always ensure we're fulfilling a certain condition.

"You have a question, Tetra?"

10.5.3 Tetra's Question

Tetra put her hand down. "I do." She started speaking slowly and carefully. "I understand how we're adjoining radicals to the field of coefficients to create a field extension. I want to make sure I have something straight, though. The reason why we want to keep extending and extending is so we end up with a field big enough to allow us to factor, right?"

"That's right."

"Well in that case, why don't we just start out with some super enormous field? The field of complex numbers $\mathbb{C}$, for example? Wouldn't that be big enough to do the factorization?"

$$f(x) = (x - \alpha_1)(x - \alpha_2)(x - \alpha_3) \cdots (x - \alpha_m)$$
$$\text{where } \alpha_1, \alpha_2, \alpha_3, \ldots, \alpha_m \in \mathbb{C}$$

"Of course," Miruka said, nodding. "If we start out in $\mathbb{C}$ we can easily factor a polynomial into linear expressions, so you're right there. But we won't know if we can solve it using *only radicals*."

"Ah."

"Gauss proved that all equations have roots if we're working in $\mathbb{C}$, but his proof only shows the *existence* of a solution. What Lagrange and Ruffini and Abel and Galois and all the other mathematicians of the time were struggling to discover was not a way to show the existence of a solution, but the conditions for representing a solution as radicals. Doing so—in other words, algebraically solving the equation—is what requires these repeated field extensions."

Tetra nodded. "Okay, I get it. Sorry, I was confused for a bit there."

"Hey, Miruka!" Yuri said. "This is kind of like the angle trisection problem!"

"How's that?" Miruka said, smiling.

"Like this!"

- A trisecting angle exists. However, there may not be a finite number of straightedge and compass operations that allow you to construct that angle.

- A solution to the fifth-degree equation exists. However, there may not be a finite number of arithmetic operations and radicals that allow you to express that solution.

"See! They're exactly the same!"

"That's a good understanding of things," Miruka said, clearly happy. "You have a knack for seeing the logical structure of a problem. In fact, we can make several analogies between the angle trisection problem and solvability of equations. You've discovered one."

Analogies... I recalled the letter Miruka had given me.

"So Miruka," Yuri said. "After all that, what *are* the necessary and sufficient conditions for solving an equation? And what are these 'certain conditions' we have to fulfill when we reduce a Galois group?"

Miruka smiled again. "That's exactly what we're going to talk about next. Galois has led us across his bridge from the world of fields to the world of groups, but if we can't solve the problem there, we've made a pointless trip. So, back to the matter at hand."

10.5.4 Adjoining pth Roots

"So we're going to adjoin some radicals to the equation's field of coefficients, extending that field," Miruka began. "But what kind of radicals should we adjoin? Square roots? Cube roots? Fourth roots? Galois's first paper says we only have to think of pth roots when considering elements to adjoin, with p prime. For example, if you want to adjoin a sixth root, you can just adjoin a square root and a cube root in turn, since $\sqrt[6]{} = \sqrt[3]{\sqrt{}}$. Let's call this adjoined element r.

"Okay, so this means we'll be adjoining to K some r where $r^p \in K$, creating the field extension $K(r)$. So did this adjunction reduce the

Galois group for the equation? Well, maybe it did, maybe it didn't. If the Galois group does not reduce no matter what r we adjoin, then its equation is not algebraically solvable.

"Let's say the primitive pth root of unity ζ_p starts out as an element in the coefficient field K."

$$\zeta_p \in K$$

"In that case, then just by adjoining one radical we can automatically adjoin all radicals. For example, when we adjoin $r = \sqrt[3]{2}$ to create $K(\sqrt[3]{2})$, say we've started with the third root of unity $\zeta_3 = \omega$ as an element in the coefficient field K from the start. In other words, $\omega \in K$. Then, $\sqrt[3]{2}\omega \in K(\sqrt[3]{2})$ and $\sqrt[3]{2}\omega^2 \in K(\sqrt[3]{2})$. So just by adjoining $\sqrt[3]{2}$, we've adjoined three roots, $\sqrt[3]{2}$, $\sqrt[3]{2}\omega$, and $\sqrt[3]{2}\omega^2$."

"Oh! Look!" Tetra said. "$\sqrt[3]{2}$ isn't all alone any more!"

"It's in good company," Miruka agreed. "Anyway, we can obtain ζ_p from a radical that's smaller than p—Gauss proved this before Galois got involved—so there's no problem assuming $\zeta_p \in K$.

"Now it's time to use Theorems 2 and 3. From Theorem 2,[13] we know that when adjoining one root, the pth root, of the auxiliary equation, we can divide G up into p parts. From Theorem 3,[14] we know that by adjoining all roots of the auxiliary equation, the Galois group will reduce to a normal subgroup. The auxiliary equation we're thinking of here is this."

$$x^p - r^p = 0 \qquad (r^p \in K)$$

"In other words, an auxiliary equation for finding the pth root of r^p. Since $\zeta_p \in K$, just by adjoining one root r of this auxiliary equation, we've adjoined all p roots. Namely, $K(r)/K$ is a normal extension. Now the Galois group G has reduced to the normal subgroup H, and the order of the quotient group G/H is the prime p. In other words..."

> If we can <u>adjoin radicals to the coefficient field</u> of an equation so that its Galois group is reduced, the <u>quotient group produced by the reduced Galois group has prime order.</u>

---

[13] p. 358
[14] p. 361

"This works the other way, too."

> If there exists a normal subgroup such that <u>the order of the quotient group is prime</u>, we can <u>adjoin a radical to the field of coefficients</u> such that the Galois group reduces.

"This bit about the order of the quotient group being prime, *that's* the 'certain condition' we need to maintain. Then we start repeating. I'm going to number these Galois groups to help keep track."

- Refer to the Galois group G for a given equation as G_0.
- Find a normal subgroup G_1 of G_0 such that the order of the quotient group G_0/G_1 is prime.
- Find a normal subgroup G_2 of G_1 such that the order of the quotient group G_1/G_2 is prime.
- Find a normal subgroup G_3 of G_2 such that the order of the quotient group G_2/G_3 is prime.
- ... Keep repeating this until the normal subgroup is the trivial group E.

"So we're creating this chain of normal subgroups such that the order of the quotient group G_k/G_{k+1} is prime."

$$G = G_0 \triangleright G_1 \triangleright G_2 \triangleright G_3 \triangleright \ldots \triangleright G_n = E$$

"If we can create a chain like that, the equation we started with is algebraically solvable. If we can't, in other words if there is no radical we should adjoin and no auxiliary equation to solve, then there is no algebraic solution to the equation. All of this is summarized in Galois's Theorem 5, his solvability theorem."

Galois Theory

> **Theorem 5 (Necessary and sufficient conditions for algebraic solvability of an equation)**
>
> An equation can be solved if and only if the Galois group G for the equation has a sequence of normal subgroups
>
> $$G = G_0 \triangleright G_1 \triangleright G_2 \triangleright G_3 \triangleright \cdots \triangleright G_n = E,$$
>
> where
>
> - $G_k \triangleright G_{k+1} \iff G_{k+1}$ is a normal subgroup of G_k,
> - the order of the quotient group G_k/G_{k+1} is prime, and
> - E is the trivial group.
>
> In this case, group G is called a "solvable group."

"So now we can say that the necessary and sufficient conditions for algebraically solving an equation are that the Galois group for the equation is solvable."

> **Theorem 5 (Necessary and sufficient conditions for algebraic solvability of an equation, reworded)**
>
> Equation is solvable $\iff$ Its Galois group is solvable

"Finally, we have a satisfactory answer."

> Question: Under what conditions can we algebraically solve an equation?
> Answer: When the Galois group for the equation is solvable!

"We've finally arrived!" I said.

10.5.5 Galois's Adjunction Elements

I sat back, intending to savor the feeling of arriving at our destination. However, both Yuri and Tetra's hands shot up.

"Miruka! Does this really work both ways?" Yuri asked.

Tetra immediately followed that up with, "What kind of elements are we adjoining, specifically?"

"Actually, Galois answered both of those questions at once," Miruka said. "In his first paper, Galois showed that if the order of the quotient group is a prime p, we can concretely construct the pth root we need to adjoin. First, we define a rational expression θ created using the roots of a given equation $f(x) = 0$. Let G be the Galois group for the equation, and let H be a reduced normal subgroup. Then, we give θ these properties."

- If a substitution σ is in both G and H ($\sigma \in G$ and $\sigma \in H$), the value of θ is invariant ($\sigma(\theta) = \theta$).

- If instead $\sigma \in G$ and $\sigma \notin H$, the value of θ varies ($\sigma(\theta) \neq \theta$).

"We can construct the rational expression θ by creating a symmetric expression that is invariant under the substitutions in H. We use a substitution σ, for which $\sigma \in G$ and $\sigma \notin H$, to permute the order of the roots. We'll represent this new rational expression as $\sigma(\theta)$. Then we can use this notation."

$$\theta_0 = \theta, \ \theta_1 = \sigma(\theta_0), \ \theta_2 = \sigma(\theta_1), \ \theta_3 = \sigma(\theta_2), \ \ldots$$

"Here, θ_k is the result of k applications of substitution σ to θ. Note that $\theta_0, \theta_1, \theta_2, \theta_3, \ldots$ is not an infinite sequence. The number of elements in the quotient group G/H will be a prime number p. So p applications of σ brings us back where we started."

$$\theta_p = \theta_0 = \theta$$

"Galois proposed this r to use as the element to adjoin to the field K, with ζ_p being the pth primitive root of unity."

$$r = \zeta_p^1 \theta_1 + \zeta_p^2 \theta_2 + \zeta_p^3 \theta_3 + \cdots + \zeta_p^{p-1} \theta_{p-1} + \zeta_p^p \theta_p$$

Tetra nervously raised her hand again. "Isn't that r kind of—"

"Convoluted, yes. But it should look familiar. A sum of products of pth roots of unity is called a 'Lagrange resolvent.' Galois actively studied the works of Lagrange, and proposed using this Lagrange resolvent as the adjunction element. Taking a step back, we want to know what element r we should adjoin to the field K to reduce the Galois group from G to H. But there's one thing that worries us— does r^p really belong to the field K? This is important because we want to build our tower by adjoining known pth-root elements, so we're in trouble if we can't be sure $r^p \in K$. In his first paper, Galois says *il est clair* that this is the case, but let's do a little extra work to make sure this really is clear."

Problem 10-1 (Galois's adjunction elements)

Defining Galois's adjunction element r as

$$r = \zeta_p^1 \theta_1 + \zeta_p^2 \theta_2 + \zeta_p^3 \theta_3 + \cdots + \zeta_p^{p-1}\theta_{p-1} + \zeta_p^p \theta_p,$$

where

- p is prime and the order of the quotient group G/H,

- ζ_p is a primitive pth root of unity,

- K is the field of coefficients, to which ζ_p has already been adjoined,

- σ is a substitution satisfying $\sigma \in G$ and $\sigma \notin H$,

- θ is a rational expression in the roots, invariant under the normal subgroup H but varying under other substitutions in the Galois group G for the equation, and

- θ_k is the rational expression resulting from k applications of substitution σ to θ ($\theta_{k+1} = \sigma(\theta_k)$, $\theta_p = \theta_0 = \theta$),

does $r^p \in K$ hold?

"Okay, let's prove that $r^p \in K$. We want to say that $r^p \in K$, in other words that the value r^p is known, and to that end we just need to be able to say the value of r^p is invariant under substitutions belonging to the Galois group G. In other words, we want to be able to say this equation holds."

$$\sigma(r^p) = r^p \qquad \text{the value of } r^p \text{ is invariant under } \sigma$$

"Our proof starts with the definition of r."

$$r = \zeta_p^1 \theta_1 + \zeta_p^2 \theta_2 + \zeta_p^3 \theta_3 + \cdots + \zeta_p^{p-1} \theta_{p-1} + \zeta_p^p \theta_p$$

"Let's apply the substitution σ to both sides."

$$\sigma(r) = \sigma\big(\zeta_p^1 \theta_1 + \zeta_p^2 \theta_2 + \zeta_p^3 \theta_3 + \cdots + \zeta_p^{p-1} \theta_{p-1} + \zeta_p^p \theta_p\big)$$

"ζ_p is known, so the substitution σ will not change the value of ζ_p. That means this equation holds."

$$\sigma(r) = \zeta_p^1 \sigma(\theta_1) + \zeta_p^2 \sigma(\theta_2) + + \cdots + \zeta_p^{p-1} \sigma(\theta_{p-1}) + \zeta_p^p \sigma(\theta_p)$$

"Using $\theta_{k+1} = \sigma(\theta_k)$, we get this."

$$\sigma(r) = \zeta_p^1 \theta_2 + \zeta_p^2 \theta_3 + \zeta_p^3 \theta_4 + \cdots + \zeta_p^{p-1} \theta_p + \zeta_p^p \theta_{p+1}$$

"$\zeta_p^p \theta_{p+1} = \theta_1$ holds for the final term."

$$\sigma(r) = \zeta_p^1 \theta_2 + \zeta_p^2 \theta_3 + \zeta_p^3 \theta_4 + \cdots + \zeta_p^{p-1} \theta_p + \underset{\sim}{\theta_1}$$

"Move the final term up front."

$$\sigma(r) = \underset{\sim}{\theta_1} + \zeta_p^1 \theta_2 + \zeta_p^2 \theta_3 + \zeta_p^3 \theta_4 + \cdots + \zeta_p^{p-1} \theta_p$$

"Multiply both sides by ζ_p^1 and clean up."

$$\zeta_p^1 \sigma(r) = \zeta_p^1 \theta_1 + \zeta_p^2 \theta_2 + \zeta_p^3 \theta_3 + \zeta_p^4 \theta_4 + \cdots + \zeta_p^p \theta_p$$

"See how the the right side equals r?"

$$\zeta_p^1 \sigma(r) = r$$

"Now we can divide both sides by ζ_p^1 to get this."

$$\sigma(r) = \frac{r}{\zeta_p^1}$$

"Raise both sides to the pth power."

$$\bigl(\sigma(r)\bigr)^p = \frac{r^p}{\zeta_p^p}$$

"Since $\zeta_p^p = 1$ ..."

$$\bigl(\sigma(r)\bigr)^p = r^p$$

"... and since $\bigl(\sigma(r)\bigr)^p = \sigma(r^p)$ ..."

$$\sigma(r^p) = r^p$$

Tetra raised her hand. "This $\bigl(\sigma(r)\bigr)^p = \sigma(r^p)$ you used in the last step. Where does that come from?"

"Because raising r to the pth power after substituting roots is the same as substituting roots after raising r to the pth power."

"Ah, makes sense. Thanks!"

"Anyway, now we know that $\sigma(r^p) = r^p$. This shows that the value of r^p is invariant under a substitution σ in the Galois group for the equation, and since it's invariant, it's known. Thus, $r^p \in K$, which is what we wanted to show. End of proof."

"*Quod erat demonstrandum!*" Tetra chimed.

"Since $r^p \in K$, r is a solution to an auxiliary equation $x^p - r^p = 0$ over K. So if we adjoin r to K to create the extension field $K(r)$, we can use Theorems 3 and 4 to reduce the Galois group G to the normal subgroup H, and the quotient group G/H will have prime order p. This r is the adjunction element Galois was thinking of. And there you go. Done and done."

Answer 10-1 (Galois's adjunction element)

Defining Galois's adjunction element r as

$$r = \zeta_p^1 \theta_1 + \zeta_p^2 \theta_2 + \zeta_p^3 \theta_3 + \cdots + \zeta_p^{p-1} \theta_{p-1} + \zeta_p^p \theta_p,$$

$r^p \in K$ holds.

10.5.6 Yuri's Response

"Miruka!" Yuri moaned, fidgeting nervously. "I just don't get it."
"Still struggling?" Miruka asked.
"I am," she admitted, uncharacteristically serious. "I'm sorry."
"No need to apologize. Want to see it all in more detail?"
Yuri grimaced. "Not the whole thing, but... Well, can you clear up this thing about the order of the quotient group being prime?"
"Ah, right. Lisa, where's that Cayley graph from yesterday?"
"This way," Lisa said, leading us to the next... *Er, hallway?*

10.6 Two Towers

10.6.1 The General Cubic Equation

Several conference posters adorned the hallway's walls.
"Hey!" Yuri shouted. "That's the poster we made yesterday!"
Sure enough, hanging there was the graph of S_3 and its quotient group we'd made the day before.

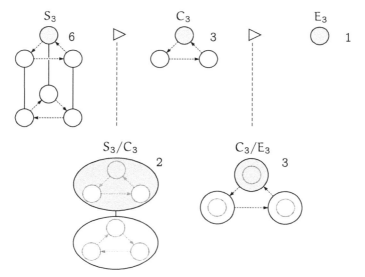

Decomposition of the Galois group S_3 for the general cubic equation ($S_3 \triangleright C_3 \triangleright E_3$).

Miruka pointed at the figure as she spoke.

"From left to right we have a sequence of normal subgroups."

$$S_3 \triangleright C_3 \triangleright E_3$$

"Below that, the two quotient groups S_3/C_3 and C_3/E_3. The orders of the quotient groups are 2 and 3, both primes. So that's one example."

"Neat!" Yuri said.

"The Galois group for the general cubic formula is the symmetric group S_3, a sequence of normal subgroups that lead to the trivial group. Further, the order of each quotient group is prime. Therefore, S_3 is solvable."

"Is this the kind of image Galois had in mind when he was working on all this?" Tetra asked.

"I imagine he saw a far richer structure than this." Miruka paused. "Not that I can say for certain, of course. Anyway, these primes 2 and 3 that appear here also show up in the formula for roots of cubic equations. After solving a second-degree equation, we found a third root. We did that so we could adjoin second and third roots."

"The 2 and 3, right."

"So let's see what happens when we create our tower of fields and tower of groups for a general cubic equation, $ax^3 + bx^2 + cx + d = 0$. Here's the figure when we let $K = \mathbb{Q}(a, b, c, d, \zeta_2, \zeta_3)$, explicitly writing square roots as $\sqrt[2]{}$."

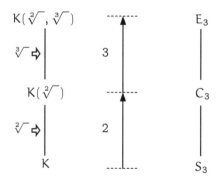

Tower of fields and tower of groups for the general cubic equation.

Looking at the figure, Tetra and Yuri spoke at the same time.

"The tower of fields extends fields until we get the minimal splitting field at the top!"

"The tower of groups reduces until we get the trivial group at the top!"

10.6.2 The General Quartic Equation

"Let's try to do the same thing for the general quartic equation," Miruka said.

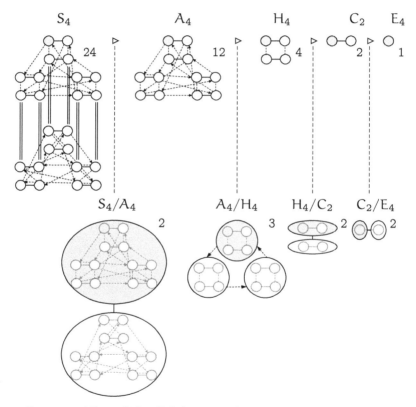

Decomposition of the Galois group S_4 for the general quartic equation ($S_4 \triangleright A_4 \triangleright H_4 \triangleright C_2 \triangleright E_4$).

"Let's see," Tetra said. "The orders of $S_4/A_4, A_4/H_4, H_4/C_2$, and C_2/E_4 are... $2, 3, 2$, and 2."

"All prime!" Yuri added.

"The Galois group for the general quartic equation is the symmetric group S_4." Miruka said. "Reduction of this group goes like this."

$$S_4 \triangleright A_4 \triangleright H_4 \triangleright C_2 \triangleright E_4$$

"The order changes like this."

$$24 \xrightarrow{\frac{1}{2}} 12 \xrightarrow{\frac{1}{3}} 4 \xrightarrow{\frac{1}{2}} 2 \xrightarrow{\frac{1}{2}} 1$$

"So when solving the general quartic equation, we adjoin pth roots in the order square root→cube root→square root→square root. See how the adjunction of pth roots and the change in order $\frac{1}{p}$ correspond so well?"

"They do!" Tetra said.

"With this, we can construct a tower of fields and tower of groups for the general quartic equation $ax^4 + bx^3 + cx^2 + dx + e = 0$. A picture of these towers looks like this."

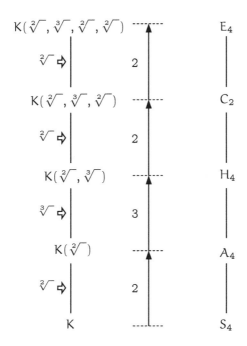

Tower of fields and tower of groups for the general quartic equation.

"In his first paper, Galois wrote about how to reduce Galois groups," Miruka said. "He didn't draw pictures of towers, though. Instead, he represented Galois group reductions in a table as arrangements of roots. I'll use $\alpha_1, \alpha_2, \alpha_3, \alpha_4$ to represent the roots, and show reductions of $\frac{1}{2} \to \frac{1}{3} \to \frac{1}{2} \to \frac{1}{2}$ in a row."

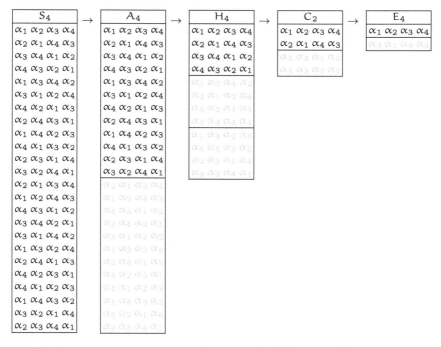

"So these are arrangements of roots, right?" Tetra said.

Miruka nodded. "Yes, in Galois's notation, which gathers permutations of roots as substitution groups. Let me explain. For example, here's the set of arrangements for four items, representing the substitution group formed from four elements."

$\alpha_1\ \alpha_2\ \alpha_3\ \alpha_4$
$\alpha_2\ \alpha_1\ \alpha_4\ \alpha_3$
$\alpha_3\ \alpha_4\ \alpha_1\ \alpha_2$
$\alpha_4\ \alpha_3\ \alpha_2\ \alpha_1$

"If you gather together the substitutions that create these four permutations from the $\alpha_1\ \alpha_2\ \alpha_3\ \alpha_4$ in the first row here, you've created the substitution group."

- [1 2 3 4] is the substitution that creates $\alpha_1\,\alpha_2\,\alpha_3\,\alpha_4$ from $\alpha_1\,\alpha_2\,\alpha_3\,\alpha_4$.

- [2 1 4 3] is the substitution that creates $\alpha_2\,\alpha_1\,\alpha_4\,\alpha_3$ from $\alpha_1\,\alpha_2\,\alpha_3\,\alpha_4$.

- [3 4 1 2] is the substitution that creates $\alpha_3\,\alpha_4\,\alpha_1\,\alpha_2$ from $\alpha_1\,\alpha_2\,\alpha_3\,\alpha_4$.

- [4 3 2 1] is the substitution that creates $\alpha_4\,\alpha_3\,\alpha_2\,\alpha_1$ from $\alpha_1\,\alpha_2\,\alpha_3\,\alpha_4$.

"You'll get the same substitution group regardless of which row you start with. For example, if you collect the substitutions that create these permutations from the $\alpha_2\,\alpha_1\,\alpha_4\,\alpha_3$ in the second row, you'll end up with the same substitution group, just in a different order. This is what Galois meant when he wrote 'the original arrangement does not matter' when defining substitution groups.[15]"

- [2 1 4 3] is the substitution that creates $\alpha_1\,\alpha_2\,\alpha_3\,\alpha_4$ from $\alpha_2\,\alpha_1\,\alpha_4\,\alpha_3$.

- [1 2 3 4] is the substitution that creates $\alpha_2\,\alpha_1\,\alpha_4\,\alpha_3$ from $\alpha_2\,\alpha_1\,\alpha_4\,\alpha_3$.

- [4 3 2 1] is the substitution that creates $\alpha_3\,\alpha_4\,\alpha_1\,\alpha_2$ from $\alpha_2\,\alpha_1\,\alpha_4\,\alpha_3$.

- [3 4 1 2] is the substitution that creates $\alpha_4\,\alpha_3\,\alpha_2\,\alpha_1$ from $\alpha_2\,\alpha_1\,\alpha_4\,\alpha_3$.

"This should also give you a sense of how Galois felt when, considering the set of permutations of roots as substitution groups, he wrote that 'the partition will be into p <u>groups</u>' in his Theorem 2.[16] Because the p cosets of the partition can each be regarded as a substitution group."

---

[15]p. 365
[16]p. 358

10.6.3 General Quadratic Equations

"Can you draw these towers of fields and groups for a general quadratic equation too?" Yuri asked.

"Of course," Miruka replied. "They're pretty short towers, but once again you can confirm that the order of the quotient group is 2, another prime."

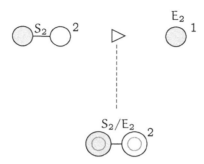

Decomposition of the Galois group S_2 for the general quadratic equation ($S_2 \triangleright E_2$).

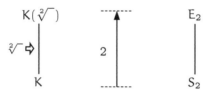

Tower of fields and tower of groups for the general quadratic equation.

"Such cute little towers!" Tetra said. "Just two floors!"

"So here's problem for you, Yuri," Miruka said. "What specifically is the $\sqrt[2]{}$ we're adjoining here?"

"Who, me?" Yuri said, but paused to think. "Oh, you told us this before. That's the square root of the discriminant?"

"Well done. In other words, the roots for the general quadratic equation belong to the extension field resulting from adjoining the square root of the discriminant to the coefficient field."

$$\frac{-b \pm \sqrt{b^2 - 4ac}}{2a} \quad \in \quad \mathbb{Q}(a, b, c, \zeta_2, \sqrt{b^2 - 4ac})$$

"Wow, Miruka!" Yuri said, clearly excited. "This totally changes how I think about the quadratic formula!"

"Oh yeah? How's that?" Miruka asked.

"Like, up until now it's been pretty much just a chant I'd memorized, 'negative b plus or minus root b squared minus 4ac divided by 2a.' But now I know formulas like this aren't just a random arrangement of letters and symbols. Like, they really... *mean* something, I guess? I'm not sure how to put it..."

"You mean that now you're friends with the quadratic formula!" Tetra said. "You've received the message in the math!"

With that, Yuri showed us a very happy smile.

10.6.4 No Formula for the Roots of Quintic Equations

"Ruffini and Abel gave proofs that there exists no general formula for the roots of a fifth-degree equation," Miruka said, "but we can also use Galois theory to show the same thing. To do so, we just have to show that the Galois group for the general quintic equation is an unsolvable group. That Galois group is the symmetric group S_5, so let's create a table of normal subgroups in S_5 and look at each in turn. Here's a table, with numbers above the $\triangleright$ symbols so we can easily see the group index, with prime indices circled.[17]"

$$\text{General quadratic equation} \quad S_2 \overset{(2)}{\triangleright} E_2$$

$$\text{General cubic equation} \quad S_3 \overset{(2)}{\triangleright} A_3 \overset{(3)}{\triangleright} E_3$$

$$\text{General quartic equation} \quad S_4 \overset{(2)}{\triangleright} A_4 \overset{(3)}{\triangleright} H_4 \overset{(2)}{\triangleright} C_2 \overset{(2)}{\triangleright} E_4$$

$$\text{General quintic equation} \quad S_5 \overset{(2)}{\triangleright} A_5 \overset{60}{\triangleright} E_5$$

$$\text{General sextic equation} \quad S_6 \overset{(2)}{\triangleright} A_6 \overset{360}{\triangleright} E_6$$

$$\text{General septic equation} \quad S_7 \overset{(2)}{\triangleright} A_7 \overset{2520}{\triangleright} E_7$$

$$\text{General octic equation} \quad S_8 \overset{(2)}{\triangleright} A_8 \overset{20160}{\triangleright} E_8$$

$$\vdots$$

"As the table shows, the Galois group S_n for a general nth-degree equation is solvable only when $n = 2, 3, 4$, since when $n \geqslant 5$, the normal subgroups look like this."

$$S_n \overset{(2)}{\triangleright} A_n \overset{\frac{n!}{2}}{\triangleright} E_n$$

"So if $n \geqslant 5$, the group A_n will only have itself and the trivial group as normal subgroups, with non-prime orders. So the symmetric group S_n isn't solvable when $n \geqslant 5$."

Yuri spoke up. "There's still plenty I don't understand, but one thing's for sure—there's a whole lot of interesting stuff here. But, still..."

Miruka raised an eyebrow. "Yes?"

"Can we *please* eat lunch now? I'm starving!"

10.7 SUMMER'S END

10.7.1 The Fundamental Theorem of Galois Theory

"Wow, that's some tasty sherbet," Yuri said, now fully recovered from her trial by math.

---

[17]Here, the C_3 on p. 379 is represented as A_3.

"Cassis sorbet," Lisa muttered.

We were in Oxygen, the cafe on the Narabikura Library's third floor, enjoying some after-lunch dessert.

"Bet it doesn't beat this chocolate cake though," Miruka said, licking her lips.

"*Gâteau au chocolat*," Lisa corrected.

"Sorbet... *Gâteau*..." Tetra said. "Does the Oxygen menu have a French theme today, in honor of Galois?"

Lisa silently nodded.

Tetra turned to Miruka. "By the way, would I be right in saying that, in essence, Galois theory is all about the relation between field extensions and group reductions?"

"You would, and you'd have summarized what we now call the Fundamental Theorem of Galois Theory."

She pointed to a poster conveniently located behind our table.

The Fundamental Theorem of Galois Theory

Let L be the splitting field of a polynomial f(x) with no repeated roots over field K, and let G be the Galois group for the equation f(x) = 0 over K. Further, define mappings ϕ (phi) and Ψ (psi) as follows:

ϕ is a mapping from <u>the set of all subfields of L</u> to <u>the set of all subgroups of G</u>, with

ϕ(M) = the subgroup of G under which elements in M are invariant
   = $\{\sigma \in G \mid \sigma(a) = a \text{ for all elements } a \text{ in } M\}$,

(M is a subfield of L), and

Ψ is a mapping from <u>the set of all subgroups of G</u> to <u>the set of all subfields of L</u>, where

Ψ(H) = the subfield of L that is invariant under H
   = $\{a \in L \mid \sigma(a) = a \text{ for all elements } \sigma \text{ in } H\}$

(H is a subgroup of G).

Then, ϕ and Ψ are both bijections and inverse mappings of each other. In other words,

$$\Psi(\phi(M)) = M, \qquad \phi(\Psi(H)) = H$$

The relation between the subfields and subgroups by ϕ and Ψ is called a "Galois correspondence."

"A Galois correspondence shows that the tower of fields and the tower of groups stand side-by-side as a perfect pairing," Miruka said. "You can think of it as a correspondence between fields which are invariant under groups and groups under which fields are invariant. Invariance is important, and invariant things are worth naming."

"Wow, I wish I was anywhere near as smart as Galois," Tetra said.

"Galois was a genius," Miruka said, "but he too stood on the shoulders of giants. The cyclotomic polynomials that Gauss studied, Lagrange's root substitutions and his resolvent—Galois's work is based on all that and more. And without his teacher and mentor Richard, he may have never been introduced to any of it."

"Good point," Tetra said, nodding.

Miruka spoke in a singsong voice: "Through abstraction, discoveries become theories. Through verbalization, theories become papers."

Tetra added: "Through metaphor, discoveries become stories. Through melodies, stories become songs."

Yuri sat there, staring at both of them.

"So sure, Galois was a genius," Miruka continued. "More importantly, he left us his findings in the form of papers. He passed his discoveries on to the world and the future. He did this in several papers, and in a letter to his friend Chevalier he wrote the night before his duel. That letter was published after his death, but few people took note. It was fourteen years after his death before his complete works, including his first paper, were finally published by Joseph Liouville."

"Fourteen years!" Yuri said.

"After that, Galois's ideas finally started to spread to other mathematicians. Camille Jordan published his *Traité des substitutions et des équations algebraique*, which explicitly calculated Galois groups, and Dedekind started giving the first lectures on Galois theory. About a century later, Emil Artin reformulated Galois groups, setting aside the equations in favor of field automorphisms. Galois groups arose from the problem of solvability of equations, but today that problem has fallen under the Galois theory of algebraic equations, so it's been reduced to an application of Galois theory. A theory's range of application broadens with increased abstraction. Applying the fruits of group theory by introducing group structures as mathematical objects is a natural approach, so mathematicians found Galois theory to be a useful tool. Galois theory even has connections to Wiles's proof of Fermat's Last Theorem. It marks the end of a long history of theories of equations, and the start of new branchings for mathematics. In that sense, it's kind of similar to how Gödel's incompleteness theorems deflected the course of number theory."

"That's wonderful," I said.

"Galois yearned for someone who could take up his ideas and interpret them, and thankfully later mathematicians obliged. Mathematics is built up by a variety of people—those who think, those who convey, those who learn, those who teach, those who popularize..."

We sat silently for a moment, considering the broad web of people supporting mathematics.

"But that's enough from me for today," Miruka said.

"Aw! But I still have some questions!" Tetra said.

"And I want to hear more about Galois," Yuri added.

"Thankfully there are books that will answer your questions, Tetra, and tell you more about Galois, Yuri. There are texts on groups and Galois theory, from primers to advanced works, and biographies of Galois himself. You've only just begun learning. Definitions of fields and groups, vector spaces and degrees of extension, quotient groups and group indices, fields and subfields, groups and subgroups, correspondences between groups and fields, field extensions and group reductions, normal extensions and normal subgroups, cosets and equivalence classes... They're all out there, just waiting for you. All in a sea of books."

I couldn't wait to dive in.

10.7.2 Browsing the Exhibits

After lunch, the number of visitors to the Narabikura Library's Galois Festival increased. I heard discussions of all things mathematical from sofas and around tables. The Cayley graph for the symmetric group S_4 that we helped Yuri create attracted quite a crowd. I also noticed quite a few people in the room dedicated to the angle trisection problem—the room planned by Yuri's boyfriend. He had set up a table with straightedges and compasses where visitors could try their hand at geometric constructions, at which both children and adults were clearly enjoying themselves.

Yuri stopped inside and spoke to some people who were milling about.

"It's easy to misunderstand what the angle trisection problem is about," she said. "First off, be sure not to confuse 'existence' with 'constructibility'..."

The festival was over before I knew it.

10.7.3 Oxygen at Night

After the festival, I found myself alone, drinking coffee in the Oxygen cafe.

I was thinking about separate worlds, and how we could bridge them. How problems that are difficult in one world can be carried across those bridges, allowing them to be easily solved. How mathematics can sometimes feel like a fantasy novel, and how Galois theory is one of the most beautiful mathematical fantasies of all.

I looked up and noticed it was already dark outside. I was glad Miruka had dragged me to the Galois Festival, giving me the opportunity to delve deep into Galois's first paper. But the Festival had come to an end, just as my summer vacation soon would. I felt like everything around me was ending.

My mind turned back to Galois, a mathematician who took on a serious problem and defeated it in a paper he wrote when he was seventeen years old. *Seventeen! Compared to him, what have I actually*—

"Ouch!"

I spun around holding my ear, which had been suddenly yanked from behind me.

"Found you," Miruka said, sitting next to me.

"Yeah," I said, rubbing my ear. "Just kind of zoning out."

"Hmph."

Miruka drank the coffee remaining in my cup. I sat there absorbing her presence, wondering how much longer I could count on her being near.

"You sure seem to enjoy moping," she said, bringing her face near mine.

I involuntarily pulled back. "I don't enjoy it. I just—"

"Always 'I just,' 'I just' with you..."

"I was just reflecting on how little I've actually accomplished."

"Comparing yourself to Galois, were you?"

"As if. *Yowch!*"

An elbow to the ribs this time.

"You're you," Miruka said in a near whisper. "Don't try to be someone else."

"Even if this me can't accomplish anything?"

"You exist. That's good enough."

"But existence alone isn't sufficient for—"

I stopped talking. The fact that Galois never saw his twenty-first birthday flashed through my mind. How he must have been screaming "I have no time!" as he scribbled that last letter, his final message to mathematicians of the future.

I have no time. But still... Even so...

10.7.4 Irreplaceable

Boom! I heard something that sounded like an explosion from outside the window. *What on earth?*

Miruka and I hurried to the cafe terrace, which in daytime provided an ocean view, but now showed only darkness.

Boom!

Another explosion, this time accompanied by a ring of light expanding across the night sky. Cheers and applause arose from the direction of the beach.

"Fireworks," Miruka said. "How lovely."

Another rocket launched, painting the sky in blue and red radii that shone bright, only to immediately fade.

"Such big circles they make," I said.

"Spheres, actually."

"Ah, of course."

A long silence as we stared into the darkness. "They sure are taking their time before launching the next one," she said.

I turned to look at her profile, recalling Yuri asking me what I considered irreplaceable. What was so important to me that I could never replace it.

Miruka turned to me. "What are you staring at?"

I kept looking at her, she at me.

Eventually, unable to bear the silence, I said, "Summer's almost over."

"All thing must end," she replied. "But still..."
She waved fingers at me.

$$1 \quad 1 \quad 2 \quad 3 \ldots$$

"But still?" I said, holding up an empty hand to complete the Fibonacci sign.

$$\ldots 5$$

"Endings are new beginnings."
She smiled.
There it is. The thing I can never replace.

> ... I hope there will be some who profit by deciphering all this mess.
>
> ÉVARISTE GALOIS
> in a letter to Auguste Chevalier
> 29 May 1832 [17]

Epilogue

"No use hiding, I know you're there," the teacher said, speaking into an apparently empty classroom.

Several students crept out from behind the displays that had been arranged there.

"Just a little more time, *please*?" the girl said. "We're almost done!"

"Time's up. The school's been closed for some time now."

"Just one more hour!" She said, the others in her group chiming in. *Just a little more time! Today's our last day! The exhibit's tomorrow!*

"Everything seems ready to me," he said, scanning the room, taking in the assortment of posters and displays.

"We've still got some finishing touches." *We're almost done! We're so close! Just a little more time!*

"So Math Club's theme for the school festival is Galois Theory?"

"It is! And we're putting together the best exhibits this school has ever seen!" she proudly stated.

"The best ever, huh? Then maybe as club president you can answer some questions about what you're showing."

"Bring it on! Ask me anything!" *There ya go! Show him your stuff! Buy us some extra time!*

"So what's this equation all about?" he said, pointing at one of the posters.

$$\cos\frac{2\pi}{17} = -\frac{1}{16} + \frac{1}{16}\sqrt{17} + \frac{1}{16}\sqrt{2(17-\sqrt{17})}$$
$$+ \frac{2}{16}\sqrt{17 + 3\sqrt{17} - \sqrt{2(17-\sqrt{17})} - 2\sqrt{2(17+\sqrt{17})}}$$

"It shows that a regular 17-gon is constructible using a straight-edge and compass," she immediately replied. "The only operators on the right side are the basic arithmetic operations and square roots, so the figure is constructible. Gauss's discovery of this construction method is what set him on the path of mathematics. That was in March 1796, when he was 18 years old. We wrote this equation using sixteens and seventeens because that's how old we are."

"Hmph. But can't you just say a 17-gon is constructible because 17 is prime?"

"Not quite. It's constructible because it's a prime that looks like this." She pointed to a line further down on the poster.

$$17 = 2^{2^2} + 1$$

"Interesting."

"Specifically, the necessary and sufficient condition for a regular n-gon to be constructible is that n can be expressed in this form."

$$n = 2^r p_1 p_2 p_3 \cdots p_s$$

"Here, r is a nonnegative integer, and $p_1, p_2, p_3, \cdots, p_s$ are zero or more different Fermat primes, namely primes that can be written in the form $2^{2^m} + 1$, where m is a nonnegative integer."

"Well done. So what's with this model?" he said, pointing to a three-dimensional geometric form on display in the center of the classroom.

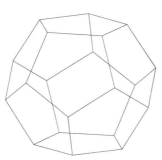

Orientations of a regular dodecahedron are isomorphic with the alternating group A_5.

She picked up the model and spun it around in her hands. "This is a regular dodecahedron, which has sixty possible orientations. There are twelve faces, each with five edges you could put in front, and $12 \times 5 = 60$, so sixty possibilities. Here's how you could count that as a cyclic group," she said, pointing to the poster next to the display.

- Count the initial orientation as 1 possibility.

- There are 6 axes of rotation passing through the centers of opposing faces. There are 5 orientations resulting from revolutions about those axes, but one of those gives the initial orientation, so there's 4 left. So 4 orientations for each of 6 axes of rotation, and $6 \times 4 = 24$ possibilities.

- There are 10 axes of rotation passing through opposing vertices. There are 3 orientations resulting from revolutions around those, but one of those gives the initial orientation, so there's 2 left. So 2 orientations for each of 10 axes of rotation, and $10 \times 2 = 20$ possibilities.

- There are 15 axes of rotation passing through the centers of opposing edges. There are 2 orientations resulting from revolutions around those, but one of those gives the initial orientation, so there's 1 left. So 1 orientation for each of 15 axes of rotation, and $15 \times 1 = 15$ possibilities.

"Add them all up, and you get $1+24+20+15 = 60$ possibilities."

"Interesting, but what does this have to do with Galois theory?"

"The group describing the possible orientations of a regular dodecahedron, in other words the symmetry group for the dodecahedron, is isomorphic to the alternating group A_5," she said. "A_5 is a normal subgroup of the fifth-order symmetric group S_5, and a permutation group generated from an even number of swaps. There's a correspondence between the 60 orientations of a regular dodecahedron and the 60 elements of A_5. The only normal subgroups of A_5 are itself and the trivial group, so there exists no normal subgroup of A_5 that can create a quotient group with prime order. In other words, A_5 is not a solvable group. Using that fact along with Galois theory, we can prove that there exists no general formula that gives the roots for a fifth-degree equation." She held up two thumbs. "And *that's* how we do it in Math Club." *Snap! You go girl! Schoolin' the teacher!*

"Impressive. Okay, tell me what's up with the soccer ball?"

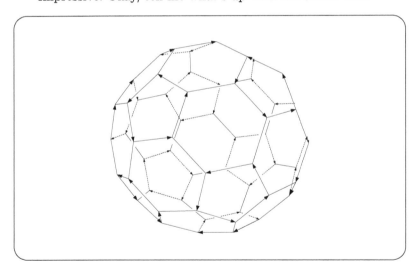

The girl beamed. "That's a Cayley graph! One form of the alternating group A_5. A soccer ball has 60 vertices, corresponding to the 60 elements in A_5. The solid lines correspond to the original operations that generate the second-order cyclic group, and the dashed arrows correspond to the original operations that generate the fifth-order cyclic group. So these two elements, the lines and arrows, generate the alternating group A_5."

"Hmm..."

The girl lifted an eyebrow. "Well?"

"All right, all right. A little more time. But I'll expect you in my office in one hour to announce that you're done."

Way to go, prez! Thanks, teach! C'mon, let's get cracking!

One hour later on the dot, the girl appeared in his office.

"All done! Thanks for the extra time!"

"It's dark already. You going to be able to get home?"

"I'm good. We're all heading back together."

"Good to see the Math Club is so supportive."

"We're all great friends... Hey, what's wrong?"

"Nothing, nothing. Just reminiscing."

"Were you involved in any clubs when you were in school?"

"An unofficial math club, I guess you could call it. We used to meet in the library after school."

"Oh, that girl from the picture the other day! Was she a member?"

"So you still remember that, huh. By the way, good presentation back there."

"Really? You think?"

"I hope you'll keep studying and tell me more."

She laughed. "Me the teacher and you the student? Yeah, right."

"Why not? A role reversal. A swapper."

"A swapper?"

"Hey, isn't everyone waiting for you?"

"Oh, right! See you at our presentation tomorrow!"

She flicked her fingers and headed out the door.

He sat back and fell into his thoughts.

He spent his last night alone, writing a message. Today, countless math lovers are receiving that message, connecting them to him. He created a new branch of mathematics at such a young age, and left this world at such a young age. But he will always live on, because we'll never forget him.

Please remember me since fate did not give me enough of a life to be remembered by my country. I die your friend.

ÉVARISTE GALOIS

Afterword

> "When we cry without order, we cannot share our sadness with others. This is because people find it difficult to feel our sorrow well. When we are within sorrow we cannot find happiness, but we can bring order to our sadness. When we are within sadness, we should find ways to save it. It's like a song."
>
> HIDEO KOBAYASHI
> *Words*

Thank you for reading *Math Girls 5: Galois Theory*. This is the fifth volume in the *Math Girls* series, following *Math Girls* (2007), *Math Girls 2: Fermat's Last Theorem* (2008), *Math Girls 3: Gödel's Incompleteness Theorems* (2009), and *Math Girls 4: Randomized Algorithms* (2011). As in the previous books, this is a story in which Miruka, Tetra, Yuri, Lisa, and the narrator explore mathematics and youth. This story also brought back Ay-Ay, who was absent in the fourth book.

By far, the hardest part of creating this book was writing Chapter 10. I asked Miruka to introduce us to Galois's first paper, but she went on to show and discuss all the complementary proofs that Galois omitted there. In doing so, she made that chapter far too long and far too difficult. I therefore asked her once again to limit her discussion to just that first paper, so she decided to just take us on a tour of the Galois Festival to trace through what Galois wrote.

This book presented a hefty load of mathematical concepts, including quotient groups, field extensions, group indices, orders of extension, normal subgroups, normal extensions, solvability, and the Galois correspondence. I hope Miruka was able to elucidate these topics for you, and that you enjoyed the journey. If you would like to look deeper into Galois theory and its related proofs, the Recommended Reading section suggests some books you might want to read.

As with the other books in the *Math Girls* series, this book was created using $\LaTeX 2_\varepsilon$ and the AMS Euler font. Also as before, Haruhiko Okumura's book *Introduction to Creating Beautiful Documents with $\LaTeX 2_\varepsilon$* was an invaluable aid during layout, and I thank him for it. I created the diagrams using Microsoft Visio and the elementary mathematics handout macro 'emath,' created by Kazuhiro Okuma (a.k.a. tDB).

While I was writing this book, Media Factory released manga editions of *Math Girls 2: Fermat's Last Theorem* and *Math Girls 3: Gödel's Incompleteness Theorems*. I thank artists Shun Kasuga and Miyuki Matsuzaki, along with the entire editorial staff at Media Factory, for helping to broaden the world of *Math Girls*. Further, Bento Books published an English translation of *Math Girls*, so I wish to thank translator Tony Gonzalez and the editorial staff for that.

I would also like to thank the following persons for proofreading and giving me invaluable feedback while I was writing this book. Of course, any remaining mathematical errors are solely the responsibility of the author.

> Ryo Akazawa, Tatsuya Igarashi, Mirai Ikebuchi, Kazuhiro Inaba, Ryuhei Uehara, Ken Okada, Hiromichi Kagami, Toshiki Kawashima, Iwao Kimura, Jun Kudo, Kazuhiro Kezuka, Kayo Kotaki, Hiroaki Hanada, Aya Hayashi, Yoichi Hirai, Hiroshi Fujita, Masahide Maehara, Kiyoshi Miyake, Yusuke Muraoka, and Kenji Yamaguchi

I would like to thank my readers and the visitors to my website, and my friends for their constant prayers.

I thank Tetsujiro Kamei of Kameshobo for long supporting the *Math Girls* series and for his suggestion that it should take on Galois theory.

I would like to thank my editor, Kimio Nozawa, for his continuous support during the long process of creating these books.

I thank the many readers of the *Math Girls* series for their support. Your encouragement is a treasure beyond all others.

I thank my dearest wife and my two sons.

I dedicate this book to my mother, who has always sent me encouraging messages.

Finally, thank *you* for reading my book. I hope that someday, somewhere, our paths shall cross again.

Hiroshi Yuki
2012
https://www.hyuki.com/girl/

Recommended Reading

> "So do you study this group theory stuff in high school?"
> "You might get the very basics in high school ... But to get the full story you have to study on your own, in math books."
>
> *Math Girls 5: Galois Theory*

To Those Who will Continue Studying

Following Galois' approach, this book defined Galois groups as groups of root substitutions, but most modern texts define them as automorphism groups of fields. Further, this book did not address finite fields (Galois fields), but such fields will be addressed in general discussions of Galois theory. In such descriptions, not only normal extensions but "separable extensions" are an important concept. Specifically, a field extension that is both normal and separable is called a "Galois extension." Please keep these points in mind as you continue your studies of Galois theory.

> [Note: The following references include all items that were listed in the original Japanese version of *Math Girls 5: Galois Theory*. Most of those references were to Japanese sources. Where an English version of a reference exists, it is included in the entry.]

THE *Math Girls* SERIES

[1] Yuki, H. (2011). *Math Girls*. Bento Books. Published in Japan as *Sūgaku gāru* (Softbank Creative, 2007).

> The story of two girls and a boy who meet in high school and work together after school on mathematics unlike anything they find in class. In the school library, classrooms, and cafes, they investigate topics like prime numbers, absolute values, the Fibonacci sequence, the relation between arithmetic and geometric means, convolutions, harmonic numbers, the zeta function, Taylor expansions, generating functions, the binomial theorem, Catalan numbers, and numbers of partitions.

[2] Yuki, H. (2012). *Math Girls 2: Fermat's Last Theorem*. Bento Books. Published in Japan as *Sūgaku gāru / Ferumā no saishū teiri* (Softbank Creative, 2008).

> The second book in the *Math Girls* series, where new "math girl" Yuri joins the others on a quest to understand "the true form" of the integers. Presents groups, rings, and fields among other topics, building to a tour of Fermat's last theorem. Other topics include prime numbers, the Pythagorean theorem, Pythagorean triples, prime factorization, greatest common divisors, least common multiples, proof by contradiction, the pigeonhole principle, the definition of groups, abelian groups, integer remainders, congruence, and Euler's formula.

[3] Yuki, H. (2016). *Math Girls 3: Gödel's Incompleteness Theorems*. Bento Books. Published in Japan as *Sūgaku gāru / Gēderu no saishūteiri* (Softbank Creative, 2009).

The third book in the *Math Girls* series, where the math girls (and boy) use formal systems to learn "the mathematics of mathematics" and learn about Gödel's oft-misunderstood theorems. Topics include the Peano axioms, mathematical induction, basic set theory, Russell's paradox, mappings, limits, why $0.999\cdots = 1$, the basics of mathematical logic, the ϵ–δ definition of limits, diagonalization, equivalence relations, radians, the sine and cosine functions, Hilbert's program, and a proof of Gödel's incompleteness theorems.

[4] Yuki, H. (2019). *Math Girls 4: Randomized Algorithms*. Bento Books. Published in Japan as *Sūgaku gāru / Rantaku arugorizumu* (Softbank Creative, 2011).

The fourth book in the *Math Girls* series, in which everyone moves ahead one grade in school and are joined by computer whiz Lisa. Together, they explore applications of probability to create algorithms that employ randomization and learn how to quantitatively analyze algorithms. Other topics include the Monty Hall problem, permutations and combinations, Pascal's triangle, the definition of probability, sample spaces, probability distributions, random variables, expected values, indicator random variables, order, big-O notation, matrices, linear transformations, matrix diagonalization, random walks, the 3-SAT problem, the P versus NP problem, linear searches, binary searches, and algorithms for bubble sort, quicksort, and randomized quicksort.

GENERAL READING

[5] Yamashita, J. (1986). *Garoa e no rekuiemu* [A Requiem for Galois]. Gendai Sugakusha.

> A well-balanced book presenting a wide range of Galois-related topics. Also describes Galois' studies from before the paper he submitted to the Academy and his explorations other than those related to the theory of equations. (Note: As of 2012, this book has become difficult to obtain.)

[6] Nakamura, A. (2010). *Garoa no gunron* [Galois' Theory of Groups]. Kodansha (Blue Backs B-1684).

> A book that explains Galois groups using substitution groups as Galois did. Describes building up Galois theory through use of what the author calls "unit Galois theory," which limits operations to a single adjunction of nth roots and reduction of the Galois group. Contains a lot of content, so perseverance is required.

[7] Kojima, H. (2010). *Tensai garoa no hassō-ryoku* [The Imagination of Genius Galois]. Gijutsu Hyoron-sha.

> A book that uses automorphism groups of fields to explain Galois groups. Presents an intuitive and easy-to-understand description of the Galois correspondence using rectangles. Also presents three versions of Galois theory: a "super simplified version," a "simplified version," and a "more-or-less full version."

[8] Kim, J. (2011). *Jūsan-sai no musume ni kataru garoa no sūgaku* [An Explanation of Galois Theory for My Thirteen-Year-Old Daughter]. Iwanami Shoten.

As the title suggests, this book is written as an explanation of Galois theory to the author's thirteen-year-old daughter. Discusses formulas for solutions to equations, permutation groups, residue groups, and normal subgroups, and explains Galois theory focusing on solutions to generalized nth-order equations. An easy-to-read book with many illustrations, including a fun discussion of applying group theory to Rubik's cubes.

[9] Yano, K. (2006). *Kaku no santōbun* [Angle Trisections]. Chikuma Shobo.

The first edition of this book was a historical publication regarding the "Angle Trisection Problem" published during the Second World War. In addition to conventional methods using a straightedge and compass, the book discusses other solutions using special devices. In a second part, mathematician Shin Hitotsumatsu provides a detailed explanation of the proof for why arbitrary angle trisections are impossible. The final part of this book is an essay by Tetsujiro Kamei about "Trisectionists," his term for amateur mathematicians making claims to have found methods for trisecting angles.

[10] Iyanaga, S. (1999). *Garoa no jidai garoa no sūgaku, dai-ichibu, jidai-hen* [Galois' Era, Galois' Mathematics, Vol. 1: Era]. Springer–Verlag Tokyo.

This book describes Galois' life against the political and mathematical backgrounds of his time. Mathematically detailed topics are covered in volume 2 of this series [28].

[11] Kato, F. (2010). *Garoa* [Galois]. Chuo Koronshinsha.

> A brief book describing Galois' life. Very easy to read, and a vibrant description of the social atmosphere in Galois' time. Also provides a full translation of a preface to the papers Galois hoped to write but was unable to finish.

ALGEBRA AND GALOIS THEORY

[12] Shiga, K. (1988). *Senkei daisū sanjū-kō* [Thirty Lectures on Linear Algebra]. Asakura Shoten.

> A text that provides a step-by-step course on linear algebra. Starting with simultaneous equations, it explains vector spaces, linear transformations, determinants, and eigenvalue problems.

[13] Shiga, K. (1989). *Gunron he no sanjū-kō* [Thirty Lectures on Group Theory]. Asakura Shoten.

> A text that provides a step-by-step course on group theory. A well-structured approach showing how to use groups to capture mathematical forms.

[14] Shiga, K. (1994). *Hōteishiki: Sūgaku ga sodatte iku monogatari 5* [Stories Developed by Mathematics, Vol. 5: Equations]. Iwanami Shoten.

> A book that discusses algebra without straying too far from the issue of solving equations. Explains how to solve various fourth-degree equations, and uses Kronecker's proof to show the impossibility of algebraically solving fifth- and higher-degree equations. Sandwiched between chapters are interludes with titles like "Tides of the Time," "A Conversation with the Teacher," and "Teatime"

that help to convey the quiet joy of mathematics. Also provides a simple development of Galois theory.

[15] Yakabe, I. (1976). *Sūsan hōshiki garoa no rion* [Galois Theory for High School Seniors]. Gendai Sugakusha.

> An ambitious work that attempts to explain Galois theory in a conversational format. Provides an easy-to-understand explanation using specific equations, but care must be taken not to become overwhelmed by the large volume of equations. (Note: As of 2012, this book has become difficult to obtain.)

[16] Ueno, K. (2011). "*Garoa no yume—Garoa no kangaeta koto*" [Galois' Dreams—Galois' Thoughts]. *Sugaku Bunka, Vol. 15.* Nihon Hyoronsha.

> An article in a mathematics publication that succinctly summarizes the highlights of Galois' accomplishments.

[17] Stewart, I. (2003). *Galois Theory, Third Edition.* Translated by Namiki, M. and Suzuki, J. as *Meikai garoa riron* (Kodansha, 2008).

> A textbook that explains Galois theory using many examples and problems.

[18] Cox, D. (2004). *Galois Theory.* Translated by Kajiwara, T. as *Meikai garoa riron* Vols. 1 and 2 (Kodansha, 2008, 2011).

> Another Galois theory textbook. Takes care to provide informative notes allowing the reader to trace the historical development of the theory. Starting with a discussion of polynomials and field theory, the book goes on to discuss solvability via radicals, cyclotomic polynomials, and construction problems (including origami).

[19] Yukie, A. (2010). *Daisūgaku 1 gunron nyūmon* [Algebra, Vol. 1: Introduction to Group Theory]. Nihon Hyoronsha.

> As the title states, this is a textbook that introduces group theory. Provides detailed supplementary notes regarding "well-defined" and other topics that can easily trip up budding mathematicians.

[20] Yukie, A. (2010). *Daisūgaku 2 kan to tai to garoa riron* [Algebra, Vol. 2: Rings, Fields, and Galois Theory]. Nihon Hyoronsha.

> As the title states, this is a textbook that explains rings, fields, and Galois theory.

[21] Artin, E. (1966). *Galois Theory*. Translated by Terada, F. as *Garoa riron nyūmon* (Chikuma Shobo, 2010).

> A text that uses linear algebra to reorganize Galois theory. This translation is good for self-learners, due to the addition of by-chapter summaries and practice problems with answers. Concludes with an afterword by Ichiro Satake that summarizes the key points of Galois theory in just a few pages.

[22] *Iwanami sūgaku nyūmon jiten* [The Iwanami Dictionary of Introductory Mathematics]. (2005) Iwanami Shoten.

> A dictionary of mathematical terms with easy-to-understand definitions.

[23] Harada, K. (2001). *Gun no hakken* [Discovering Groups]. Iwanami Shoten.

> A mathematics text that starts with symmetry in geometric figures, leading to the solubility of equations and Galois theory.

[24] Nakajima, S. (2006). *Daisū hōteishiki to garoa riron* [Algebraic Equations and Galois Theory]. Kyoritsu Shuppan.

> A mathematics textbook that covers polynomials, field extensions, linear spaces, and Galois theory. Formulas for solutions of algebraic equations are treated as an application of Galois theory. Includes a discussion of notation for mathematical descriptions, making this a good read for beginners unfamiliar with mathematical writing.

[25] Rotman, J. (1990). *Galois Theory*. Translated by Sekiguchi, J. as *Kaitei shinpan garoa riron* (Springer Japan, 2000).

> A brief textbook describing Galois theory. Includes an appendix that explains Galois theory using substitution groups in what the author calls "sepia Galois theory."

GALOIS' FIRST PAPER

[26] Galois, E. (1831). "Mémoire sur les conditions de résolubilité des équations par radicaux" [Memoir on the conditions for solvability of equations by radicals]. *Journal des Mathématiques Pures et Appliquées*, pp. 417–433.

> Galois' first paper. Discusses one principle, "the necessary and sufficient conditions for algebraically solving equations," and one application, "the necessary and sufficient conditions for algebraically solving a certain kind of prime order equation."

[27] Kurata, R. (1987). *Garoa o yomu—Dai-ichi ronbun kenkyū* [Reading Galois: His First Memoir]. Nihon Hyoronsha.

> A stoic presentation of Galois' first paper, excising the more dramatic elements. Provides

> an analysis of that paper that maintains a good balance between the modern mathematics and the mathematics of Galois' time. Discusses two very different theories regarding how Lagrange influenced Galois.

[28] Iyanaga, S. (2002). *Garoa no jidai garoa no sūgaku, dainibu, sūgaku-hen* [Galois' Era, Galois' Mathematics, Vol. 2: Mathematics]. Springer–Verlag Tokyo.

> Continuing from Vol. 1 [10], this book describes the mathematical aspects of Galois' work. Provides a translation and analysis of Galois' first paper in Chapter 3, "Galois' Writings."

[29] Masada, K. and Yoshida, Y. (Eds.), Moriya, M. (Trans.) (1975). *Gun to daisū hōteishiki (gendai sūgaku no keifu 11)* [Development of Modern Mathematics, Vol. 11: Groups and Algebraic Equations]. Kyoritsu Shuppan.

> Provides translations and analyses of Galois' first paper and Abel's proof of the insolubility of general fourth- and higher-degree algebraic equations.

[30] Edwards, H. (1984). *Galois Theory*. Springer.

> A text that provides an analysis of Galois theory and an English translation of his first paper. I referenced this book when writing Tetra's description of the word "cyclotomic" (p. 121).

WEBSITES

[31] Kobayashi, K.: *Garoa riron no sūchi jikken* [Numeric Experiments using Galois Theory] http://www.nasuinfo.or.jp/FreeSpace/kenji/sf/perm/glgrp.htm.

Uses Python sf and Maxima to show and describe concrete examples from Galois theory (creation of subgroups and subfields).

[32] Kobayashi, K.: *Garoa riron no sūchi jikken* [Numeric Experiments using Galois Theory] http://hooktail.sub.jp/.

A community website with many members studying physics, mathematics, and computer science. This site's "algebra" page is extremely helpful for understanding Galois theory.

[33] Yuki, H.: *Math Girls* http://www.hyuki.com/girl/en.html.

The English version of the author's *Math Girls* web site, listing titles in and news related to the *Math Girls* series.

"Together, we were attempting to tackle the great truths of mathematics. I would need all the allies I could find for that endeavor..."

HIROSHI YUKI
Math Girls 5: Galois Theory

Index

A
Abel, Niels Henrik, 90, 240, 366, 386
abelian group, 90, 367
adjunction, 51
alternating group, 398
angle trisection, 133
argument (of a complex number), 104
arithmetic operations, 29
Artin, Emil, 349, 390
ascending chain of fields, 248
associativity, 74
automorphism, 390
automorphism group, 349
automorphism groups of fields, 405
auxiliary equation, 356

B
basis, 192

C
$\mathbb{C}$, 29

Cardano, Gerolamo, 227, 356
Cauchy, Augustin-Louis, 321
Cayley graph, 398
chain rule, 254
classification, 285
closure, 72
coefficient, 31
commutative group, 90
complex conjugate, 124
complex plane, 30, 103, 187
constructible number, 139
coordinate plane, 187
coprime, 121, 127
coset, 288
counterexample, 171, 266, 267, 270
cube roots, 102
cyclic group, 85, 86
cyclotomic polynomial, 120

D
de Moivre's formula, 106, 111
Dedekind, Richard, 325, 390
definition, 73

degree, 263
degree (of a field extension), 249
degree of extension, 201
del Ferro, Scipione, 227
dimension, 192
discriminant, 53, 238

E
elementary symmetric polynomial, 44
equation, 33, 97
Euler, Leonhard, 356
extension field, 188

F
factor, 98
factor group, 296
Fermat prime, 396
field, 48
field extension, 52, 163
field of coefficients, 50
finite field, 405
Fiore, Antonio, 227
Fourier, Joseph, 321
Fundamental Theorem of Galois Theory, 388
fundamental theorem of symmetric polynomials, 45

G
G ▷ H (H is a normal subgroup of G), 305
(G : H) (group index), 304, 362
Galois correspondence, 48, 276, 389

Galois extension, 405
Galois field, 405
Galois group, 340, 341
Galois theory, 48
Galois, Évariste, 48, 240, 321
Gauss, Carl Friedrich, 366, 369, 371
general quadratic equation, 345
generator, 89
group, 327
    axioms, 71
    operation, 72
    order, 86

I
identity, 75
identity group, 85
imaginary axis, 29
imaginary unit, 28
indeterminate, 32
index
    of a subgroup, 304
invariance, 44, 199
invariant, 341
inverse, 76
irreducibility, 325
irreducible polynomial, 273, 361
isomorphism, 92

J
Jordan, Camille, 390

K
known, 341

Index

L
[L : K] (degree of a field extension), 249, 362
Lacroix, Sylvestre, 321
Lagrange resolvent, 209, 375
Lagrange, Joseph-Louis, 47, 228, 303, 356, 375
Lagrange's theorem, 303
Legendre, Adrien-Marie, 245
linear combination, 183
linear dependence, 196, 198
linear equation, 31
linear independence, 196, 198
Liouville, Joseph, 390

M
mapping, 389
minimal polynomial, 258
monic, 119
multiplicativity formula for degrees, 254

N
normal extension, 274, 361
normal subgroup, 297, 305, 361
number line, 29

O
omega waltz, 229

P
permutation, 299, 327
$\Phi_k(x)$, 119
plopper, 18
Poisson, Siméon, 321–323, 340
polynomial, 33, 97
polynomial division, 99
primitive nth root of unity, 114

Q
$\mathbb{Q}$ (rational numbers), 50
quadratic equation, 31, 36
quadratic extension, 250
quotient group, 296

R
$\mathbb{R}$, 29
rational expression, 189
rational number, 145, 189
rationalization, 51, 190
real axis, 29
real numbers, 28
reducibility, 325
reduction (of a group), 357
relation between roots and coefficients, 39, 169
relatively prime, 121, 127
repeated roots, 32
Richard, Louis, 245
root, 32, 97
root of unity, 101, 114
Ruffini, Paolo, 240, 366, 386

S
S_3, 71
scalar, 180
scrambler, 18
separable extension, 405
sequence of field extensions, 248
skew, 177
solution, 97
solution (to a quadratic equation), 32

solvable group, 373
space, 180
span, 193
splitting field, 234, 273, 336
square root, 102
structure, 71, 81, 180, 192, 221, 283, 295
subgroups, 82
subset, 82
substitution, 327
substitution group, 327
swapper, 18
symmetric group, 71, 328
symmetric polynomial, 43, 346

T
Tartaglia, Niccolò, 227
tower of fields, 248
tower rule, 254
trivial group, 85, 343
Tschirnhaus transformation, 204, 205
Tschirnhaus, Ehrenfried, 204

U
unknown variable, 32
unsolvable, 30

V
variable, 32
$\varphi(n)$ (Euler's totient function), 121
vector space, 180

W
well-defined, 312
Wiles, Andrew, 241, 390

X
$X \subset Y$ (X is a subset of Y), 82

Other works by Hiroshi Yuki

(in English)

- *Math Girls*, Bento Books, 2011
- *Math Girls 2: Fermat's Last Theorem*, Bento Books, 2012
- *Math Girls 3: Gödel's Incompleteness Theorems*, Bento Books, 2016
- *Math Girls 4: Randomized Algorithms*, Bento Books, 2019
- *Math Girls Manga, Vol. 1*, Bento Books, 2013
- *Math Girls Manga, Vol. 2*, Bento Books, 2016
- *Math Girls Talk About Equations & Graphs*, Bento Books, 2014
- *Math Girls Talk About the Integers*, Bento Books, 2014
- *Math Girls Talk About Trigonometry*, Bento Books, 2014

    (in Japanese)

- *The Essence of C Programming*, Softbank, 1993 (revised 1996)

- *C Programming Lessons, Introduction*, Softbank, 1994 (Second edition, 1998)
- *C Programming Lessons, Grammar*, Softbank, 1995
- *An Introduction to CGI with Perl, Basics*, Softbank Publishing, 1998
- *An Introduction to CGI with Perl, Applications*, Softbank Publishing, 1998
- *Java Programming Lessons (Vols. I & II)*, Softbank Publishing, 1999 (revised 2003)
- *Perl Programming Lessons, Basics*, Softbank Publishing, 2001
- *Learning Design Patterns with Java*, Softbank Publishing, 2001 (revised and expanded, 2004)
- *Learning Design Patterns with Java, Multithreading Edition*, Softbank Publishing, 2002
- *Hiroshi Yuki's Perl Quizzes*, Softbank Publishing, 2002
- *Introduction to Cryptography Technology*, Softbank Publishing, 2003
- *Hiroshi Yuki's Introduction to Wikis*, Impress, 2004
- *Math for Programmers*, Softbank Publishing, 2005
- *Java Programming Lessons, Revised and Expanded (Vols. I & II)*, Softbank Creative, 2005
- *Learning Design Patterns with Java, Multithreading Edition, Revised Second Edition*, Softbank Creative, 2006
- *Revised C Programming Lessons, Introduction*, Softbank Creative, 2006
- *Revised C Programming Lessons, Grammar*, Softbank Creative, 2006

- *Revised Perl Programming Lessons, Basics*, Softbank Creative, 2006
- *Introduction to Refactoring with Java*, Softbank Creative, 2007
- *Math Girls / Fermat's Last Theorem*, Softbank Creative, 2008
- *Revised Introduction to Cryptography Technology*, Softbank Creative, 2008
- *Math Girls Comic (Vols. I & II)*, Media Factory, 2009
- *Math Girls / Gödel's Incompleteness Theorems*, Softbank Creative, 2009
- *Math Girls / Randomized Algorithms*, Softbank Creative, 2011
- *Math Girls / Galois Theory*, Softbank Creative, 2012
- *Java Programming Lessons, Third Edition (Vols. I & II)*, Softbank Creative, 2012
- *Etiquette in Writing Mathematical Statements: Fundamentals*, Chikuma Shobo, 2013
- *Math Girls Secret Notebook / Equations & Graphs*, Softbank Creative, 2013
- *Math Girls Secret Notebook / Let's Play with the Integers*, Softbank Creative, 2013
- *The Birth of Math Girls,* Softbank Creative, 2013
- *Math Girls Secret Notebook / Round Trigonometric Functions*, Softbank Creative, 2014
- *Mathematical Writing, Refinement Edition*, Chikuma Shobo, 2014
- *Math Girls Secret Notebook / Chasing Derivatives*, Softbank Creative, 2015